Sonar Signal Processing

The Artech House Acoustics Library

Acoustics and Electroacoustics by Mario Rossi

Acousto-optic Signal Processing: Fundamentals and Applications by P.K. Das and C. DeCusatis

Bulk Acoustic Wave Theory and Devices by Joel F. Rosenbaum

Engineering Applications of Acoustics by John A. Kleppe

Signal Detection and Estimation by M. Barkat

Sonar Signal Processing by Richard O. Nielsen

Surface Acoustic Waves for Signal Processing by M. Feldmann and J. Henaff

Sonar Signal Processing

Richard O. Nielsen
Interstate Electronics Corporation
Anaheim, California

Artech House
Boston • London

Library of Congress Cataloging-in-Publication Data

Nielsen, Richard O.
Sonar signal processing / Richard O. Nielsen
p. cm.
Includes bibliographical references and index.
ISBN 0-89006-453-9
1. Signal processing--Digital techniques. 2. Sonar. I. Title.
TK5102.5.N515 1991 91-3102
621.389'5--dc20 CIP

British Library Cataloguing in Publication Data

Nielsen, Richard O.
Sonar signal processing.
1. Signals. Processing
I. Title.
621.3822

ISBN 0-89006-453-9

685 Canton Street
Norwood, MA 02062

International Standard Book Number: 0-89006-453-9
Library of Congress Catalog Card Number: 91-3102

10 9 8 7 6 5 4 3 2 1

To Dana
to my parents,
Dina and Torben,
and to my wife, Vicki

Contents

Preface

This book is the outgrowth of the author's short course, Acoustic Signal Processing. It is intended to provide the reader with a current discussion of sonar signal processing. Such a discussion by necessity bridges a number of related fields such as acoustic propagation in the medium, detection, and estimation theory; filter theory; sensor array processing and spectral analysis (both conventional and high resolution); fast transforms; matrix analysis; and digital signal processing.

The author hopes that this book will be useful to practicing hardware, software, and systems engineers, as well as engineering and project managers, working in sonar, radar, and seismic processing. In addition, the book can serve as a reference text for an upper division or graduate level college course in digital signal processing.

Until recently, practical sonar systems have employed conventional processing algorithms for the detection of signals and the estimation of their parameters. By conventional processing, we mean that no knowledge of the environment, such as spatial or temporal correlation, is included in the signal processing algorithms. In the opinion of this author, we are entering an era in which sufficient processor and memory resources will be available to permit matrix operations, required by high-resolution beamforming and spectral analysis, to be performed on platforms in real time.

In recognition of the fact that sonar signals are typically digitized at the sensors, detectors and estimators are developed using discrete, rather than continuous, variables.

To limit this book to a reasonable length, a large amount of introductory material is not included. We expect that the reader has a working understanding of the concepts of linear systems theory, probability theory, and sampled systems. Chapter 1 introduces the topics of analog signal conditioning, digital filtering, and the calculation of the discrete Fourier transform. Included in this treatment are analog filters and analog-to-digital conversion. Finite impulse and infinite impulse response digital filters are covered, as are multirate processing techniques. Next, the book discusses conventional beamforming in the time and frequency domains, as well as the effects of quantized delays and other errors.

The topic of Chapter 3 is the calculation of optimum detectors of broadband and narrowband signals and estimators of signal parameters such as bearing, power, frequency, and relative delay. These optimum detectors and estimators serve as the framework for the material in Chapter 4. Here, we discuss how these optimum structures are approximated by systems in the field and the associated losses in performance.

Chapter 5 treats active sonar processing. An overview of the characteristics of the sonar channel is followed by an investigation of sonar pulse types that reduce the impact of the channel. The optimum detector for range-velocity processing is discussed in a background noise-limited environment and its performance is assessed in the presence of reverberation noise. Finally, fast Fourier transform-based approximations of this detector are presented.

A comprehensive set of passive sonar processing requirements is proposed and, using the realizable detectors developed in Chapter 4, the overall computational throughput and data bandwidth requirements are determined. A generic signal processing architecture is introduced that can take advantage of the parallelism inherent in these system requirements.

To this point, only conventional sonar signal processing has been discussed. Chapter 7 presents an overview of high-resolution beamforming and spectral estimation as they apply to sonar. One motivation for this overview is the identification of the matrix operations that a modern sonar processor must be able to execute. The use of numerically stable matrix operations, such as QR decomposition and singular value decomposition, for solving least-squares and eigensystem problems, is discussed.

It is becoming feasible to implement matrix operations in fielded sonar systems, due in large part to the advent of parallel processors, in which operations of order n^3 can be executed in time n^2 by using an array of n processors. Each of these processors operates on a portion of the matrix problem in parallel. Chapter 8, written by Doctors Jaime H. Moreno and Thomás Lang, presents a technique for mapping matrix operations onto arrays of processors. This technique is illustrated with the LU decomposition algorithm.

Finally, Chapter 9 briefly treats three recent developments in sonar signal processing. These include processing nonstationary signals by using time-frequency distributions, correcting for wavefront curvature with focused beamforming, and differentiating between Gaussian and nonGaussian broadband noise sources using the bispectrum.

The author wishes to acknowledge his fruitful associations with Thomas Becker, William Burdic, Mauro Dentino, Philip Hallenborg, Robert Jones, Douglas Sherwood, Jeffrey Strauss, and Alan Willson, Jr.

RICHARD O. NIELSEN
ORANGE, CALIFORNIA
NOVEMBER, 1990

Chapter 1
Overview of Digital Signal Processing

This chapter is intended to present the basic signal processing concepts of analog and digital filtering, analog-to-digital conversion, multirate processing, and the calculation of the discrete Fourier transform. This information, which is covered in great detail in numerous texts, is used throughout much of the remainder of the book.

A short, and therefore far from exhaustive, list of such texts is presented next. Collections of papers dealing with *digital signal processing* (DSP) were published in 1972 [1] and 1975 [2] by the IEEE Press. The first of these was edited by Rabiner and Rader and the second by the Digital Signal Processing Committee of the IEEE Acoustics, Speech, and Signal Processing (ASSP) Society. Also appearing at this time were the general DSP texts by Oppenheim and Schafer [3], Rabiner and Gold [4], and Tretter [5]. A collection of DSP programs [6] was published in 1979, also edited by the Digital Signal Processing Committee of the ASSP. The topic of digital filters was treated by Antoniou [7] in 1979, and multirate DSP was the subject of the 1983 book by Crochiere and Rabiner [8]. Burrus and Parks [9] discuss the calculation of the discrete Fourier transform and convolution in their 1985 text.

Three volumes of papers dealing with *very large scale integration* (VLSI) and signal processing appeared in 1985, 1986, and 1988. These were edited by Kung, Whitehouse, and Kailath [10]; Kung, Owen, and Nash [11]; and Brodersen and Moscovitz [12]. Finally, a text that takes a system design approach to DSP, by DeFatta, Lucas, and Hodgkiss [13], appeared in 1988.

1.1 ANALOG SIGNAL CONDITIONING

This section discusses the analog antialiasing filter and analog-to-digital conversion steps required to generate discrete-time sequences for further processing. We start by assuming the existence of a transducer or sensor that converts sound pressure in the medium to analog voltage signals. The electromechanical considerations that enter into the analysis, design, and construction of such devices are beyond the scope, as well as the intent, of this book. The first function, that of protecting against aliasing in the digitized data, is provided by an analog low-pass filter. This topic is treated next.

1.1.1 Analog Filtering

For a given signal source, there is an upper frequency of interest, f_p. Therefore, to satisfy the Nyquist criterion, one needs to sample the analog data at a frequency f_s that is at least two times f_p. A convenient value to relate these two frequencies is 2.56. As we shall see, this results in a reasonable compromise between analog filter length and digital computation rate. (The exact value is associated with the popular power-of-two length fast Fourier transform; see Section 4.4.1.)

A well-known relationship between the Fourier transform, $X_a(f)$, of an analog signal, $x_a(t)$, and the Fourier transform, $X(f)$, of the corresponding sampled sequence, $x(n) = x_a(nT)$, is [4, eq. (2.65)]

$$X(f) = \frac{1}{T} \sum_{k=-\infty}^{\infty} X_a(f + kf_s) \tag{1.1}$$

where $T = 1/f_s$.

The effect of the antialiasing filter is illustrated in Figure 1.1, where an analog line spectrum is shown in Figure 1.1(a), the digital spectrum without antialiasing filtering is shown in Figure 1.1(b), and the correct digital spectrum with antialiasing filtering is found in Figure 1.1(c). With proper analog domain filtering, the digital spectrum (1.1) is given by

$$X(f) = \frac{1}{T} X_a(f), \quad \text{for} \quad -f_s/2 \leq f \leq f_s/2 \tag{1.2}$$

The voltage or magnitude specification for a low-pass filter is provided in Figure 1.2. Here, we are interested in passing signals in the frequency range 0 to f_p (the passband) with no attenuation while providing zero response for frequencies above f_a (the stopband). To obtain realizable filters, we allow a $\pm\delta_1$ deviation from unity response in the passband and tolerate a nonzero gain, δ_2, in the stopband. The passband ripple, R, and stopband attenuation, A, both in decibels (dB), are defined by

$$R = 20 \log_{10} \frac{1 + \delta_1}{1 - \delta_1} \quad \text{and} \quad A = -20 \log_{10} \delta_2$$

We frequently are interested in processing the output of an acoustic array of sensors.

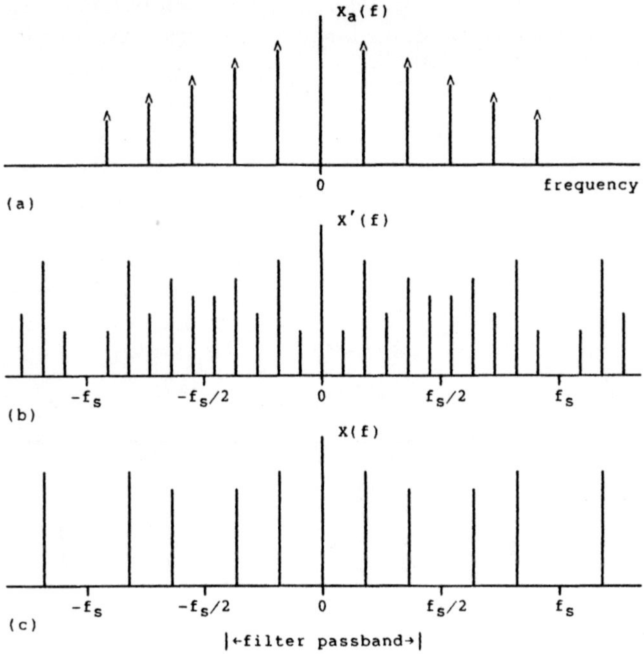

Figure 1.1 Effect of an analog antialiasing filter: (a) analog line spectrum, $X_a(f)$; (b) digital spectrum, $X'(f)$, without an antialiasing filter; (c) digital spectrum, $X(f)$, with an ideal antialiasing filter.

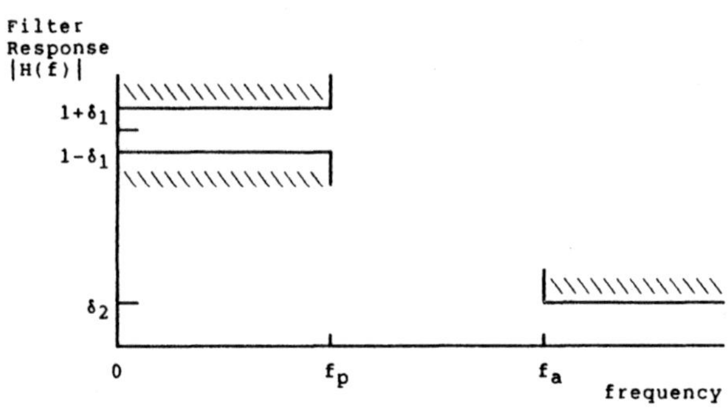

Figure 1.2 Magnitude of the frequency response for a low-pass filter.

For coherent combination (or *beamforming*) of these sensor channels, it is important that the variation in phase response of the analog filters from channel to channel be small. (Section 2.5.2 discusses the effects of uncertainties in sensor phase and position on the formation of beams.) In general, filter phase stability, gain stability, and reliability are inversely proportional to filter order. The performance of a low-order filter is less sensitive to analog-component tolerance and drift. It therefore is of interest to minimize the order of the analog filter, where filter order is equal to the number of poles in the s-domain filter transfer function. Popular analog filters include Butterworth, Chebyschev, and elliptic filters. Figure 1.3 illustrates the filter order for these three types of filters as a function of the transition ratio. This ratio, which is an indication of the "steepness" of the filter response between the passband

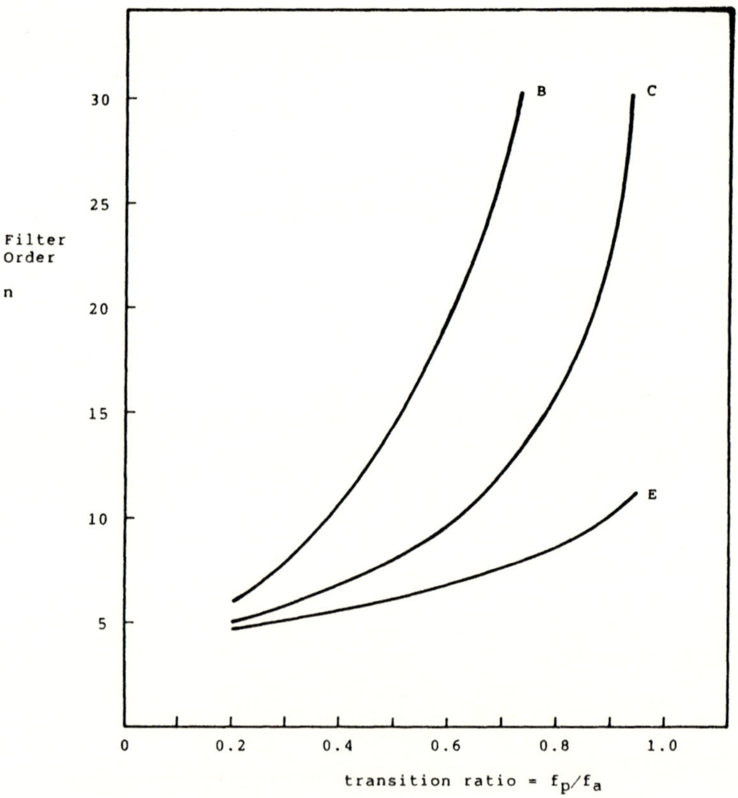

Figure 1.3 Analog filter order, n, *versus* transition ratio (B = Butterworth, C = Chebyschev, E = elliptic).

and the stopband, is given by f_p/f_a. Values for passband ripple of 0.5 dB and stopband attenuation of 70 dB were used to generate Figure 1.3.

As we can see from Figure 1.3, the order of the analog antialiasing filter is proportional to the transition ratio. The value of this ratio, and therefore the order of the filter, can be reduced if we are willing to accept aliasing in the transition region (the region from f_p to f_a). This situation is shown in Figure 1.4. With no aliasing in the transition region, the transition ratio is equal to $2f_p/f_s$. With aliasing, this ratio is reduced to $f_p/(f_s - f_p)$.

In the interest of reducing the complexity of the analog filter, we might consider the sampling implementation given in Figure 1.5. Here, the *analog-to-digital* (A/D) converter is operated at a frequency that is an integer, D, times the required frequency, f_s. The order of the analog filter is reduced because of the decrease in transition ratio. As an example of this approach, we consider the two cases given in Table 1.1. (Decimation and phase linearity are treated in Section 1.2.) We may conclude that even though (1) the *sample-and-hold* (S/H) and the A/D converters operate at twice the required sampling frequency, and (2) a digital filter is required, case 2 is preferable because of the reduction in analog filter hardware. This technique of oversampling the data has utility in the following context. As we shall see in Chapter 2, the process of simple time-domain beam formation often requires the availability of finely spaced time samples. (That is, the required sampling frequency $\hat{f}_s$ is many times the data sampling frequency f_s.) For this application, a reasonable approach is to oversample the data, process it with the beamforming operation, and then filter and reduce the sampling frequency to the final value f_s.

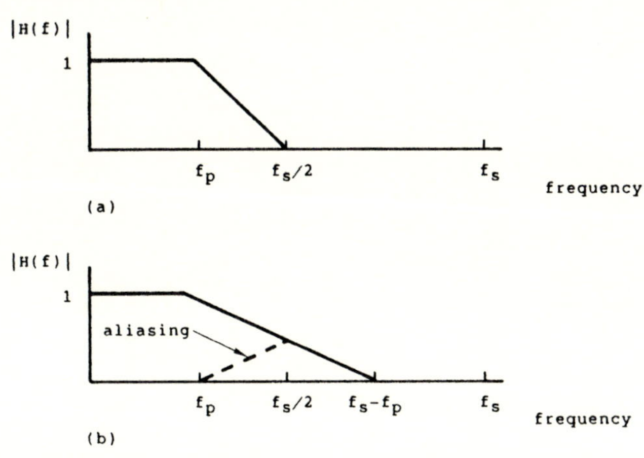

Figure 1.4 Low-pass filter response with (a) no aliasing in the transition region; (b) aliasing in the transition region.

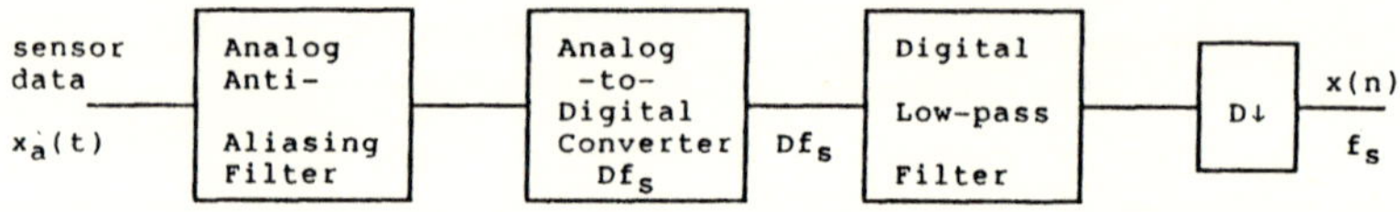

Figure 1.5 Sampling strategy that reduces the order of the analog antialiasing filter by sampling the input data at D times the required frequency.

The approach of decreasing the order of the analog antialiasing filter by increasing the sampling frequency of the A/D converter recently has been implemented in an integrated circuit. This device uses the sigma-delta A/D conversion technique, wherein the analog signal is sampled by a one-bit A/D converter at a frequency in excess of fifty times the desired frequency f_s. Because of this high sampling frequency, the transition ratio in Figure 1.3 is extremely small and a very simple analog antialiasing filter is required (one or two poles).

Figure 1.6 provides a block diagram of a sigma-delta A/D converter. The modulator converts the analog input signal to a one-bit pulse-density modulated data sequence, $y'(n)$. A multiple-bit representation of the input signal is obtained after digital low-pass filtering. The modulator acts as a high-pass filter, or noise shaper, for the large quantization noise that results from one-bit A/D conversion. This predominantly high-frequency noise at the modulator output is then filtered by the digital low-pass filter. The response of an nth order noise shaper is shown in Figure 1.7.

Because a large portion of the quantization noise is effectively "moved" out of the frequency range of interest, it is possible to realize a compromise between

Table 1.1
Comparison of Two Implementations for A/D Conversion

Case	*Analog Filter*	*A/D Sampling Frequency*	*Digital Filter*	*Decimation*
1	8-pole elliptic	f_s	none	none
2	4-pole elliptic	$2f_s$	53-stage	2

Notes:
Case 1 requires two adjustments; has poor channel-to-channel phase tracking, poor phase linearity, and moderate reliability.
Case 2 requires no adjustments; has good channel-to-channel phase tracking, good phase linearity, and good reliability.

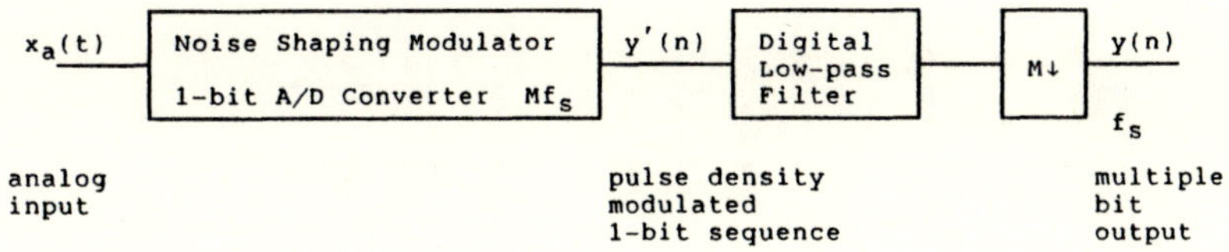

Figure 1.6 Simple block diagram of a sigma-delta A/D converter.

word length and sampling frequency at the output of the device. That is, the higher the final sampling frequency, the more quantization noise that is present in the output and the shorter is the useable word length. A device is commercially available [14] that offers such a compromise. Effectively using a third-order noise shaping circuit, it provides 12-bit resolution at a 400 kHz sampling frequency or 16-bit resolution at 100 kHz. Similar approaches have been discussed in the literature [15, 16].

1.1.2 Analog-to-Digital Conversion

The type of time-domain sampling that we have discussed to this point has not taken advantage of any special features of the analog signal in question. That is, we have simply arranged for the sampling frequency, f_s, to be some factor times the highest frequency of interest in the data. This often is referred to as *low-pass sampling*. If we consider a signal with the spectrum shown in Figure 1.8, the use of a different sampling scheme can result in a lower sampling frequency. The type of signal shown

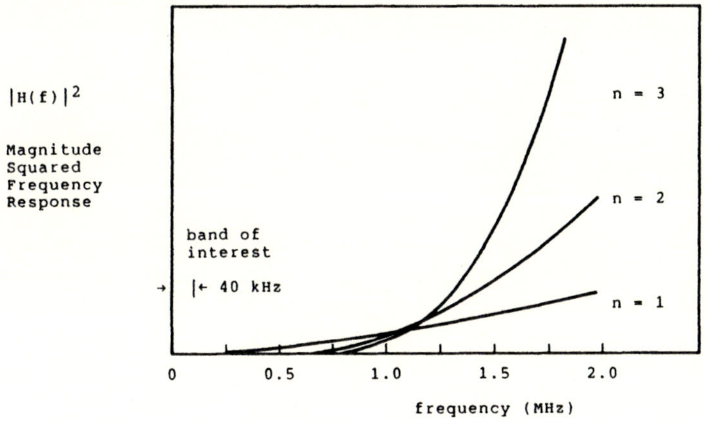

Figure 1.7 Frequency response of an *n*th order noise shaper (one-bit A/D converter sampling frequency = 6.4 MHz).

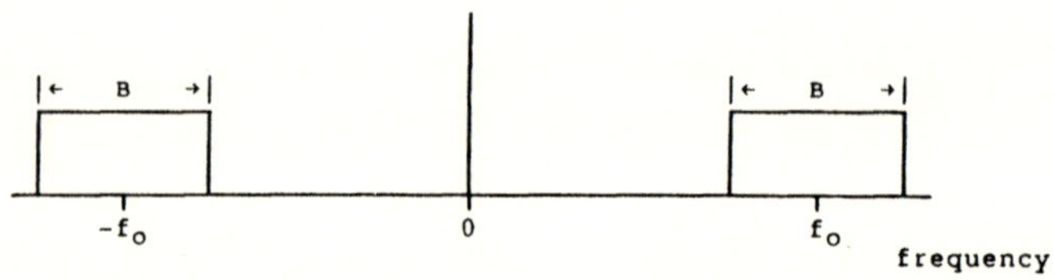

Figure 1.8 Spectrum of a bandpass signal (the width of the spectrum, B, is small compared to frequency, f_0).

in this figure is found in active sonar and communications, referred to as a *narrowband,* or *bandpass, signal*. The extent of the spectrum is confined to a narrow frequency region, B, centered around a high frequency, f_0. Low-pass sampling would require a sampling frequency f_s that is greater than twice the highest frequency in the data, which is $f_0 + B/2$ Hz. By using a representation of the signal known as the *complex envelope,* however, a significantly lower sampling frequency can be used.

The complex envelope, $\tilde{x}(t)$, of a time-domain waveform can be defined as

$$\begin{aligned} \tilde{x}(t) &= [x(t)\, e^{-j2\pi f_0 t}]_{LP} \\ x(t) &= 2 \operatorname{Re}[\tilde{x}(t)\, e^{j2\pi f_0 t}] \end{aligned} \tag{1.3}$$

Here LP denotes a low-pass filter operation, Re $[\cdot]$ denotes "real part of," and $j = \sqrt{-1}$. When dealing with the detection of signals in narrowband noise, as in Section 3.2.3, we frequently find the factor of 2 in (1.3) appearing in the first equation in the literature.

Quadrature Sampling

Because all the information of a bandpass signal is contained in the complex envelope, $\tilde{x}(t)$ can be sampled at a frequency f_s that is greater than only twice the highest frequency of $B/2$ Hz (rather than greater than twice $f_0 + B/2$ Hz). The implementation of this approach, often referred to as *quadrature sampling,* is shown in Figure 1.9. The complex envelope is calculated by shifting the spectrum of $x(t)$ by f_0 Hz to zero frequency, low-pass filtering, and sampling. The output of the A/D converters is termed the in-phase and quadrature components ($x_I(n)$ and $x_Q(n)$, respectively). The desired complex envelope is then $\tilde{x}(n) = x_I(n) + jx_Q(n)$. This calculation is also known as *complex demodulation*.

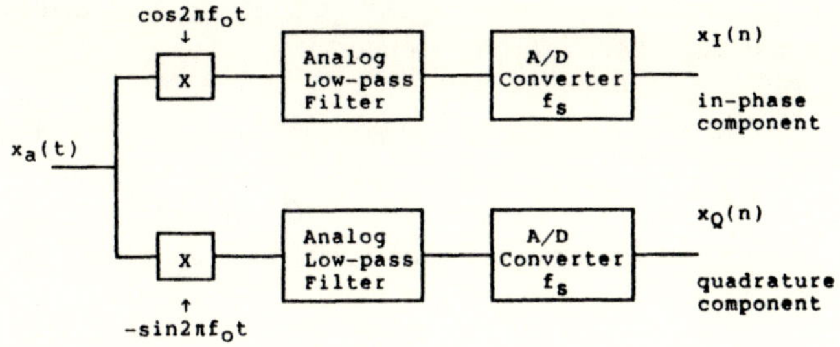

Figure 1.9 Calculation of the complex envelope.

Discrete Random Processes

At this point, we introduce a few definitions pertaining to discrete random processes, which are treated, for instance, in [3, Chapter 8] and in Marple's text [17, Section 4.3]. A discrete random process is a collection, or ensemble, of time sequences that may result from an experiment, such as measuring the noise sequences from a number of identical receivers. It is possible to describe this ensemble in terms of a mean, correlation functions, and power spectral density functions. Frequently, however, we have available only one sample sequence, $x(n)$, of the ensemble. It often is assumed that the random process is stationary, so that the process statistics are independent of time, and ergodic, so that the ensemble averages can be replaced by time averages based on the single sequence $x(n)$.

A random process is wide-sense stationary if the mean and autocorrelation function do not vary over time. A pair of random processes is jointly wide-sense stationary if the cross-correlation function does not vary over time. Thus wide-sense stationarity is a weaker condition than stationarity.

The expected value of $x(n)$ is

$$E\{x(n)\} = \lim_{M\to\infty} \frac{1}{2M + 1} \sum_{n=-M}^{M} x(n)$$

where $E\{\cdot\}$ is the expectation operator. The autocorrelation function is defined by

$$R_{xx}(m) = E\{x(n)x^*(n + m)\} = \lim_{M\to\infty} \frac{1}{2M + 1} \sum_{n=-M}^{M} x(n)x^*(n + m)$$

where * denotes complex conjugation. If $x(n)$ and $y(n)$ are sample sequences of two jointly stationary processes, the cross-correlation function is $R_{xy}(m) = E\{x(n)y^*(n + m)\}$. The zero mean sequences are uncorrelated if

$$E\{x(n)y^*(n + m)\} = E\{x(n)\}E\{y^*(n)\} = 0, \quad \text{for all } m$$

The *power spectral density* (psd) or spectrum, $S_{xx}(f)$, which is assumed to be band-limited to $\pm f_s/2$ Hz, and the autocorrelation function are related by

$$S_{xx}(f) = \frac{1}{f_s} \sum_{m=-\infty}^{\infty} R_{xx}(m) e^{-j2\pi f m/f_s}$$

$$R_{xx}(m) = \int_{-f_s/2}^{f_s/2} S_{xx}(f)\, e^{j2\pi f m/f_s}\, df$$

In particular, the variance for a zero mean sequence is equal to

$$\sigma^2 = R(0) = \int_{-f_s/2}^{f_s/2} S(f)\, df$$

where $S(f)$ is used to represent $S_{xx}(f)$. A zero mean white noise sequence is one for which $R(m) = \sigma^2\delta(m)$, and $S(f) = \sigma^2/f_s$, for $|f| \leq f_s/2$. (The function, $\delta(m)$, is the discrete delta function and equals unity for $m = 0$ and zero for $m \neq 0$.) The *cross-power spectral density* (cpsd) or cross spectrum, $S_{xy}(f)$, is defined by

$$S_{xy}(f) = \frac{1}{f_s} \sum_{m=-\infty}^{\infty} R_{xy}(m)\, e^{-j2\pi f m/f_s}$$

We will be dealing primarily with sample sequences taken from zero mean wide-sense stationary random processes.

Effect of A/D Converter Word Length

Because the A/D converter uses fixed word length, or number of levels, to represent the analog signal, distortions occur both in terms of quantization noise (the input is either rounded or truncated to the appropriate A/D converter level) and saturation noise (the input exceeds the maximum A/D converter level and is clipped to that level). In terms of quantization noise and assuming that rounding is performed, the largest error that can occur between the input signal and the output of the A/D converter is equal to $\pm q/2$, where q is the quantization step size. Therefore an error sequence, $e(n)$, due to rounding is generated where

$$e(n) = x(n) - \text{int}[x(n)]$$

The function, int[u], denotes "rounding to nearest A/D converter level," given in Figure 1.10. When the input signal is complicated and the A/D converter word length is long, so that $x_a(t)$ can be expected to vary over many quantization steps, we can assume a uniform *probability density function* (pdf) for $e(n)$, as shown in Figure 1.11. In addition, we can assume that samples of the error sequence $e(n)$ are not correlated; $e(n)$ is not correlated with the exact sequence $x(n)$; and $e(n)$ is a sample sequence of a stationary random process [3, p. 415]. Clearly these assumptions are invalid when, for example, the input is at a dc level.

It is desirable to increase the input signal level to reduce the effect of quantization noise. On the other hand, this may result in an unacceptable level of saturation distortion. Based on the assumption that the input signal is a Gaussian-distributed random process, Gray and Zeoli [18] have obtained values for a scalar, k, that minimizes total noise power (quantization and saturation noise) as a function of A/D

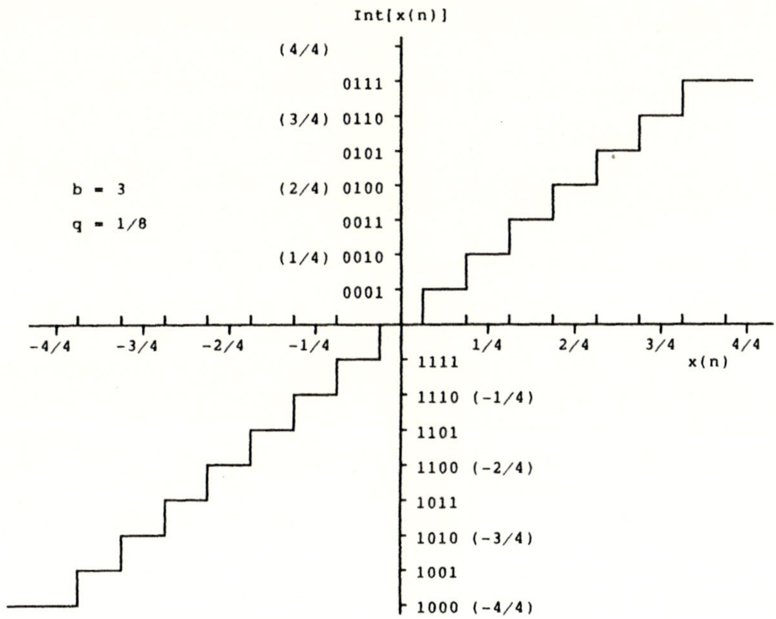

Figure 1.10 Transfer function for a four-bit A/D converter employing rounding and two's-complement arithmetic.

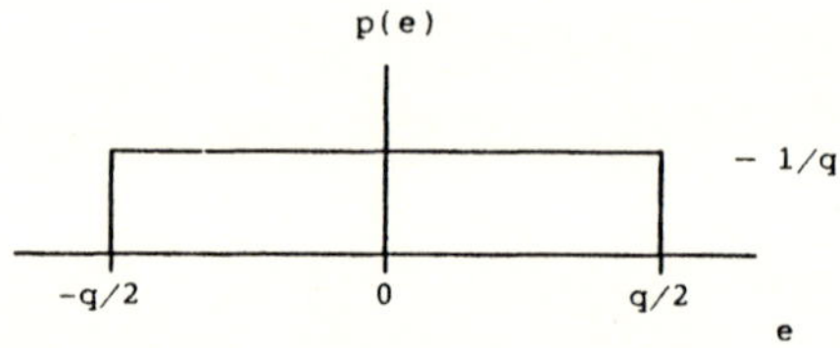

Figure 1.11 Probability density function for the A/D converter quantization noise, assuming rounding.

converter word length. The input signal is scaled so that its standard deviation, σ, is $1/k$ times the maximum level that can be represented by the A/D converter.

Often k is simply set equal to 4, so that saturation occurs only for those values of the input signal that exceed 4σ in magnitude. The probability of saturation for signals of interest therefore is considered acceptable.

The *signal-to-quantization noise* (S/QN) ratio is determined as follows: We assume a word length of $b + 1$ bits. The quantization step size, q, is then equal to 2^{-b}, and, from the assumption of a uniform pdf (Figure 1.11), the quantization noise power is equal to $q^2/12$. The input signal is scaled so that its standard deviation is equal to $1/k$ times the maximum A/D converter level, which is $(2^b - 1)/2^b$, or approximately unity. Therefore,

$$
\begin{aligned}
S/QN &= 10 \log_{10} \frac{\text{signal power}}{\text{quantization noise power}} = 10 \log_{10} \frac{1/k^2}{2^{-2b}/12} \\
&= 20\, b \log_{10} 2 + 10 \log_{10} 12 - 20 \log_{10} k
\end{aligned}
$$

With k set equal to 4, the S/QN ratio is given by $6.02b - 1.25$ dB. Therefore, a word length of $b + 1 = 13$ bits is required if we desire an S/QN value of 70 dB.

A control point, $1/k$, thus is selected to minimize the overall noise contribution from the A/D converter. A block diagram of the analog processing steps discussed in this section is shown in Figure 1.12. An amplifier, with gain selectable in steps a fraction of a dB, is used to position the standard deviation of the input signal at the A/D converter control point. This procedure is called *automatic gain control*

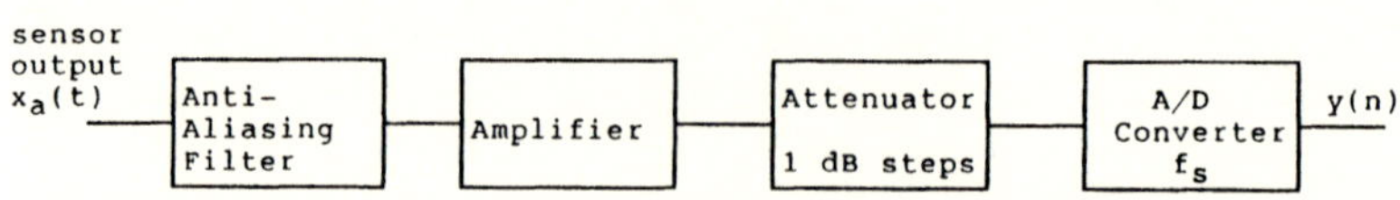

Figure 1.12 Block diagram of typical analog signal conditioning (an attenuation step size of 1 dB is shown).

(AGC). An estimate of signal power is obtained over some time interval, t_{agc}. If this interval is excessively long, the input signal can either saturate or fall below a reasonable level in the A/D converter word during the calculation. If t_{agc} is too short, the estimate of signal power is noisy, and the A/D converter control point responds to short-term variations in power.

1.2 DIGITAL FILTERING

This section deals with low-pass filters for two reasons. First, the filters most commonly encountered in sonar signal processing are the low-pass and the bandpass types. As we shall see in various sections of this book, a bandpass filter frequently is implemented by a mixing function followed by a low-pass filter (actually, a pair of identical filters because the mixing operation generates complex data). The mixing function shifts the center frequency, f_m, of the band of interest to zero frequency. This equivalent bandpass filter is shown in Figure 1.13. Second, a set of transformations from Constantinides [7] is available whereby we can map a prototype low-pass filter into a bandpass, a bandstop, or a high-pass filter. Table 1.2 summarizes the features of two major classes of digital filters: the recursive, or *infinite impulse response* (IIR) filter; and the nonrecursive, or *finite impulse response* (FIR) filter.

Referring to the low-pass filter response plot (Figure 1.2), typical values for the passband ripple parameter, R, are 0.2 dB to 1.0 dB, and the stopband attenuation parameter, A, often is specified in the range 50 dB to 80 dB.

We mentioned in Section 1.1 that the order of the analog antialiasing filter can be reduced if we allow aliasing in the transition region. This situation also is true for the case in which a digital low-pass filter is followed by a reduction in sampling frequency (decimation). By permitting aliasing in the transition region, the width of this region is $f_a - f_p = f_s - 2f_p$, rather than $f_a - f_p = f_s/2 - f_p$ without aliasing. The quantity f_s is the sampling frequency at the output of the filter after decimation. These relationships were again illustrated in Figure 1.4. Assuming that $f_s = 2.56 f_p$, the transition region is $0.56 f_p$ and the transition ratio is $1/1.56 = 0.64$ if aliasing is permitted in the transition region. Without aliasing, the transition region and transition ratio are $0.28 f_p$ and $1/1.28 = 0.78$, respectively. The lower transition ratio

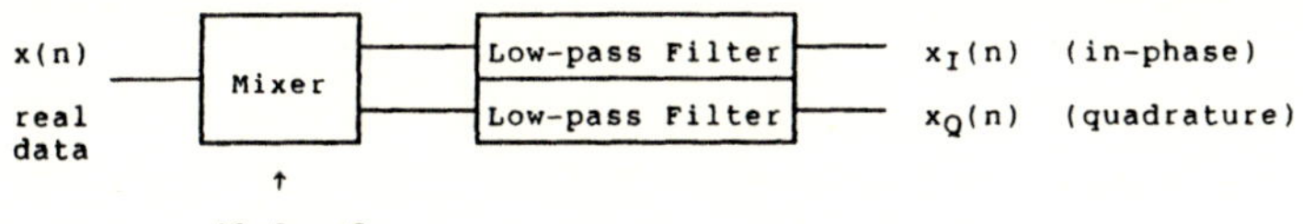

Figure 1.13 Equivalent bandpass filter implemented with low-pass filters.

Table 1.2
Brief Comparison of IIR and FIR Filters

IIR Filters	*FIR Filters*
Specified by poles and zeros	Specified by zeros
Recursive implementation	Nonrecursive implementation
Calculations performed at input sampling frequency	Calculations performed at output sampling frequency
Usually nonlinear phase	Exact linear phase with symmetric impulse response
Designs often based on analog filters	Unique digital design techniques

associated with transition region aliasing results in a reduction in the order of both FIR and IIR filters.

1.2.1 Finite Impulse Response Filters

This type of filter is defined in terms of the z-transform filter transfer function, $H(z)$:

$$H(z) = \sum_{n=0}^{N-1} h(n)\, z^{-n}$$

and the difference equation:

$$y(n) = \sum_{k=0}^{N-1} h(k)\, x(n-k) \tag{1.4}$$

where $h(n)$ is the filter impulse response, N is the length of the filter, and $x(n)$ and $y(n)$ are the input and output sequences, respectively. For the FIR filter, the impulse response sequence $h(n)$ and the filter coefficients are identical.

A simple formula for estimating the length of a FIR filter is, from [19],

$$N = \frac{-10 \log_{10}(\delta_1 \delta_2) - 13}{14.6\,(f_a - f_p)/f_s} + 1 \tag{1.5}$$

where f_s is the sampling frequency at the input to the filter. As can be seen from (1.5), the length of the FIR filter is inversely proportional to the width of the transition region, normalized by the input sampling frequency.

The McClellan-Parks algorithm [6, 20] often is used for designing FIR filters. This algorithm generates linear phase FIR filters with minimum weighted Chebyschev error in approximating a desired frequency response.

Linear Phase Considerations

Table 1.2 indicates that the phase of the FIR filter is linear with respect to frequency if the impulse response, $h(n)$, is symmetric. IIR filters, in general, do not possess a linear phase function (the Bessel filter achieves approximately linear phase in the passband [7]). The effect of linear phase is discussed next.

The frequency response of a filter can be expressed as $H(f) = |H(f)|e^{j\theta(f)}$. The group delay, $\tau(f)$, of the filter is defined as $\tau(f) = -d\theta(f)/df$ and is independent of frequency if $\theta(f) = \alpha f$, where α is a constant (that is, if the phase is a linear function of frequency). We consider a signal, $x(t)$, as consisting of two sinusoids, where

$$x(t) = A_1 \cos(2\pi f_1 t) + A_2 \cos(2\pi f_2 t)$$

We assume an ideal low-pass filter so that

$$|H(f)| = \begin{cases} 1, & |f| \leq f_p \\ 0, & \text{otherwise} \end{cases}$$

and also assume that the frequencies contained in the input signal, $x(t)$, are in the filter passband (that is, $0 \leq f_1 \leq f_p$ and $0 \leq f_2 \leq f_p$). Then the output of the filter, $y(t)$, is

$$y(t) = A_1 \cos(2\pi f_1 t + \alpha f_1) + A_2 \cos(2\pi f_2 t + \alpha f_2)$$

or, equivalently,

$$y(t) = A_1 \cos[2\pi f_1 (t + \alpha/2\pi)] + A_2 \cos[2\pi f_2 (t + \alpha/2\pi)]$$

That is, each sinusoid is shifted by the same quantity $t_0 = \alpha/2\pi$. The only effect of this filter is the insertion of a constant time delay, t_0, for each frequency component of $x(t)$ in the filter passband. Therefore, a filter with a linear phase characteristic is nondispersive. The time delay of a signal as a function of frequency is constant. A linear phase characteristic may be desirable in replica correlation (Section 5.3.3).

FIR Filter Computational Considerations

As we shall see in Chapter 6, a major portion of the computational load imposed by a sonar processing system results from digital filtering. Therefore, it is desirable to reduce the number of computations required to implement a filter. When the FIR filter is followed by decimation, this can be accomplished by (1) calculating the filter output only at the required rate, (2) using multiple low-pass filters, and (3) using half-band filters.

Because the difference equation for the FIR filter (1.4) involves only present and past values of the input sequence $x(n)$, clearly we need calculate only those output samples $y(n)$ of interest. That is, if D is the decimation value,

$$y(n) = \sum_{k=0}^{N-1} h(k)\, x(Dn - k)$$

For the case in which the desired decimation value is a composite number $D = D_1D_2 \ldots$, the filter can be realized as a cascade connection of filters, each followed by a decimation factor. The resulting multistage filter implementation, shown in Figure 1.14 for $D = D_1D_2$, often can result in a considerable reduction in the overall computation rate, when compared to that required for a single filter implementation. Qualitatively, this reduction is due to the fact that the first filters in such a multistage structure generally have few stages (the length, N, is small), because the widths of the transition regions in (1.5) are large. The last filter may be relatively long but still shorter than the single filter implementation. This is because the sampling frequency at the input to the last filter is lower than the original sampling frequency. Crochiere and Rabiner [8] provide a set of curves for optimizing the factorization of the desired decimation, D, with respect to either the computation rate or the storage required for filter coefficients. We often can dramatically reduce the total coefficient storage and computational rate by using this multistage structure. Frequently little is to be gained, however, by factoring D into more than three or four factors [8].

We point out that the multistage structure can be implemented with IIR filters as well. Because of the recursive nature of this type of filter, the output of each filter has to be calculated at its input sampling frequency. However, only those sam-

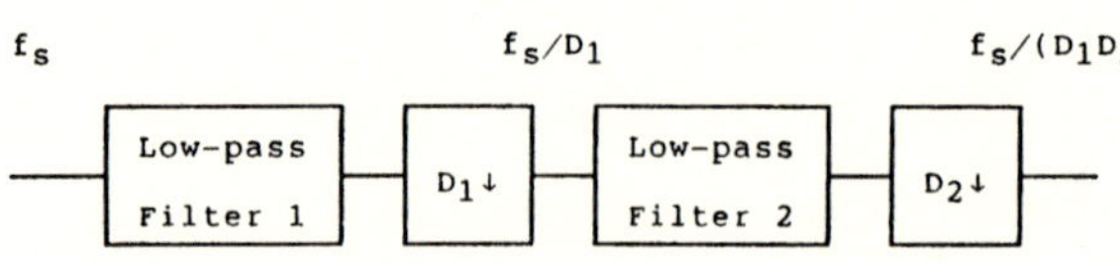

Figure 1.14 A two-filter implementation of a large decimation factor, $D = D_1D_2$.

ples corresponding to the lower output sampling frequency are processed by the following filter.

We can show that if we impose the following constraints on the amplitude response of a FIR low-pass filter,

$$\delta_1 = \delta_2 = \delta, \quad \text{and} \quad f_p + f_a = f_s/2 \tag{1.6}$$

then, with the exception of the center term, the odd-indexed terms of the impulse response are zero. That is,

$$h(n) = \begin{cases} 0, & n \text{ odd}, n \neq (N-1)/2 \\ 1/2, & n = (N-1)/2 \end{cases}$$

for $n = 0, \ldots, N-1$.

The impulse response, $h(n)$, and the magnitude frequency response, $|H(f)|$, of such a filter are shown in Figure 1.15. Due to the constraints in (1.6), this response

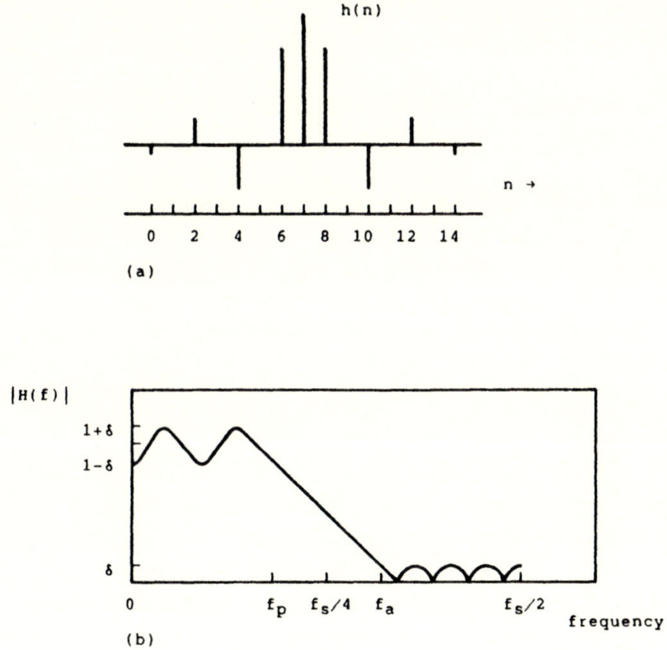

Figure 1.15 Half-band FIR filter: (a) impulse response, (b) magnitude frequency response ($f_p + f_a = f_s/2$).

is equal to 1/2 at frequency $f_s/4$ and the length, N, is odd and equal to $4k + 7$, where k is a nonnegative integer.

This type of FIR filter is called a *half-band filter*. It is ideally suited for the case wherein filtering is followed by decimation by two. (If this is not the desired decimation, D, then perhaps D can be factored into one or more values of two in Figure 1.14.)

The properties of the impulse response of the half-band filter suggest a reduction in computations by about one-half relative to a conventional FIR filter. This reduction is not realized because of the constraint that the passband and stopband parameters (δ_1 and δ_2) must be equal results in a larger value for N for the half-band filter than for a filter in which δ_1 and δ_2 can be specified separately.

Example 1.2.1

Assume the following set of design parameters:

$$\begin{aligned} f_s &= 5000 \text{ Hz} && \text{(input)} \\ R &= 0.2 \text{ dB} && (\delta_1 = 0.0115) \\ A &= 60 \text{ dB} && (\delta_2 = 0.001) \\ f_p &= 1000 \text{ Hz} && (\text{thus } f_a = f_s/2 - f_p = 1500 \text{ Hz}) \end{aligned}$$

Using (1.5), the conventional filter has length $N = 26$. With a symmetric impulse response, this is equivalent to $N - 1$ additions and $N/2$ multiplications, or 38 operations per output.

Using (1.5) with $\delta = \min\{\delta_1, \delta_2\} = 0.001$, a value of $N = 34$ is obtained. Because the length of a half-band filter is $4k + 7$, N is set equal to 35. This is equivalent to $(N + 1)/2$ additions and $(N + 1)/4 + 1$ multiplications, or 28 operations per output, which corresponds to 74 percent of the operations required for the standard filter. A more dramatic reduction in computations is obtained when δ_1 is closer in value to δ_2.

Example 1.2.2

Assume the same parameter values as in example 1.2.1, except let $R = 0.05$ dB ($\delta_1 = 0.00288$). The conventional filter now requires 31 stages for a total of 46 operations per output. The length of the half-band filter remains the same, because δ_1 is still larger than δ_2. Therefore the half-band filter requires 61 percent of the calculations of the standard filter.

A variation of the McClellan-Parks program [20] provides an exact design for a half-band filter [21]. (Using the McClellan-Parks method directly to design a half-band filter leads to an impulse response in which the odd-indexed elements of $h(n)$,

$n \neq (N-1)/2$, are small but not identically zero.) Here a prototype linear phase filter, with transfer function $G(z)$ and length $(N-1)/2$, is designed with a passband edge of $2f_p$, a transition region from $2f_p$ to $f_s/2$, and a passband parameter equal to 2δ. The half-band filter transfer function $H(z)$ and impulse response $h(n)$ are then obtained by

$$H(z) = \frac{G(z^2) + z^{-(N-1)/2}}{2}$$

$$h(n) = \begin{cases} g(n/2)/2, & n \text{ even} \\ 0, & n \text{ odd} \neq (N-1)/2 \\ 1/2, & n = (N-1)/2 \end{cases}$$

In addition to calculating the impulse response exactly, the filter design program requires less execution time because the prototype filter is of length $(N-1)/2$, rather than N.

1.2.2 Infinite Impulse Response Filters

This type of filter is defined in terms of the z-domain transfer function, $H(z)$:

$$H(z) = \left[\sum_{k=0}^{M} a(k)z^{-k}\right] \Big/ \left[1 - \sum_{k=1}^{N} b(k)z^{-k}\right] \tag{1.7}$$

and the difference equation:

$$y(n) = \sum_{k=1}^{N} b(k)\, y(n-k) + \sum_{k=0}^{M} a(k)\, x(n-k) \tag{1.8}$$

where N and M are the number of poles and zeros, respectively. As can be seen from the difference equation (1.8), both past values of the output sequence $y(n)$ and present and past values of the input sequence $x(n)$ are required for each filter calculation. Therefore, the output of the IIR filter must be calculated at the input sampling frequency.

Digital elliptic filters often are used because, like their analog counterparts, they require the fewest number of stages of any filter type for a given transition ratio f_p/f_a. A common implementation for this filter is obtained by factoring the transfer function, $H(z)$, in (1.7) into biquadratic sections. The transfer function for the ith filter section, $H_i(z)$ is equal to the ratio of two second-order polynomials in z^{-1}:

$$H_i(z) \frac{1 + a_{1i}z^{-1} + a_{2i}z^{-2}}{1 - b_{1i}z^{-1} - b_{2i}z^{-2}}$$

An implementation is shown in Figure 1.16. In the event that the order of $H(z)$ in (1.7) is odd, a first-order section is obtained by setting a_{2i} and b_{2i} equal to zero.

A design program such as DOREDI (design and optimization of recursive digital filters) [6] can be used to design a digital elliptic filter (low-pass, bandpass, bandstop, or high-pass). Once the appropriate transfer function, $H(z)$, has been obtained, programs of this type (1) group pole pairs and zero pairs in each biquadratic section and (2) order the biquadratic sections to optimize the noise performance of the filter. These steps are especially important when the filter is implemented with fixed-point arithmetic. A poor ordering of biquadratic sections and a poor grouping of pole and zero pairs, without scaling, can result in large filter responses at various points in the filter, although the input and output sequences are well behaved. Interstage scaling is then required to avoid overflow. Such scaling can reduce the S/QN ratios at various points in the filter.

Comparing the computational requirements of FIR and IIR filters is difficult. The results of such a comparison are affected by requirements for output decimation and linear phase. (A linear phase characteristic can be approximated for the case of the elliptic filter by using delay equalization.) A comparison has been performed [4, 22] based on the number of multiplications required for each type of filter to realize a given design specification. In particular, a symmetric FIR filter of length N and an nth-order elliptic filter (n poles and n zeros) were considered. Because the zeros for the elliptic filter are located on the unit circle in the z domain, the coefficient a_{2i}, for the ith stage (Figure 1.16) is equal to unity. One multiplier preceding the elliptic filter is used to provide scaling. The required number of multiplications for the two types of filters are

FIR filter: $[(N + 1)/2]$

IIR filter: $[(3n + 3)/2]$

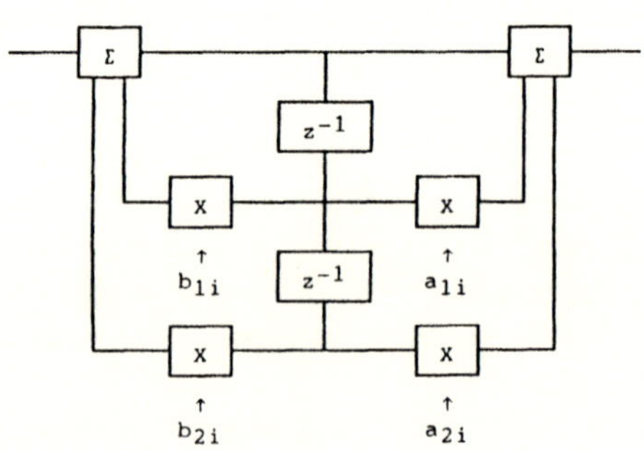

Figure 1.16 Biquadratic IIR filter section.

where [·] denotes "integer part." Using these values, it follows that the elliptic filter is more efficient in meeting a given filter specification than the FIR filter when the inequality $N/n > 3$ is satisfied. The results of a comparison are shown in Figure 1.17. Therefore, in most cases, the elliptic filter requires fewer computations.

Because of the availability of DSP devices, in which a multiplication and an addition both are calculated in one instruction cycle, the results of this comparison are slightly different. We assume that, after initialization, one stage of the FIR filter (i.e., one multiplication-accumulate) can be calculated each instruction cycle. Also, we assume that one biquadratic section (with zeros on the unit circle) requires five instruction cycles and that a first-order section requires three cycles. Floating-point arithmetic is assumed, so no scaling multiplier is used. Under these conditions, the elliptic filter is more efficient if $N/n > 2.5$. Therefore, the disparity in number of computations between FIR and elliptic filters is even more pronounced. In the com-

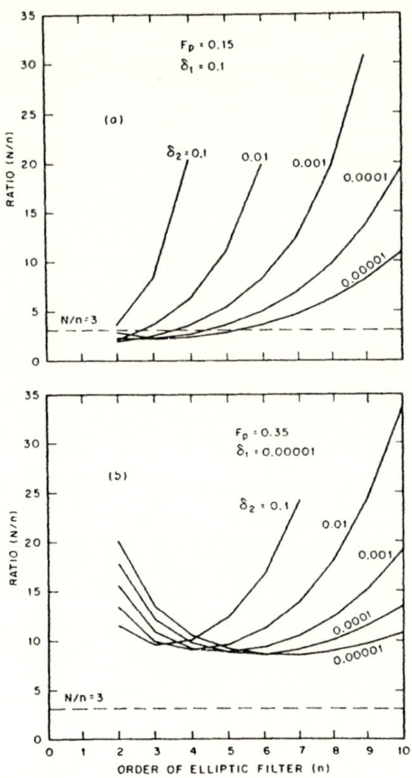

Figure 1.17 Plot of N/n as a function of n for optimum FIR filters and elliptic filters meeting identical low-pass specifications.

Source: [4] L.R. Rabiner and B. Gold, *Theory and Application of Digital Signal Processing,* Prentice-Hall, Englewood Cliffs, NJ, p. 285. (© 1975. Reprinted with permission.)

parison in [22], if an elliptic filter of order $n = 8$ satisfies a given filter specification, the FIR filter length would have to exceed 24 before it required more multiplications. With the present analysis, the length of the FIR filter need only be greater than 20 stages for it to require more execution time.

Example 1.2.3

Consider the following filter specification [4, p. 251]:

$\delta_1 = 0.01$ (passband ripple, $R = 0.174$ dB)
$\delta_2 = 0.0001$ (stopband attenuation, $A = 80$ dB)
$f_p = 480$ Hz (passband edge)
$f_a = 520$ Hz (stopband edge)
$f_s = 8000$ Hz (input and output sampling frequencies)

Results:

n = 12th-order elliptic filter (30 instruction cycles per output calculation)
N = 645 stage FIR filter (645 instruction cycles per output calculation)

Example 1.2.4

Consider the filter specification in example 1.2.3, except that the input sampling frequency now is 2000 Hz.

Results:

n = 11th-order elliptic filter (28 instruction cycles per output calculation)
N = 162 stage FIR filter (162 instruction cycles per output calculation)

These examples demonstrate the rather weak dependence of the order of the digital elliptic filter on sampling frequency. The length of the FIR filter, however, is directly related to f_s.

1.2.3 Multirate Signal Processing

Although there are numerous examples in digital signal processing where a change in sampling frequency is required, common examples in sonar signal processing are spectral analysis and beamforming. As discussed in Chapter 4, it often is of interest to perform spectral analysis using several frequency resolutions. A typical technique involves the use of a fixed-length *discrete Fourier transform* (DFT). Multiple resolutions are generated by changing the sampling frequency, f_s, of the data applied to the DFT. Because the resolution of a length-N DFT with a rectangular window is equal to f_s/N, decimation (that is, a smaller value of f_s at the input to the DFT) results in higher resolution.

Interpolation often is required in beamforming. Although the sensor data have been sampled at a frequency consistent with the Nyquist criterion, the process of accurate beam formation frequently requires the availability of samples that are more closely spaced in time (Section 2.5). An excellent treatment of multistage signal processing is provided by Crochiere and Rabiner [8].

Decimation

We now discuss the process of decimation, or a decrease in the sampling frequency by an integer factor, M. Consider a sequence $w(n)$. We are interested in forming a new sequence by taking every Mth sample so that $y(n) = w(Mn)$. It is instructive to consider the transform of a filtered and decimated sequence. We assume that the sequence $w(n)$ corresponds to the output of a filter, with transfer function $H(z)$ and impulse response $h(n)$, to which the sequence $x(n)$ has been applied. We can show [8] that the spectrum of $y(n)$ has the following form:

$$Y(f) = \frac{1}{M} \sum_{m=0}^{M-1} H(f - mf_s')\, X(f - mf_s'), \quad |f| \le f_s'/2$$

where $f_s' = f_s/M$ = sampling frequency after decimation. Expanding this equation gives

$$Y(f) = \frac{1}{M} \Big\{ \underbrace{H(f)X(f)}_{m\,=\,0} + \underbrace{H(f - f_s')X(f - f_s')}_{m\,=\,1} + \underbrace{H(f - 2f_s')\, X(f - 2f_s')}_{m\,=\,2} + \dots \Big\}, \quad |f| \le f_s'/2 \tag{1.9}$$

Figure 1.18 provides a plot of the spectrum, $X(f)$, of the input to the filter for $M = 3$. We can see that if the filter, $H(f)$, has the response shown in Figure 1.18,

$$H(f) = \begin{cases} 1, & |f| \le f_s'/2 \\ 0, & \text{otherwise} \end{cases}$$

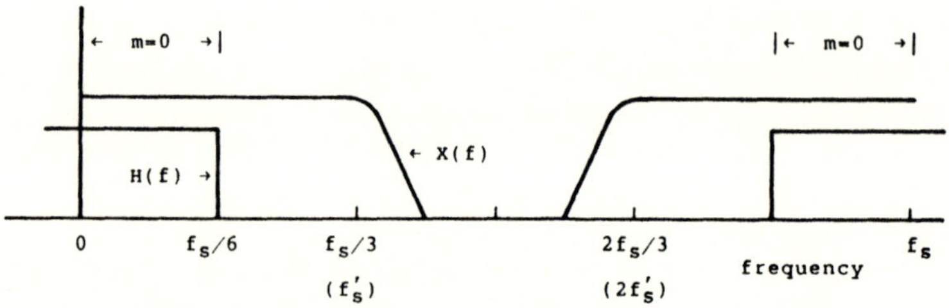

Figure 1.18 Spectrum, $X(f)$, of the input signal for decimation by $M = 3$.

then the only contribution to the sum in (1.9) is the $m = 0$ contribution and

$$Y(f) = \frac{1}{M} X(f), \quad |f| \leq f_s'/2$$

This filter thus serves as an antialiasing filter.

Interpolation

We next consider an increase in the sampling frequency by an integer, L. Therefore, the new sampling frequency is $f_s' = Lf_s$, and the time separation of samples is $T' = T/L$, where $T = 1/f_s$.

We begin with a sequence, $x(n)$, where the samples are separated by T seconds, and form

$$y(m) = \begin{cases} x(m/L), & m = 0, \pm L, \pm 2L, \ldots \\ 0, & \text{otherwise} \end{cases}$$

This operation often is referred to as *zero padding*, where $L - 1$ zeros are inserted between each pair of terms in the original sequence, $x(n)$. An operation is then performed to interpolate between the nonzero samples in the sequence $y(m)$. The process is illustrated in Figure 1.19.

Again, it is instructive to examine the transforms of the sequences in Figure 1.19. The transform of the zero-padded sequence $y(m)$ is equal to $Y[e^{j2\pi f/(Lf_s)}] = X(e^{j2\pi f/f_s})$. We note that $X(e^{j2\pi f/f_s})$ is periodic relative to the original sampling frequency f_s. That is, the spectrum of the zero-padded sequence $y(m)$ is correct for $|f| \leq f_s'/2L$.

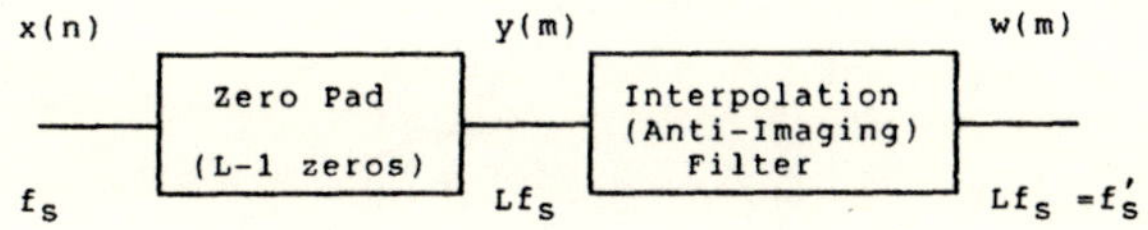

Figure 1.19 Block diagram of the interpolation process.

Undesired images of $X(e^{j2\pi f/f_s})$ are centered on integer multiples of f_s, as shown in the spectrum, $Y(f)$, in Figure 1.20. Thus an ideal low-pass filter operation is required to remove these images. The frequency response of this filter is

$$H(f) = \begin{cases} L, & |f| \leq f_s'/2L \\ 0, & \text{otherwise} \end{cases} \tag{1.10}$$

We address the issue of proper passband gain for the antialiasing and the anti-imaging filters. For an analog signal, $x_a(t)$, sampled at f_s (such that the Nyquist criterion

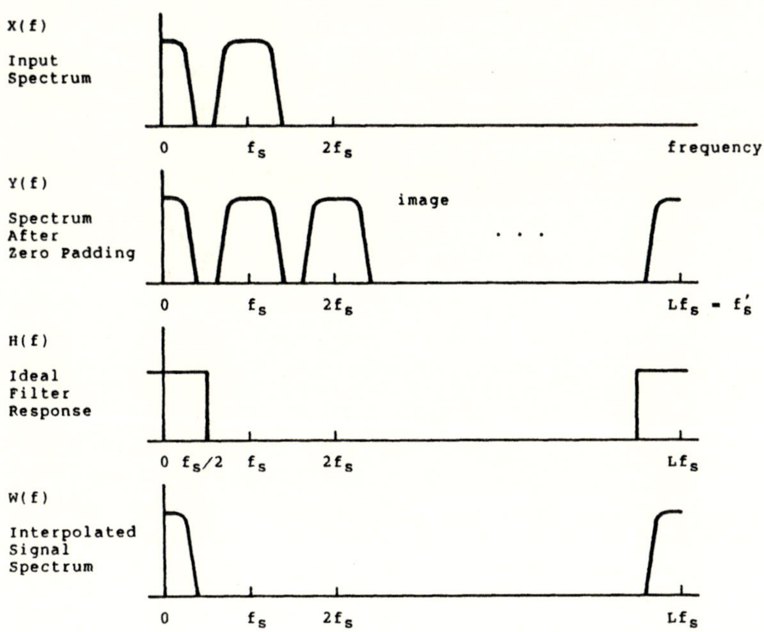

Figure 1.20 Spectra of sequences at various steps in the interpolation process.

is satisfied), the spectrum of the sampled signal repeating (1.2), is

$$X(f) = \frac{1}{T} X_a(f), \quad \text{for } -f_s/2 \leq f \leq f_s/2$$

For the case of decimation, had we originally sampled $x_a(t)$ at the lower frequency $f_s' = f_s/M$, the relationship of the digital spectrum to the analog spectrum would have been

$$Y(f) = \frac{1}{T'} X_a(f) = \frac{1}{TM} X_a(f) = \frac{1}{M} X(f)$$

Because the spectrum after decimation is $Y(f) = X(f)/M$, a passband gain of unity is correct for the antialiasing low-pass filter.

For the case of interpolation, had we originally sampled $x_a(t)$ at the higher frequency $f_s' = L f_s$, the relationship of the digital spectrum to the analog spectrum would have been

$$Y(f) = \frac{1}{T'} X_a(f) = \frac{L}{T} X_a(f) = L\, X(f)$$

Therefore, it is necessary to provide a passband gain of L for the anti-imaging filter.

It is apparent that a low-pass filter is required to generate the proper spectrum at the output of the interpolation process (Figure 1.20). We wish to determine the time-domain effect of this filter on the zero-padded sequence, $y(n)$. The impulse response of the ideal anti-imaging filter in (1.10) is

$$h(n) = \frac{\sin(\pi n/L)}{\pi n/L}, \quad -\infty \leq n \leq \infty \tag{1.11}$$

The output of the filter is

$$w(m) = \sum_{n=-\infty}^{\infty} y(n)\, h(m-n) = \sum_{n=-\infty}^{\infty} y(n) \frac{\sin[\pi(m-n)/L]}{\pi(m-n)/L}$$

Because $y(n)$ is nonzero only for values of n equal to integer multiples of L, $w(m) = y(m) = x(m/L)$ for m as an integer multiple of L. That is, the output is simply equal to the input to the interpolation process. For other values of m, the output, $w(m)$, is an interpolated result based on all elements of $y(n)$. This is shown in Figure 1.21, where the interpolation factor is $L = 4$ and we wish to calculate the interpolated output at time $n_o T'$. The sinc function (1.11) is centered on sample n_o, and the output

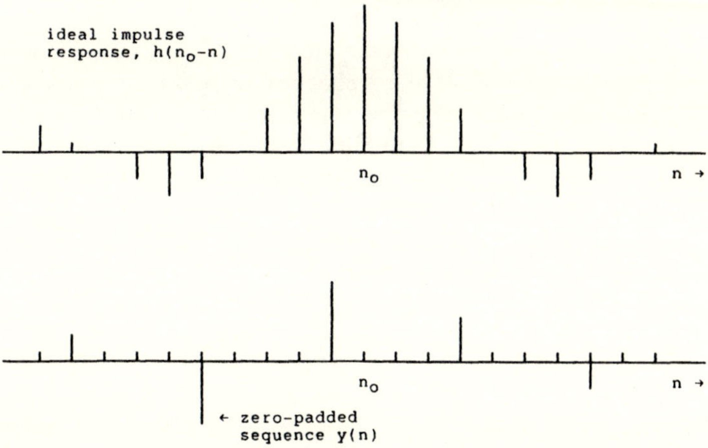

Figure 1.21 Calculation of the interpolated value for time n_0T' and $L = 4$ using the ideal interpolation filter.

is obtained by multiplying each element of $y(n)$ by the corresponding element of $h(n)$, for all n, and summing the results.

Of course, in practice, we are not able to implement the ideal impulse response in (1.11), and a finite-length filter must be designed to approximate the ideal low-pass filter response (1.10).

An intuitive approach to interpolating between pairs of samples is simply to use linear interpolation. This approach for calculating the output sample for time n_0T' is shown in Figure 1.22 for $L = 4$. Note that linear interpolation is equivalent, for this value of L, to filtering with a seven-stage symmetric FIR filter and that

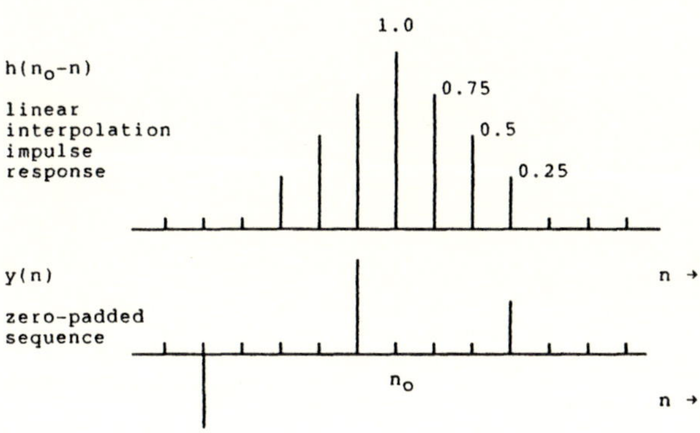

Figure 1.22 Calculation of the interpolated value for time n_0T' and $L = 4$ using linear interpolation.

exactly two input samples are used for each calculation. Intuitively, we would expect linear interpolation to perform poorly in the presence of high-frequency signals. This is supported by an examination of the frequency response of the seven-stage FIR filter. As shown in Figure 1.23, this filter attenuates the spectral images only in the vicinity of multiples of the original sampling frequency, f_s. Therefore, the original signal $x(n)$ must be considerably oversampled (that is, the highest frequency in $x(n)$ must be much smaller than $f_s/2$) for linear interpolation to provide adequate performance.

Change in Sampling Frequency by L/M

We close this section on multirate signal processing by considering changes in sampling frequency by ratios of integers, L/M. This is achieved by implementing the zero-pad operation (to increase the sampling frequency by a factor, L) and anti-imaging filter (Figure 1.19) followed by an antialiasing filter and decimation by M.

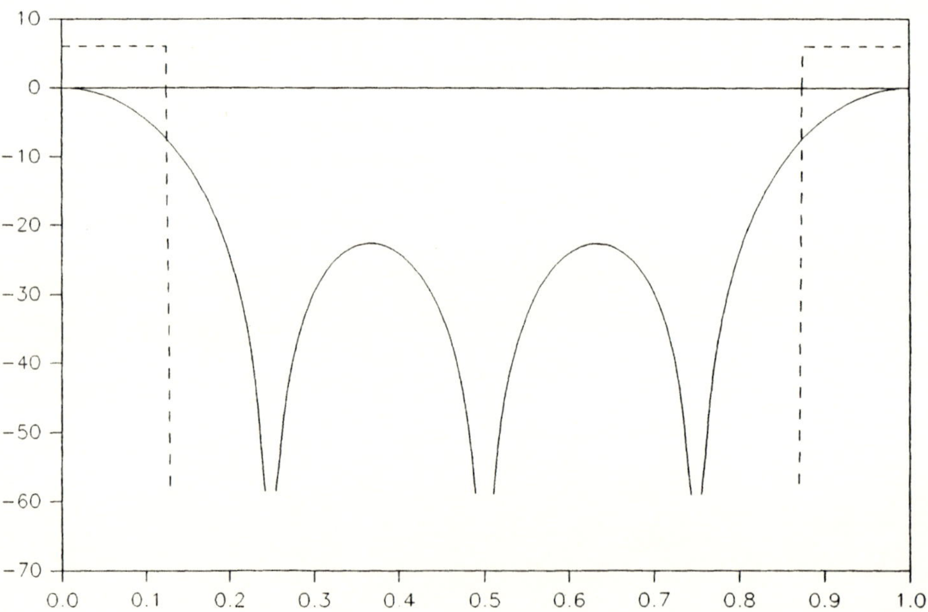

Figure 1.23 Power response (dB) of the $L = 4$ linear interpolation filter in Figure 1.22 *versus* frequency, normalized by f_s' Hz (ideal low-pass filter (1.10) is indicated with a dashed line).

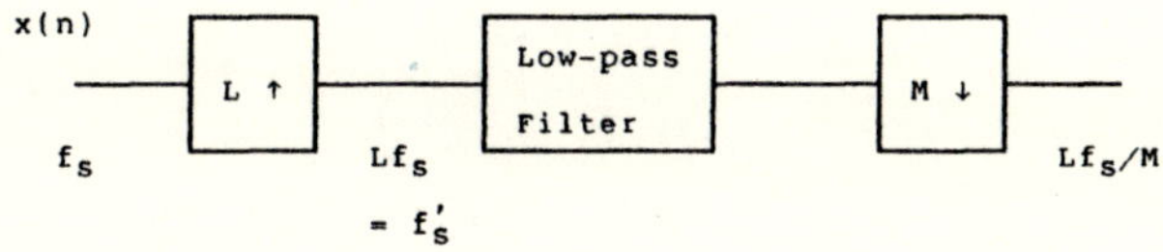

Figure 1.24 Signal processing structure to achieve an L/M change in sampling frequency.

The structure shown in Figure 1.24 is obtained by combining the two low-pass filters. The ideal low-pass filter must satisfy

$$H(f) = \begin{cases} L, & |f| \le \min\{f_s'/2L, f_s'/2M\} \\ 0, & \text{otherwise} \end{cases}$$

where $f_s' = Lf_s$, the sampling frequency after interpolation by L.

1.3 DISCRETE FOURIER TRANSFORM

The DFT pair for a sequence $x(n)$ is defined by

$$X(k) = \sum_{n=0}^{N-1} x(n)\, e^{-j2\pi kn/N}, \qquad k = 0, \ldots, N-1$$
$$x(n) = \frac{1}{N}\sum_{k=0}^{N-1} X(k)\, e^{j2\pi kn/N}, \quad n = 0, \ldots, N-1 \tag{1.12}$$

A length-N DFT provides a bank of identical contiguous filters centered at kf_s/N, where f_s is the sampling frequency at the input to the DFT. The filter centered at frequency $k_o f_s/N$ has the frequency response:

$$\exp[-j\pi(k_o/N - f/f_s)(N-1)] \cdot \frac{\sin[\pi(k_o/N - f/f_s)N]}{\sin[\pi(k_o/N - f/f_s)]} \tag{1.13}$$

The magnitude of this response is sketched in Figure 1.25, where the frequency variable is in terms of bins. To verify (1.13), we apply a complex sinusoid:

$$x(n) = e^{j2\pi fn/f_s} \tag{1.14}$$

to the DFT and determine the response of the k_oth bin as the frequency, f, is varied.

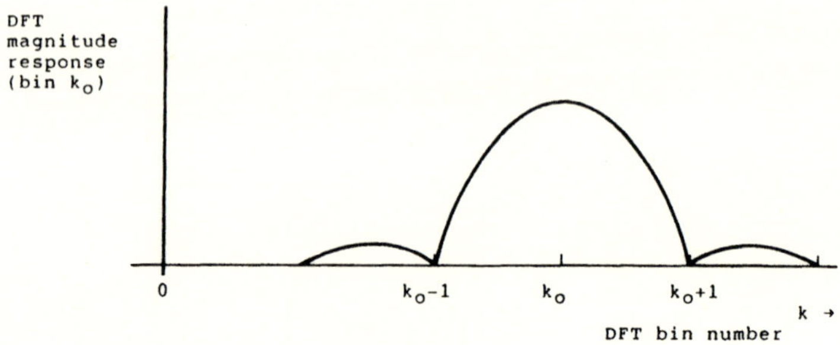

Figure 1.25 Magnitude response of the DFT filter for $k = k_o$.

With the input sequence given in (1.14), the output for this DFT bin, from (1.12), is

$$X(k_o) = \sum_{n=0}^{N-1} e^{j2\pi fn/f_s} e^{-j2\pi k_o n/N}$$

or, equivalently,

$$X(k_o) = \sum_{n=0}^{N-1} \exp[-j2\pi n(k_o/N - f/f_s)]$$

The summation of the geometric progression to M terms is

$$\sum_{n=0}^{M-1} r^n = \frac{r^M - 1}{r - 1} \tag{1.15}$$

The output of the k_oth DFT bin is found by substituting $r = \exp[-j2\pi(k_o/N - f/f_s)]$ into (1.15). The result is

$$X(k_o) = \frac{\exp[-j2\pi N(k_o/N - f/f_s)] - 1}{\exp[-j2\pi(k_o/N - f/f_s)] - 1}$$

After some manipulation, we obtain

$$X(k_o) = \frac{\exp[-j\pi N(k_o/N - f/f_s)] \sin[\pi(k_o/N - f/f_s)N]}{\exp[-j\pi(k_o/N - f/f_s)] \sin[\pi(k_o/N - f/f_s)]}$$

which is equivalent to (1.13).

From (1.13), we can see that each filter in the DFT repeats every N frequency bins and that there are $N - 1$ uniformly spaced zero crossings within an N-bin interval. The main lobe of a filter is located at the first null of its neighbor. For a complex sinusoid centered in a bin so that its frequency is $f = kf_s/N$, for some k $\varepsilon[0, N - 1]$, the output of the DFT consists only of the response of that bin. For a sinusoid not bin centered (that is, its frequency is not equal to any DFT frequency), (1.13) indicates that each bin has a nonzero output. This is called *spectral leakage*.

The response given in (1.13) results from applying a rectangular window function of length N to the data. Other window functions are used to obtain a compromise between a narrow main lobe (for high resolution) and low sidelobes (for low spectral leakage). High resolution provides accurate estimation of the frequency of a sinusoid and results in the separation of two sinusoids that are closely spaced in frequency. Low spectral leakage improves the detectability of a weak sinusoid in the presence of a strong sinusoid that is not bin centered.

1.3.1 Window Functions

An excellent tutorial paper [23] discusses many different functions. Examples of common window functions, $w(n)$, include the following.

Rectangular

$$w(n) = 1, \quad n = 0, \dots, N - 1$$

Hann (cosine squared or raised cosine)

$$w(n) = 0.5 - 0.5 \cos(2\pi n/N), \quad n = 0, \dots, N - 1$$

Hamming

$$w(n) = 0.54 - 0.46 \cos(2\pi n/N), \quad n = 0, \dots, N - 1$$

Gaussian

This window achieves the minimum time-bandwidth product of any window function. The time-domain function is obtained by truncating

$$w(n) = \exp\left[-\frac{1}{2}\left(\frac{\alpha n}{N/2}\right)^2\right]$$

to N terms. The mainlobe width can be adjusted with the parameter, α.

Dolph-Chebyschev

This window function minimizes the mainlobe width for a given sidelobe level. A window parameter allows the specification of the desired sidelobe level, which, for this window, is constant with frequency.

These window functions, and the squared magnitude of their frequency response, $W(k)$, are given in Figure 1.26. A comment is in order here concerning computations. In the past, when a multiplication required considerably more time to execute then an addition, a window with a simple frequency domain representation was desirable. For instance, the result of applying the Hann window in the frequency domain to the DFT output sequence, $X(k)$, is

$$Y_R(k) = -X_R(k-1)/4 + X_R(k)/2 - X_R(k+1)/4$$

$$Y_I(k) = -X_I(k-1)/4 + X_I(k)/2 - X_I(k+1)/4$$

where $X_R(k)$ and $X_I(k)$ denote "real" and "imaginary" parts of $X(k)$, respectively. Therefore, the Hann window, using fixed-point arithmetic, requires four real additions and six shifts per frequency bin. The corresponding operation in the time domain is

$$y_R(n) = w(n)x_R(n)$$

$$y_I(n) = w(n)x_I(n)$$

which requires two real multiplications per time sample. Neglecting the shifts, or multiplications, in the frequency domain implementation, the time-domain approach requires half the number of real operations (either a real addition or a real multiplication). Therefore, we would expect to execute the time-domain window operation in approximately half the time required for the frequency domain version with a modern DSP device (in which a multiplication and an addition are performed in one instruction cycle). A window function that does not possess a short-frequency domain expression, $W(k)$, therefore can be specified. That is, *any* window function can be implemented in the time domain with $2N$ real operations per length-N DFT calculation. Window functions, other than the Hann or Hamming windows, often require many operations per frequency bin when implemented as a convolution in the frequency domain.

We next consider some issues associated with the selection of a window function.

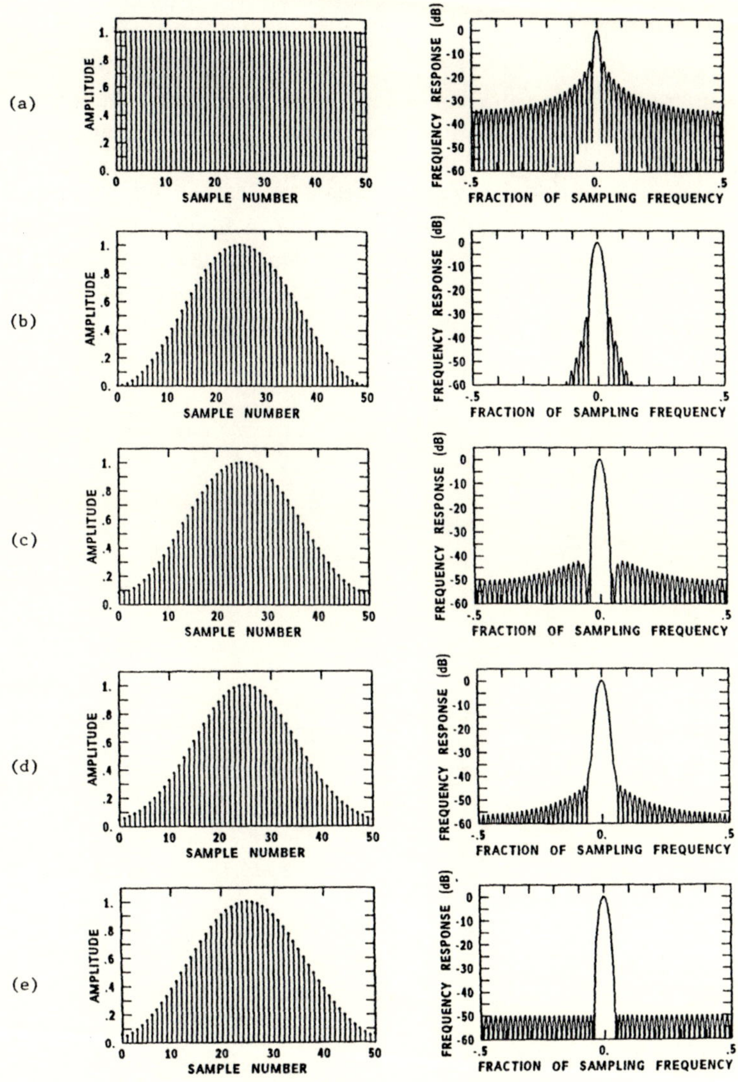

Figure 1.26 Common window functions (left) and their magnitude squared frequency responses (right): (a) Rectangular, (b) Hann or raised cosine, (c) Hamming, (d) Gaussian (α = 2.5), (e) Dolph-Chebyschev (α = 2.5, or a 50 dB sidelobe level).

Source: [17] S.L. Marple, Jr., *Digital Spectral Analysis with Applications,* Prentice-Hall, Englewood Cliffs, NJ, pp. 141–142. (© Marple 1987. Reprinted with permission.)

Sidelobe Level

The first consideration concerning the selection of a window function is the compromise between narrow mainlobe and low sidelobe level. The effect of sidelobe levels on the ability to detect a weak sinusoidal signal in the presence of a strong sinusoidal signal that is not bin centered is shown in Figure 1.27. The input consists of two sinusoids. The first has unity amplitude and a frequency that falls halfway between DFT bins 10 and 11. The second has an amplitude of 0.01 and a frequency that equals that of DFT bin 16. Figure 1.27 indicates the ability of various window functions to discern the presence of the second signal.

(a)

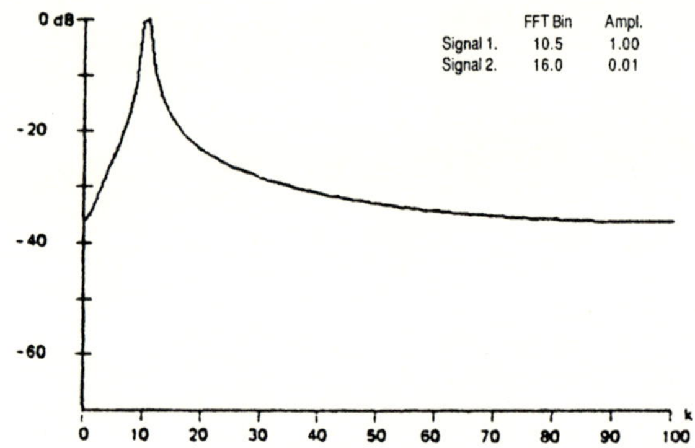

(b)

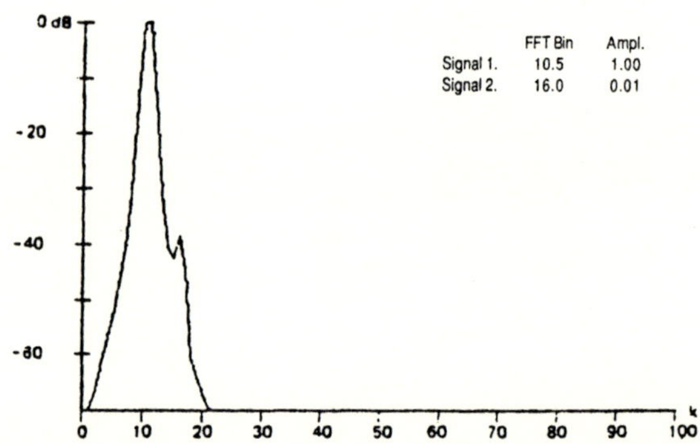

Figure 1.27 Response of various window functions to two sinusoidal signals: (a) rectangular, (b) Hann or raised cosine, (c) Hamming, (d) Gaussian (α = 3.0), (e) Dolph-Chebyschev (α = 4.0, or an 80 dB sidelobe level).

Source: [23] F.J. Harris, "On the Use of Windows for Harmonic Analysis with the Discrete Fourier Transform," *Proc. IEEE,* Vol. 66, No. 1, January 1978, pp. 51–83. (© 1978 by IEEE. Reprinted with permission.)

(c)

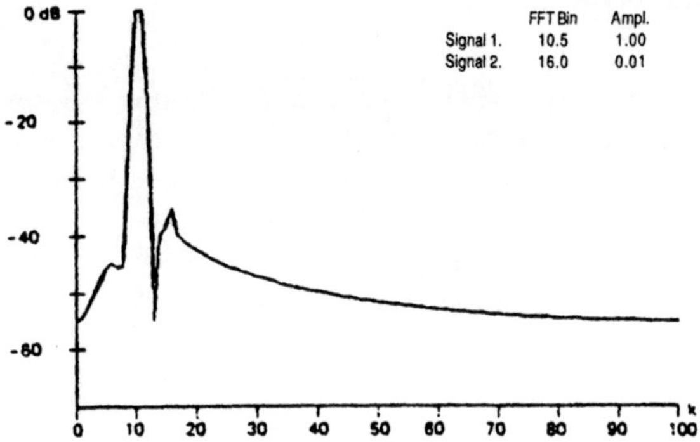

(d)

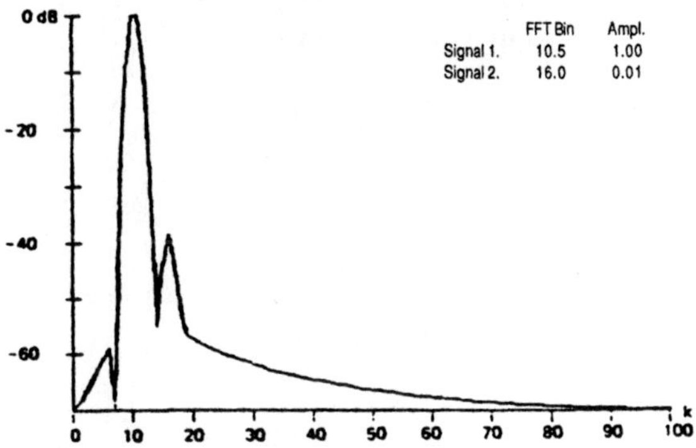

(e)

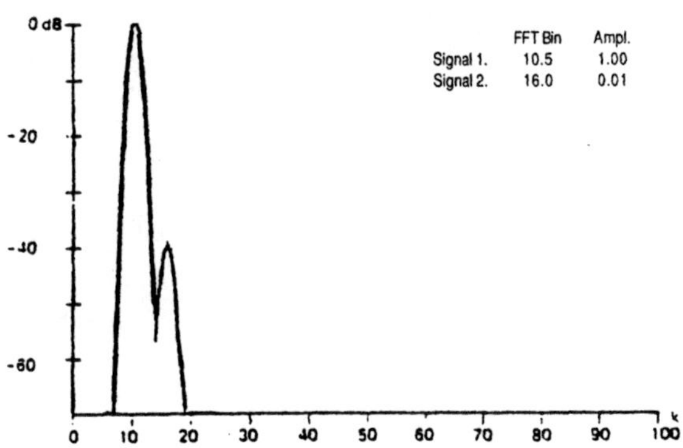

Figure 1.27 (cont'd)

Equivalent Noise Bandwidth

Another consideration in selecting a window function is the *equivalent noise bandwidth* (ENBW). This term is defined as the width of a rectangular filter with the same noise output as the window function and the same response at zero frequency. The concept is shown in Figure 1.28.

We consider a zero-mean noise sequence $g(n)$, with variance σ^2. The output noise power of a window function is given by

$$E\left\{\left|\sum_{n=0}^{N-1} w(n)g(n)\right|^2\right\} = \sum_{n=0}^{N-1}\sum_{m=0}^{N-1} w(n)w(m)E\{g(n)g^*(m)\}$$

With the assumption of white noise, $E\{g(n)g^*(m)\} = \sigma^2\delta(n - m)$, and the noise power at the output of the window function is

$$\sigma^2 \sum_{n=0}^{N-1} w^2(n) \tag{1.16}$$

We next calculate the noise power at the output of a rectangular filter with height $|W(0)|^2$, as shown in Figure 1.28. Because the noise power spectral density for the white noise sequence, $g(n)$, is σ^2/f_s, this is equal to

$$\sigma^2/f_s \int_{-\text{ENBW}/2}^{\text{ENBW}/2} |W(0)|^2\,\mathrm{d}f = \sigma^2/f_s|W(0)|^2\text{ENBW} = \sigma^2/f_s\left|\sum_{n=0}^{N-1} w(n)\right|^2 \text{ENBW} \tag{1.17}$$

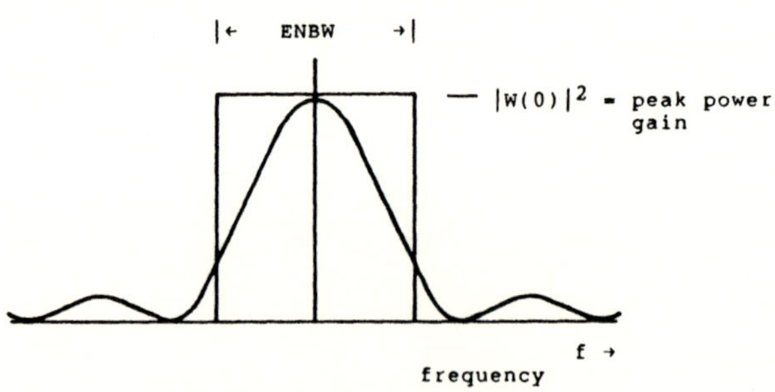

Figure 1.28 Equivalent noise bandwidth of a window function.

where the second equality in (1.17) follows from the definition of the DFT transform pair (1.12). We set the two noise power expressions (1.16 and 1.17) equal, solve for ENBW, and obtain

$$\text{ENBW} = \left[f_s \sum_{n=0}^{N-1} w^2(n)\right] \Bigg/ \left|\sum_{n=0}^{N-1} w(n)\right|^2$$

For the rectangular window, ENBW is equal to f_s/N, which is equal to the separation of the DFT bins in frequency. We can show, using the discrete Schwartz inequality, that no other window function has a smaller ENBW.

Scalloping Loss

The DFT provides a set of orthogonal functions $e^{j2\pi kn/N}$, $k = 0, \ldots, N-1$, separated in frequency by f_s/N. In the event that the frequency of a sinusoid in the data is not equal to one of these frequencies, we have the spectral leakage effect discussed earlier. In addition, there is a reduction in coherent gain if the frequency of the sinusoid differs in value from kf_s/N. This effect is termed *scalloping loss*. The origin of this term is obvious from Figure 1.29. The maximum scalloping loss occurs for a signal frequency $f = (k_o \pm 1/2)f_s/N$; that is, for a frequency that falls halfway between the frequencies of adjacent DFT bins. From (1.13), we see that the magnitude of the length-N DFT filter, with a rectangular window, for this pair of frequencies is

$$|\sin(\pi/2)|/|\sin(\pi/2N)| \approx 2N/\pi, \quad \text{for } N \text{ large}$$

After normalizing this value by the response at the center of the bin (which equals

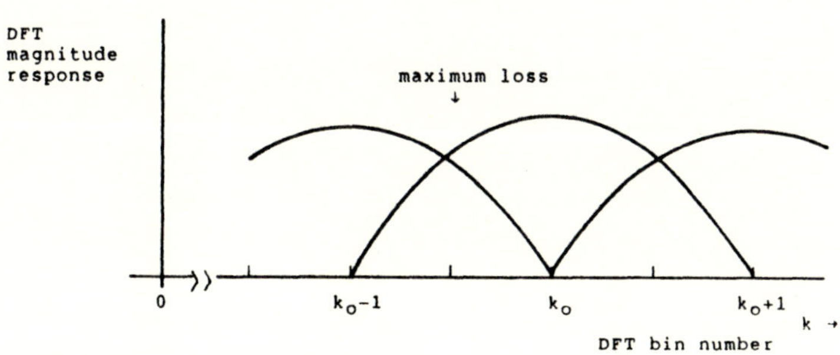

Figure 1.29 Loss in coherent gain (or scalloping loss) for a sinusoidal signal whose frequency is not equal to a DFT frequency.

N), the maximum scalloping loss for the rectangular window is $-20 \log_{10}(2/\pi)$, or about 3.9 dB.

This value and values for other window functions are presented in Table 1.3. As expected, windows other than the rectangular function have lower values of maximum scalloping loss by virtue of their broader main-lobe responses.

An average scalloping loss is defined by

$$SL_{av} = 10 \log_{10} \frac{|W(0)|^2}{\dfrac{N}{f_s} \displaystyle\int_{-f_s/2N}^{f_s/2N} |W(f)|^2 \, df}$$

and reflects the fact that in general we would expect the frequency of a sinusoid to be uniformly distributed relative to the center frequency of a bin. The average scalloping loss attains the largest value for the rectangular window.

Redundancy

We close this section with a few comments concerning the rate at which the DFT is calculated. To see that the DFT does not need to be calculated at the input sam-

Table 1.3
Window Function Parameters

Window	*Highest Sidelobe (db)*	*Sidelobe Fall-off (dB/octave)*	*ENBW*[1]	*Worst-Case Scalloping Loss (dB)*
Rectangular	−13	−6	1.00	3.92
Hann	−32	−18	1.50	1.42
Hamming	−43	−6	1.36	1.78
Gaussian				
$\alpha = 2.5$	−42	−6	1.39	1.69
$\alpha = 3.0$	−55	−6	1.64	1.25
Dolph-Chebyschev				
$\alpha = 2.5$	−50	0	1.39	1.70
$\alpha = 3.0$	−60	0	1.51	1.44
$\alpha = 3.5$	−70	0	1.62	1.25

[1]ENBW is scaled by N/f_s in this table to express this quantity in terms of DFT bin spacing.

pling frequency, we choose to view the forward transform (1.12) as having the form shown in Figure 1.30. That is, for the kth frequency bin, the DFT is equivalent to mixing the input sequence $x(n)$ with the frequency $2\pi k/N$, and then processing with a low-pass filter. This has the form of the bandpass filter structure shown in Figure 1.13. The impulse response, $h(n)$, of the two identical filters in Figure 1.30 is specified by

$$h(n) = \begin{cases} 1, & n = 0, \ldots, N-1 \\ 0, & \text{otherwise} \end{cases}$$

This type of filter is called a *comb filter*.

Because the DFT forms a set of narrow filters (the first zero crossings are separated by $2f_s/N$ Hz for the rectangular window), we can decimate the output by a large value, D. The rate at which the DFT is calculated, or, equivalently, the decimation that is applied to the bins of a length-N DFT, is related to the redundancy, R. This quantity is defined by $R = N/N_{\text{new}}$, where N_{new} is the number of new input samples used to calculate each DFT. This is shown in Figure 1.31. A redundancy of unity indicates that each DFT processes a new block of N samples. Thus, the decimation, D, in Figure 1.31 is equal to N. Similarly, with R equal to 2, each DFT processes a block of data consisting of $N/2$ new samples and $N/2$ samples from the previous DFT calculation. We shall see in Chapter 4 that a value for R greater than unity leads to an improvement in the detectability of narrowband signals embedded in noise.

1.3.2 Fast Fourier Transform Implementations

The development of the *fast Fourier transform* (FFT) algorithm made the calculation of the DFT of practical interest. Not only is the FFT used to provide rapid digital spectral estimation, but time-domain operations such as correlation and convolution, often are performed more efficiently by operating on the transforms of the sequences in question.

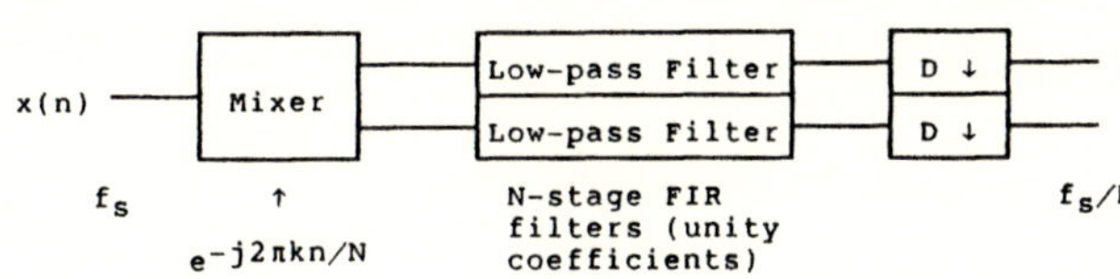

Figure 1.30 Interpretation of the DFT as a set of bandpass filters.

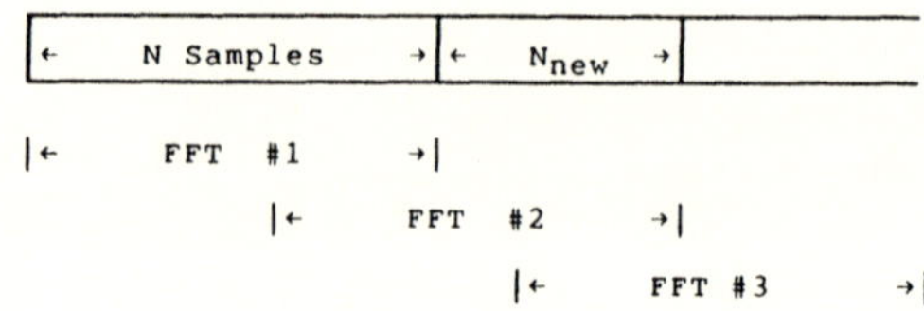

Figure 1.31 Partitioning of an input sequence for redundancy, or overlap, processing by a DFT.

Radix-2 Algorithm

Many references [3, 4] discuss in detail the derivation of the radix-2 FFT algorithm. We merely point out that an FFT algorithm can be obtained by starting with the definition of the DFT (1.12) and decomposing the input sequence into even-indexed and odd-indexed subsequences so that

$$X(k) = \sum_{n \text{ even}} x(n)\, e^{-j2\pi kn/N} + \sum_{n \text{ odd}} x(n)\, e^{-j2\pi kn/N} \tag{1.18}$$

Because n is even in the first sum, we effectively have a length-$N/2$ DFT. Similarly, by factoring the term $e^{-j2\pi k/N}$ from the second sum, this sum also corresponds to a length-$N/2$ DFT. Next, the two sums in (1.18) are decomposed as before. This procedure, when carried to its conclusion, yields the *decimation-in-time* (DIT) FFT structure. The equation for the basic DIT unit or "butterfly" is

$$X_{m+1}(p) = X_m(p) + W_N^r X_m(q), \quad m = 1, \ldots, \log_2 N$$

$$p, q \varepsilon [0, N-1]$$

$$X_{m+1}(q) = X_m(p) - W_N^r X_m(q), \quad W_N = e^{-j2\pi/N}$$

The index, m, refers to the FFT stage, whereas p and q are locations in the FFT data vector.

A second approach for deriving a radix-2 algorithm involves the decomposition of the output sequence, $X(k)$, into successively smaller subsequences. This decomposition leads to the *decimation-in-frequency* (DIF) FFT algorithm. The DIF butterfly is

$$X_{m+1}(p) = X_m(p) + X_m(q), \quad m = 1, \ldots, \log_2 N$$

$$p, q \epsilon [0, N-1]$$

$$X_{m+1}(q) = [X_m(p) - X_m(q)] W_N^r, \quad W_N = e^{-j2\pi/N}$$

An example of a length-16 FFT is shown in Figure 1.32. Here the DIT coefficients are shown to the left of the nodes, and the DIF coefficients appear to the right.

Radix-4 Algorithm

An FFT structure that requires fewer real operations (multiplications and additions) than the radix-2 FFT is one that operates on sets of four data samples. This radix-4 structure therefore is restricted to length-4^n DFTs. The basic algorithm, or butterfly, operates on the input sequence $\{x(0), x(1), x(2), x(3)\}$ to obtain the output sequence, $\{y(0), y(1), y(2), y(3)\}$, as shown:

$$\begin{bmatrix} y(0) \\ y(1) \\ y(2) \\ y(3) \end{bmatrix} = \begin{bmatrix} 1 & & & \\ & W_N^a & & \\ & & W_N^b & \\ & & & W_N^c \end{bmatrix} \begin{bmatrix} 1 & 0 & 1 & 0 \\ 0 & 1 & 0 & -j \\ 1 & 0 & -1 & 0 \\ 0 & 1 & 0 & j \end{bmatrix} \begin{bmatrix} 1 & 0 & 1 & 0 \\ 1 & 0 & -1 & 0 \\ 0 & 1 & 0 & 1 \\ 0 & 1 & 0 & -1 \end{bmatrix} \begin{bmatrix} x(0) \\ x(1) \\ x(2) \\ x(3) \end{bmatrix}$$

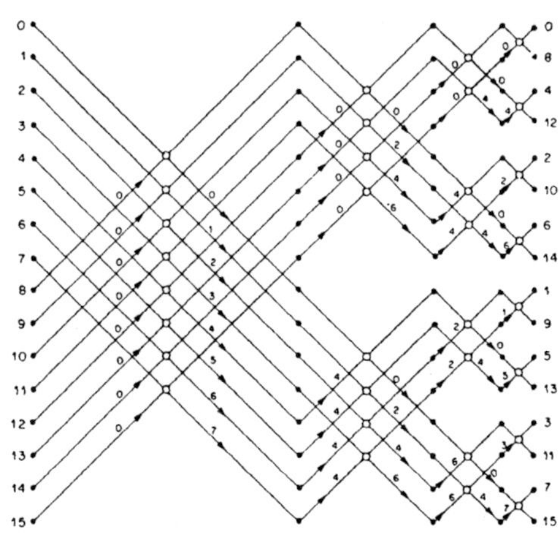

Figure 1.32 Length-16 FFT.
Source: [4] L.R. Rabiner and B. Gold, *Theory and Application of Digital Signal Processing*, Prentice-Hall, Englewood Cliffs, NJ, p. 574. (© 1975. Reprinted with permission.)

The first matrix is diagonal, and the product of the second and third matrices is equivalent to a length-4 DFT. Further reductions in computations can be realized by using higher radices.

Split-Radix Algorithm

An interesting FFT algorithm that requires fewer total real operations than either the radix-2 or the radix-4 structure, is the so-called split-radix algorithm [24, 25]. In this approach, a radix-2 structure is applied to the even-indexed terms of the input sequence, and a radix-4 structure is applied to the odd-indexed terms.

Computational Considerations

For the radix-2 FFT, each butterfly requires one complex multiplication and two complex additions. A complex multiplication can be implemented by using four real multiplications and two real additions, and a complex addition requires two real additions. Therefore, each radix-2 butterfly can be calculated by using 10 real operations. Because a length-N FFT contains $N/2 \log_2 N$ butterflies, the operations count is $5 N \log_2 N$ real operations.

The radix-4 butterfly requires eight complex additions and three complex multiplications, or 34 real operations. Multiplications by ± 1 or $\pm j$ are not counted in the calculation of the butterfly. A total of $N/4 \log_4 N$ butterflies are required for a length-N FFT and the operations count is $8.5 N \log_4 N$ real operations.

These expressions are evaluated for various values of N in Table 1.4. Note that these results correspond to the use of only one type of butterfly structure. In some butterfly stages, the term W_N^r is equal to ± 1 or $\pm j$. By using a special butterfly

Table 1.4
Real Operations Count for Radix-2, Radix-4 and Split-Radix FFTs

FFT Length	*Radix-2*	*Radix-4*	*Split-Radix*
16	320	272	236
64	1,920	1,632	1,452
256	10,240	8,704	7,852
1024	51,200	43,520	39,596
4096	245,760	208,896	191,148

to remove these trivial multiplications, the total number of computations is reduced. Radix-2 and radix-4 structures that utilize more than one butterfly structure to remove trivial multiplications are discussed by Burrus and Parks [9].

Also found in Table 1.4 are the numbers of real operations required by the split-radix FFT [25]. (Although not shown in this table, the split-radix algorithm is applicable to both length-2^n and length-4^n transforms.) As was the case with the fixed-radix algorithms, the results presented in this table are for a single butterfly structure. Additional savings in computations can be realized by using two or three different split-radix butterfly types to remove trivial multiplications.

As discussed previously in connection with filters and window functions, the time required to execute a multiplication until recently has been many times longer than the time for an addition. As a result, the goal has been to reduce the number of multiplications. One such approach, with respect to the FFT algorithm, concerns the implementation of a complex multiplication. Consider the multiplication of the FFT coefficient $W_N^k = a + jb$, and a complex number, $(c + jd)$, which requires four real multiplications and two real additions. As mentioned previously, with this form of the complex multiplication, a total of four real multiplications and six real additions are required to calculate a radix-2 butterfly.

The number of real multiplications required to calculate a complex multiplication can be reduced [9] by forming the intermediate variables:

$$x = a(c + d)$$

$$y = a + b$$

$$z = a - b$$

The real part of the product is then equal to $-yd + x$, and the imaginary part is given by $-zc + x$. Therefore, a complex multiplication has been realized with three real multiplications and five real additions. The FFT coefficients $(a + b)$ and $(a - b)$ could be stored in a table with the term, a. Then only three real additions are required. The radix-2 butterfly with this form of complex multiplication requires three real multiplications and nine real additions (or seven, if a table is used).

As mentioned previously, DSP devices can perform a multiplication and an addition in one instruction cycle. Therefore, it is desirable to minimize not only the total number of operations, but to equalize the number of real multiplications and real additions. A butterfly using a conventional complex multiplication requires at least six instruction cycles. The reduced multiplication approach requires at least nine instruction cycles without a table, or seven with a table. Thus, in the context of these DSP devices, use of the conventional complex multiplication is more efficient.

Heideman and Burrus [26] have calculated the minimum number of real multiplications required to compute a length-2^n DFT. In addition, they have developed algorithms that realize this minimum number. For a transform length of N greater than 32, however, the number of additions becomes prohibitive.

1.3.3 Real-Data Transforms

This section discusses the processing of real-input sequences using the standard FFT algorithm, which is designed to process complex data. In addition, the real-valued Hartley transform is presented.

Approach 1: Processing One Real-Data Sequence with the FFT

Frequently, we want to process sequences of real data using the FFT. Suppose we wish to perform a length-$2N$ FFT of a sequence of real numbers, $a(n)$. A straightforward approach involves forming the complex input sequence, $x(n)$, for the FFT as follows:

$$\begin{aligned} x_R(n) &= a(n) \\ & \qquad\qquad n = 0, \ldots, 2N - 1 \\ x_I(n) &= 0, \end{aligned}$$

where $x_R(n)$ and $x_I(n)$ are the real and imaginary parts of the input sequence, $x(n)$, respectively. From a computational point of view, a more attractive approach results from taking advantage of the complex conjugate symmetry of the transform, $A(k)$, of the real sequence. That is, $A(N - k) = A^*(k)$, for $k = 0, \ldots, N/2$.

The procedure is shown in Figure 1.33. The even-indexed terms of $a(n)$ are placed in $x_R(n)$, and the odd-indexed terms are placed in $x_I(n)$. A length-N FFT is calculated, and a final step recovers the desired transform of $a(n)$. Thus the resolution of a length-$2N$ transform is obtained by using a length-N FFT. Proof is found in Section 1.4.1.

This approach requires 5 $N \log_2 N$ real operations for the length-N FFT, and 3 complex additions, 1 complex multiplication, and 2 real multiplications (scaling by 1/2), or 14 real operations per output bin, for the last stage. The standard approach requires a length-$2N$ FFT. The ratio of operations of the efficient approach to the standard approach is

$$\text{Ratio} = \frac{5N \log_2 N + 14N}{10N \log_2(2N)} = 0.58 \text{ for } N = 1024$$

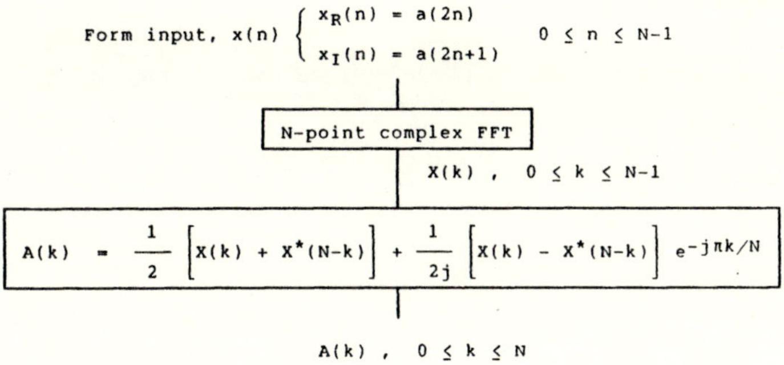

Figure 1.33 Processing one real-data sequence with a complex FFT (Approach 1).

Approach 2: Processing Two Real-Data Sequences with the FFT

A second technique provides computational savings if two length-$2N$ real sequences, $a(n)$ and $b(n)$, are to be processed. As is shown in Figure 1.34, the FFT input vector, $x(n)$, is formed by placing $a(n)$ in the real part of $x(n)$ and $b(n)$ in the imaginary part. A length-$2N$ FFT is calculated, followed by a recombination step to obtain the desired transforms, $A(k)$ and $B(k)$. Proof is provided in Section 1.4.2.

This approach requires $10\ N \log_2(2N)$ real operations for the length-$2N$ FFT and two complex additions plus four real multiplications, or eight real operations per output bin, for the last stage. The standard approach requires two length-$2N$ FFTs. The ratio of operations of the efficient approach to the standard approach is

$$\text{Ratio} = \frac{10N \log_2(2N) + 8N}{20N \log_2(2N)} = 0.54 \text{ for } N = 1024$$

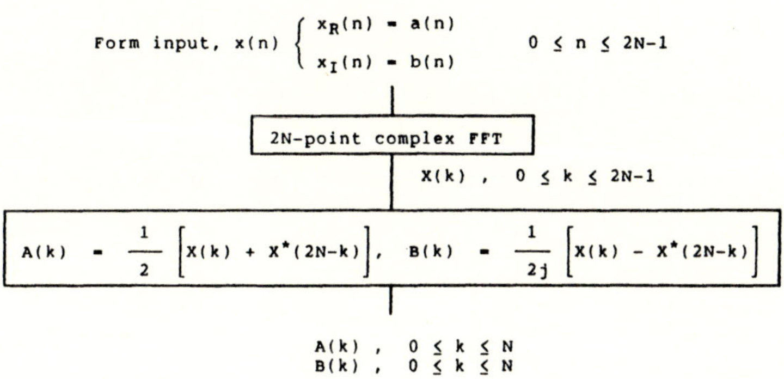

Figure 1.34 Processing two real-data sequences with a complex FFT (Approach 2).

Discrete Hartley Transform

The *discrete Hartley transform* (DHT) is defined by the transform pair:

$$H(k) = \sum_{n=0}^{N-1} x(n)\,[\cos(2\pi kn/N) + \sin(2\pi kn/N)], \quad k = 0, \ldots, N-1 \tag{1.19}$$

$$x(n) = \frac{1}{N}\sum_{n=0}^{N-1} H(k)\,[\cos(2\pi kn/N) + \sin(2\pi kn/N)], \quad n = 0, \ldots, N-1$$

where $x(n)$ is a real sequence. As can be seen from (1.19), the DHT uses real arithmetic. A set of *fast Hartley transforms* (FHTs) has been developed, including radix-2, radix-4, and split-radix structures [27]. Because these are real transforms, it has been conjectured in the literature that an FHT should require about half the computations of the corresponding FFT when processing real sequences.

This savings is not realized if Approach 1 is used to process a real-data sequence with an FFT. A length-N radix-2 FHT can be implemented with $5(N/2)\log_2 N$ real operations. Approach 1 requires $5(N/2)\log_2(N/2) + 14\,(N/2)$ real operations. For $N = 1024$, the FFT-based implementation represents only an 18 percent, rather than a 100 percent, increase in real operations relative to the FHT.

To see how the DHT and the DFT are related for a real input sequence, $x(n)$, it is instructive to express the DFT in the following form:

$$X(k) = \sum_{n=0}^{N-1} x(n)\cos(2\pi kn/N) - \mathrm{j}\sum_{n=0}^{N-1} x(n)\sin(2\pi kn/N)$$

We then can see that the output of the DHT, $H(k)$, is related to $X(k)$ by

$$H(k) = \mathrm{Re}[X(k)] - \mathrm{Im}[X(k)]$$

Using the relationship $\sin[2\pi(N-k)n/N] = -\sin(2\pi kn/N)$,

$$H(N-k) = \sum_{n=0}^{N-1} x(n)[\cos(2\pi kn/N) - \sin(2\pi kn/N)]$$

We then can obtain the following expressions relating the output of DFT to that of the DHT:

$$\mathrm{Re}[X(k)] = [H(k) + H(N-k)]/2$$

$$\mathrm{Im}[X(k)] = -[H(k) - H(N-k)]/2$$

1.4 APPENDICES

These appendices provide proofs for the two approaches given in Section 1.3.3.

1.4.1 Approach 1

This appendix verifies the expression for $A(k)$ in Figure 1.33. We are interested in calculating the length-$2N$ DFT of the real sequence $a(n)$. The DFT output, $A(k)$, can be written in this form:

$$A(k) = \sum_{\substack{n=0 \\ n \text{ even}}}^{2N-2} a(n)\, e^{-j2\pi kn/2N} + \sum_{\substack{n=1 \\ n \text{ odd}}}^{2N-1} a(n)\, e^{-j2\pi kn/2N}$$

By substituting $n = 2m$ in the first sum and $n = 2m + 1$ in the second, we obtain

$$A(k) = \sum_{m=0}^{N-1} a(2m)\, e^{-j2\pi km/N} + e^{-j\pi k/N} \sum_{m=0}^{N-1} a(2m + 1)\, e^{-j2\pi km/N} \qquad (1.20)$$

The two sums in (1.20) can be obtained with a length-N DFT. We form the sequence $x(n) = a(2n) + ja(2n + 1)$, $0 \leqslant n \leqslant N - 1$, as shown in Figure 1.33. The output of a length-N DFT is

$$X(k) = \sum_{n=0}^{N-1} a(2n)\, e^{-j2\pi kn/N} + j \sum_{n=0}^{N-1} a(2n + 1)\, e^{-j2\pi kn/N}$$

Using the identity, $e^{-j2\pi Nn/N} = 1$, $X^*(N - k)$ is equal to

$$X^*(N - k) = \sum_{n=0}^{N-1} a(2n)\, e^{-j2\pi kn/N} - j \sum_{n=0}^{N-1} a(2n + 1)\, e^{-j2\pi kn/N}$$

By combining these expressions for $X(k)$ and $X^*(N - k)$, we obtain

$$\sum_{n=0}^{N-1} a(2n)\, e^{-j2\pi kn/N} = (1/2)[X(k) + X^*(N - k)]$$

$$\sum_{n=0}^{N-1} a(2n + 1)\, e^{-j2\pi kn/N} = (1/2j)[X(k) - X^*(N - k)]$$

The desired result in Figure 1.33, which is valid for $0 \leqslant k \leqslant N$, is obtained by substituting these expressions into (1.20). In evaluating the expression in Figure

1.33 for $k = 0$ or for $k = N$, use was made of the fact that $X(N) = X(0)$. (End of proof.)

1.4.2 Approach 2

This appendix verifies the expressions for $A(k)$ and $B(k)$ in Figure 1.34. We are interested in calculating the length-$2N$ DFTs of the real sequences, $a(n)$ and $b(n)$. The DFT outputs, $A(k)$ and $B(k)$, are

$$A(k) = \sum_{n=0}^{2N-1} a(n)\, e^{-j2\pi kn/2N}$$
$$B(k) = \sum_{n=0}^{2N-1} b(n)\, e^{-j2\pi kn/2N} \qquad (1.21)$$

The two DFTs in (1.21) can be obtained with a length-$2N$ DFT. We form the sequence $x(n) = a(n) + jb(n)$, $0 \leqslant n \leqslant 2N - 1$, as shown in Figure 1.34. The output of a length-$2N$ DFT is

$$X(k) = \sum_{n=0}^{2N-1} a(n)\, e^{-j2\pi kn/2N} + j \sum_{n=0}^{2N-1} b(n)\, e^{-j2\pi kn/2N}$$

Using the identity, $e^{-j2\pi Nn/N} = 1$, $X^*(2N - k)$ is equal to

$$X^*(2N - k) = \sum_{n=0}^{2N-1} a(n)\, e^{-j2\pi kn/2N} - j \sum_{n=0}^{2N-1} b(n)\, e^{-j2\pi kn/2N}$$

By combining these expressions for $X(k)$ and $X^*(2N - k)$, we obtain the desired results (1.21):

$$A(k) = \sum_{n=0}^{2N-1} a(n)\, e^{-j2\pi kn/2N} = (1/2)[X(k) + X^*(2N - k)]$$
$$0 \leqslant k \leqslant N$$
$$B(k) = \sum_{n=0}^{2N-1} b(n)\, e^{-j2\pi kn/2N} = (1/2j)[X(k) - X^*(2N - k)]$$

In evaluating the expression in Figure 1.34 for $k = 0$ or for $k = N$, use was made of the fact that $X(2N) = X(0)$. (End of proof.)

REFERENCES

1. Rabiner, L.R., and C.M. Rader, eds., *Digital Signal Processing*, IEEE Press, New York, 1972.
2. *Selected Papers in Digital Signal Processing, II*, ed. Digital Signal Processing Committee, IEEE Acoustics, Speech, and Signal Processing Society, IEEE Press, New York, 1975.
3. Oppenheim, A.V., and R.W. Schafer, *Digital Signal Processing*, Prentice-Hall, Englewood Cliffs, NJ, 1975.
4. Rabiner, L.R., and B. Gold, *Theory and Application of Digital Signal Processing*, Prentice-Hall, Englewood Cliffs, NJ, 1975.
5. Tretter, S.A., *Introduction to Discrete-Time Signal Processing*, John Wiley & Sons, New York, 1976.
6. *Programs for Digital Signal Processing*, ed. Digital Signal Processing Committee, IEEE Acoustics, Speech, and Signal Processing Society, IEEE Press, New York, 1979.
7. Antoniou, A., *Digital Filters: Analysis and Design*, McGraw-Hill, New York, 1979.
8. Crochiere, R.E., and L.R. Rabiner, *Multirate Digital Signal Processing*, Prentice-Hall, Englewood Cliffs, NJ, 1983.
9. Burrus, C.S., and T.W. Parks, *DFT/FFT and Convolution Algorithms*, John Wiley & Sons, New York, 1985.
10. Kung, S.Y., H.J. Whitehouse, and T. Kailath, eds., *VLSI and Modern Signal Processing*, Prentice-Hall, Englewood Cliffs, NJ, 1985.
11. Kung, S.Y., R.E. Owen, and J.G. Nash, eds., *VLSI Signal Processing, II*, IEEE Press, New York, 1986.
12. Brodersen, R.W. and H.S. Moscovitz, *VLSI Signal Processing, III*, IEEE Press, New York, 1988.
13. DeFatta, D.J., J.G. Lucas, and W.S. Hodgkiss, *Digital Signal Processing: A System Design Approach*, John Wiley & Sons, New York, 1988.
14. "DSP56ADC16 16-Bit Sigma-Delta Analog-to-Digital Converter," Motorola, Inc., Data Package ADI1525, 1989.
15. Norsworthy, S.R., I.G. Post, and H.S. Fetterman, "A 14-bit 80 kHz Sigma-Delta A/D Converter: Modeling, Design, and Performance Evaluation," *IEEE J. Solid State Circuits*, Vol. 24, No. 2, April 1989, pp. 256–266.
16. Uchimura, K., T. Hayashi, and T. Kimura, "VSLI- A to D and D to A Converters with Multi-Stage Noise Shaping Modulators," *Proc. Int. Conf. Acoustics, Speech, and Signal Processing*, Tokyo, April 8–11, 1986, pp. 1545–1548.
17. Marple, S.L., Jr., *Digital Spectral Analysis with Applications*, Prentice-Hall, Englewood Cliffs, NJ, 1987.
18. Gray, G.A., and G.W. Zeoli, "Quantization and Saturation Noise Due to Analog-to-Digital Conversion," *IEEE Trans. on Aerospace and Electronics Systems*, Vol. AES-7, No. 1, January 1971, pp. 222–223.
19. Rabiner, L.R., J.H. McClellan, and T.W. Parks, "FIR Digital Filter Design Techniques Using Weighted Chebyshev Approximation," *Proc. IEEE*, Vol. 63, No. 4, April 1975, pp. 595–610.
20. McClellan, J.H., and T.W. Parks, "A Unified Approach to the Design of Optimum FIR Linear Phase Digital Filters," *IEEE Trans. Circuit Theory*, Vol. CT-20, No. 6, November 1973, pp. 697–701.
21. Vaidyanathan, P.P., and T.Q. Nguyen, "A 'TRICK' for the Design of FIR Half-Band Filters," *IEEE Trans. Circuits Syst.*, Vol. CAS-34, No. 3, March 1987, pp. 297–300.
22. Rabiner, L.R., *et al.*, "Some Comparisons between FIR and IIR Digital Filters," *Bell System Technical J.*, Vol. 53, No. 2, February 1974, pp. 305–331.

23. Harris, F.J., "On the Use of Windows for Harmonic Analysis with the Discrete Fourier Transform," *Proc. IEEE,* Vol. 66, No. 1, January 1978, pp. 51–83.
24. Duhamel, P., "Implementation of 'Split-Radix' FFT Algorithms for Complex, Real, and Real-Symmetric Data," *IEEE Trans. on Acoustics, Speech, and Signal Processing,* Vol. ASSP-34, No. 2, April 1986, pp. 285–295.
25. Sorensen, H.V., M.T. Heideman, and C.S. Burrus, "On Computing the Split-Radix FFT," *IEEE Trans. on Acoustics, Speech, and Signal Processing,* Vol. ASSP-34, No. 1, February 1986, pp. 152–156.
26. Heideman, M.T., and C.S. Burrus, "On the Number of Multiplications Necessary to Compute a Length-2^n DFT," *IEEE Trans. on Acoustics, Speech, and Signal Processing,* Vol. ASSP-34, No. 1, February 1986, pp. 91–95.
27. Sorensen, H.V., *et al.,* "On Computing the Discrete Hartley Transform," *IEEE Trans. on Acoustics, Speech, and Signal Processing,* Vol. ASSP-33, No. 5, October 1985, pp. 1231–1238.

Chapter 2
Conventional Beamforming

In this chapter, we look at arrays of acoustic sensors from the point of view of performing spatial filtering. That is, we wish to observe signals from a desired direction while attenuating the response of the array to signals from other directions. Time-domain and frequency-domain beamforming, array gain, array weighting, and quadrature beamforming are treated.

2.1 SENSOR ARRAY CHARACTERISTICS

A common form of array is the towed array, in which a set of sensors is deployed in a linear configuration some distance behind a platform. The advantages of this type of array include reduced response to the noise emanating from the platform and, by deploying long arrays, good performance at low frequencies. As we shall see, the longer the array, the lower is the frequency range for which acceptable performance can be achieved.

Modern towed arrays often consist of a set of subarrays, each designed for a specific frequency range. As is shown in Figure 2.1, a high-frequency subarray is realized by using the more closely spaced set of sensors. A low-frequency subarray can be obtained by processing the more separated set of sensors. Some of the sensors are shared by both subarrays. Sensor signals are often digitized in the array, with 8- to 12-bit resolution.

A disadvantage to this type of array is that a source can be located only relative to a conic angle. That is, without additional information, it is impossible to separate this conic angle into azimuth and elevation components. In particular, we have to contend with a left-right ambiguity. Without additional information, we cannot determine on which side of the array the source is located. Another shortcoming arises from the fact that the array is never perfectly horizontal nor straight. The resulting uncertainties in sensor positions degrade the beamforming calculation, as shown in Section 2.5.2.

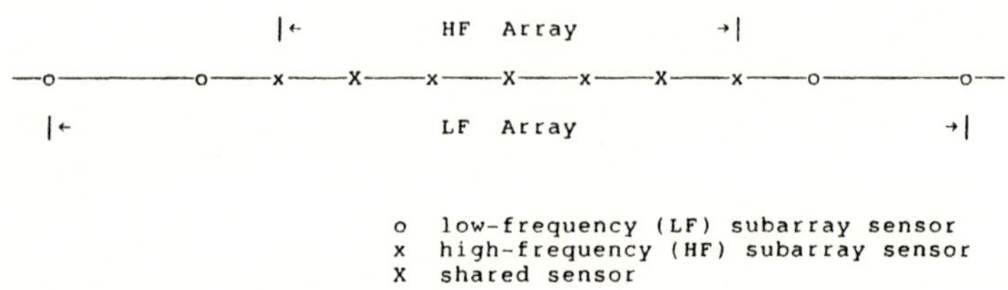

Figure 2.1 A towed array of sensors consisting of two subarrays.

A second class of array is the hull-mounted array of sensors. Such arrays are often two-dimensional, which allows the determination of the azimuth and elevation of the source. Because of restrictions on array size, these arrays in general do not have good low-frequency response. Also, because the sensors are mounted on the hull, they are exposed to a higher noise environment (both from noise due to the medium and from noise transmitted through the hull) than the sensors in the towed array.

Sonobuoys consist of one or more sensors suspended below the surface at some selectable depth. Signals often are transmitted back to an airborne platform for processing. These devices can be either active or passive, and either omnidirectional or directional. Some types deploy a vertical array of sensors that can reduce the response to surface and bottom noise.

Sonobuoys are relatively cheap and can be deployed close to a source. In practice, it is difficult, however, to coherently combine, or beamform, signals from sonobuoys. This is primarily due to uncertainties in the positions of the sonobuoy, as is shown in Section 2.5.2. The relative locations of the sonobuoys can be determined by deploying a number of active sonobuoys in the sonobuoy field. This is done at the cost of covertness. A sketch of deployments of acoustic sensors is shown in Figure 2.2.

2.2 BEAMFORMING AS SPATIAL FILTERING

We start with a three-dimensional coordinate system as shown in Figure 2.3. An arbitrary three-dimensional array of M sensors is assumed where the location of the ith sensor is denoted by the vector, $\mathbf{r}_i$. (A boldface lower-case letter denotes a vector. A matrix is indicated by a boldface upper-case letter.) We further assume that the array is in the far field for a given source so that the received signal can be modeled satisfactorily as a plane wave relative to the array. Chapter 9 discusses the situation where the dimensions of the array are an appreciable fraction of the distance to the source so that the far-field assumption is not valid.

The direction of this plane wave is defined by a unit-direction vector, $\mathbf{u}$, that is perpendicular to the wavefronts and that points to the source. As shown in Figure

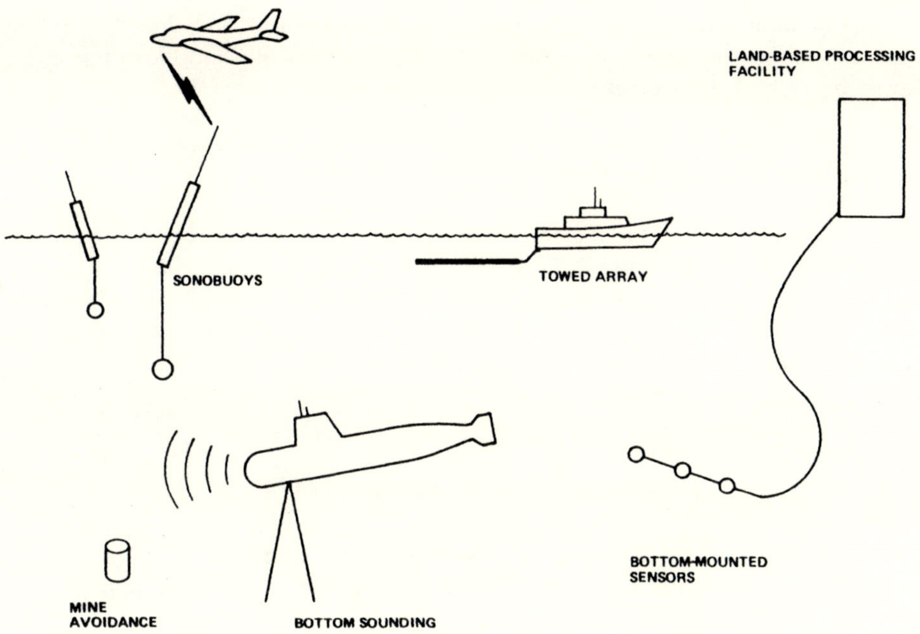

Figure 2.2 Deployments of acoustic sensors in an ocean.

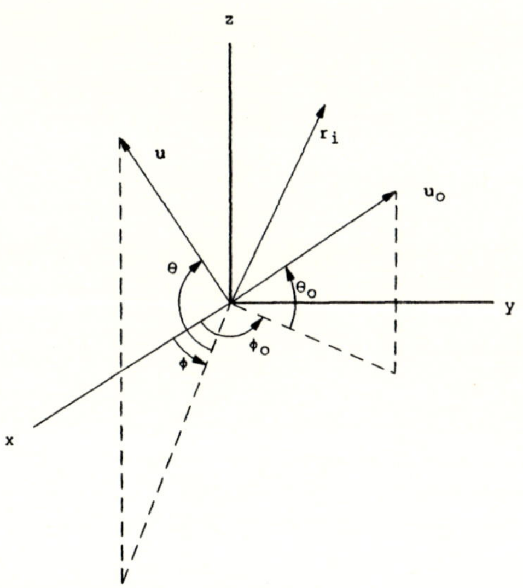

Figure 2.3 A three-dimensional coordinate system.

2.3, ϕ is the azimuth angle and θ is the elevation angle associated with $\mathbf{u}$. We will be interested in maximizing the response of our array of sensors in a direction defined by the unit-direction vector, $\mathbf{u}_0$, with corresponding azimuth and elevation angles denoted by ϕ_0 and θ_0, respectively.

2.2.1 Beamforming with a Three-Dimensional Array

We are interested in forming "a spatial filter." That is, we wish to emphasize signals from a certain direction while attenuating interferences from other directions. The result of simply summing the outputs of the M sensors in the array has the characteristics of such a spatial filter. To investigate the frequency and spatial properties of this summation, we assume that a complex sinusoid, $s(n) = e^{j2\pi fn/f_s}$ is propagating through the medium as a plane wave with a unit-direction vector, $\mathbf{u}$. The time-delayed signal at the ith sensor is

$$x_i(n) = \exp[j(2\pi fn/f_s + k\mathbf{r}_i \cdot \mathbf{u})], \quad i = 1, \ldots, M$$

where $k = 2\pi f/c$ is the wavenumber, f_s is the sampling frequency, c is the speed of sound in the medium, and $\lambda = c/f$ is the wavelength. Summing these signals yields

$$y(n) = \sum_{i=1}^{M} x_i(n) = e^{j2\pi fn/f_s} \sum_{i=1}^{M} e^{jk\mathbf{r}_i \cdot \mathbf{u}} \tag{2.1}$$

where the factor:

$$\sum_{i=1}^{M} e^{jk\mathbf{r}_i \cdot \mathbf{u}}$$

is called the *complex beam pattern*, $b(f, \mathbf{u})$. The dot product in (2.1) is equal to

$$\mathbf{r}_i \cdot \mathbf{u} = r_{xi} \cos\phi \cos\theta + r_{yi} \sin\phi \cos\theta + r_{zi} \sin\theta$$

2.2.2 Results for One-Dimensional Arrays

To gain insight into the effect of summing the outputs of the M sensors, we consider the special case of a linear array of sensors equally spaced along the x axis by a separation of d, as shown in Figure 2.4. In addition to being simple to analyze, use of this type of array is widespread. The location vector for the ith sensor is

$$\mathbf{r}_i = \{(i - 1)d, 0, 0\}, \quad i = 1, \ldots, M$$

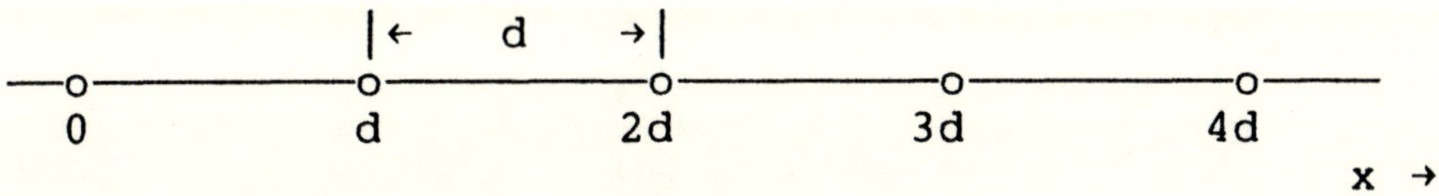

Figure 2.4 Geometry of a linear array of equally spaced sensors.

With this array geometry, (2.1) becomes

$$y(n) = e^{j2\pi fn/f_s} \sum_{i=1}^{M} \exp[jk(i-1)d\cos\phi\cos\theta] \tag{2.2}$$

The projection of the signal direction vector **u** on the x axis can be represented by a conic angle ψ, where $\sin\psi = \cos\phi\cos\theta$ as shown in Figure 2.5. The magnitude of the summation (2.2) can be written in closed form as

$$|b(f, \psi)| = \left| \frac{\sin[(\pi fMd \sin\psi)/c]}{\sin[(\pi fd \sin\psi)/c]} \right| \tag{2.3}$$

This is known as the *beam pattern* of the array. The beam power pattern, $|b(f, \psi)|^2$, is plotted in Figure 2.6, for $M = 10$ sensors and frequency $= c/(4d)$. For a given frequency, a plane wave with a conic angle ψ not equal to zero is attenuated. A plane wave with a conic angle of zero degrees arrives at all M sensors simultaneously, so the outputs of these sensors are added coherently or in phase. Therefore, the result of summing the outputs of the sensors in the array possesses the desired property of spatial selectivity.

It is instructive to examine the numerator of (2.3) in further detail. The first zero crossings occur for $(\pi fMd \sin\psi)/c = \pm\pi$, or $\psi = \pm\sin^{-1}[c/(fMd)]$. The region

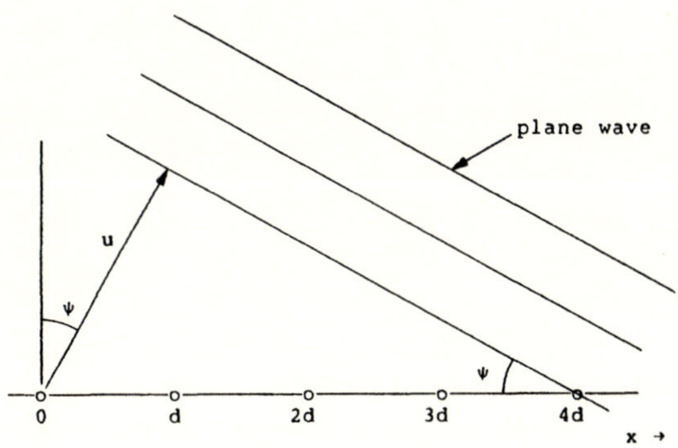

Figure 2.5 Projection of a plane wave on the linear array.

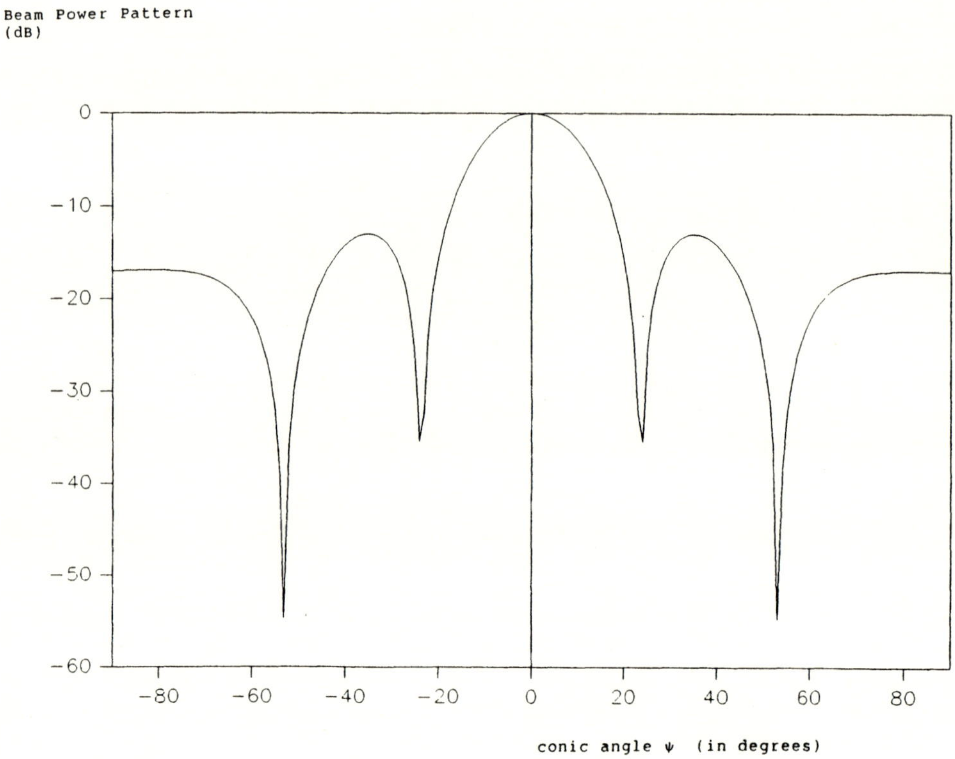

Figure 2.6 Normalized beam power pattern of a linear array consisting of 10 sensors without steering ($\psi_0 = 0°$).

of the beam pattern in Figure 2.6 between these zero crossings is called the *main lobe*. We see that the width (from zero crossing to zero crossing) in Figure 2.6 is inversely proportional to frequency, to the number of sensors, and to sensor spacing. The remaining region of the beam pattern is referred to as the *sidelobe region*.

Beam Steering

To this point, we have not delayed the sensor outputs before performing the summation in (2.1). Thus the response of the beamforming operation is maximum at "broadside" ($\psi = 0°$, or a plane wave parallel to the array). To add the sensor signals coherently (that is, in phase) for an arbitrary direction vector, $\mathbf{u}_0$, we calculate the projection of the vector, $(\mathbf{u} - \mathbf{u}_0)$, on the sensor position vectors, $\mathbf{r}_i$, $i = 1, \ldots, M$.

For the three-dimensional case, (2.1) becomes

$$y(n) = e^{j2\pi fn/f_s} \sum_{i=1}^{M} e^{jk\mathbf{r}_i \cdot (\mathbf{u} - \mathbf{u}_0)}$$

Again we are interested in the equally spaced linear array. The beam pattern is

$$|b(f, \psi)| = \left| \frac{\sin\{[\pi f M d\,(\sin\psi - \sin\psi_0)]/c\}}{\sin\{[\pi f d\,(\sin\psi - \sin\psi_0)]/c\}} \right| \tag{2.4}$$

This response is plotted in Figure 2.7, centered on $\sin\psi_0$. Parameter values are M = 10 sensors, $f = c/(4d)$, and $\psi_0 = -30°$. The process of correctly delaying each sensor output, to cause the responses to a plane wave signal with conic angle, ψ_0, to be added coherently, is termed *forming* or *steering* a beam at angle ψ_0. Such a beam often is called a *synchronous beam*.

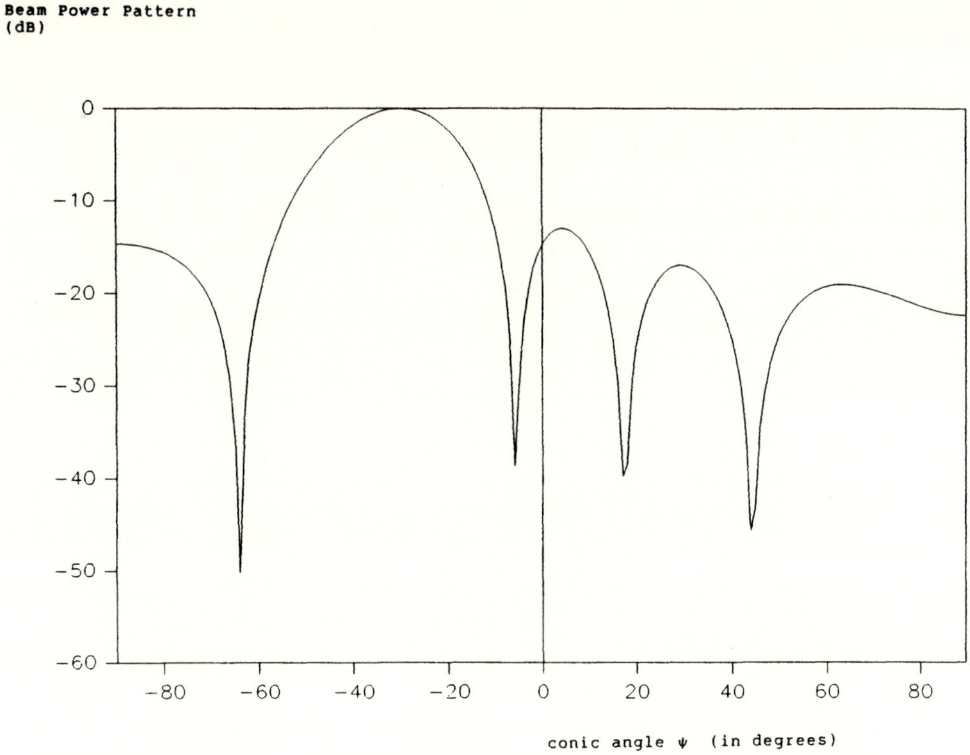

Figure 2.7 Normalized beam power pattern of a linear array consisting of 10 sensors and steered to $\psi_0 = -30°$.

Spatial Aliasing

The beam pattern (2.4) attains a maximum value of M when the denominator is zero. This occurs either for $\psi = \psi_0$ or for a value of $\psi \neq \psi_0$ such that the frequency satisfies

$$f = (mc/d)\,|\sin\psi - \sin\psi_0|^{-1} \tag{2.5}$$

where m is a positive integer. We consider first values of frequency such that $f < c/(2d)$. Because the maximum value of $|\sin\psi - \sin\psi_0|$ is 2, (2.5) has no solution and the maximum for (2.4) occurs only at angle ψ_0. If we let $f = c/(2d)$, or equivalently $d = \lambda/2$, then for the beams with angles $\psi_0 = \pm 90°$, (2.5) has a solution for $m = 1$ at $\psi = \mp 90°$. That is, a response peak (called a *grating* or *ambiguous lobe*), which is identical to the main lobe, is present in the pattern. (The beams at 90° and −90° are called *endfire beams*.) This phenomenon, known as *spatial aliasing,* occurs if the spatial domain is not sampled at a sufficiently fine spacing. We then are not able to unambiguously determine the conic angle, ψ, to the source. As the frequency is increased above $c/(2d)$, the location of the grating lobe moves in angle toward broadside, or zero degrees. This location can be determined from (2.5). For $f = c/d$, an additional grating lobe appears because a solution of (2.5) for $m = 2$ now is possible.

For beam steering angles, ψ_0, greater than zero degrees (less than zero degrees), a grating lobe appears at −90° (at 90°) at the frequency $f = (c/d)(1 + |\sin\psi_0|)^{-1}$. This frequency is given in Table 2.1, as a function of steering angle. As was the case with the endfire beams, as frequency is increased above this value, the angle of the grating lobe decreases in magnitude.

Table 2.1
Frequency at Which Grating Lobes Appear in the Beam Pattern

Beam Angle ψ_0 (degrees)	*Frequency*	*Spacing d*
±90	0.500 c/d	0.500 λ
±60	0.536 c/d	0.536 λ
±45	0.586 c/d	0.586 λ
±30	0.667 c/d	0.667 λ
±15	0.794 c/d	0.794 λ
0	1.000 c/d	1.000 λ

Beamwidth

The 3-dB beamwidth is often of interest in determining the number of beams required to provide adequate spatial coverage. This beamwidth is calculated next. Using the small angle approximation for the denominator of (2.4) (valid because we are interested in values of ψ close to ψ_0), we obtain

$$|b(f, \psi)| = \left|\frac{\sin \pi M x}{\pi x}\right|$$

where $x = (fd/c)(\sin\psi - \sin\psi_0)$. This beam pattern attains a maximum value of M for $\psi = \psi_0$. It is straightforward to verify that the beam pattern is approximately equal to $M/\sqrt{2}$ for $x = \pm 0.44/M$. Using the definition of x, we solve for the corresponding values of $\sin\psi$. These are $\sin\psi_{3\text{dB}} = \sin\psi_0 \pm 0.44c/(Mfd)$. The desired result, obtained by forming the difference of the two 3-dB angles, is

$$\Delta\psi_{3\text{dB}} \simeq \sin^{-1}[\sin\psi_0 + 0.44c/(Mfd)] - \sin^{-1}[\sin\psi_0 - 0.44c/(Mfd)] \quad (2.6)$$

We see from (2.6) that the 3-dB beamwidth in $\sin\psi$ space is not a function of the steering angle, ψ_0. It is equal to $0.88c/(Mfd)$, or $1.76/M$ for a sensor spacing of $d = \lambda/2$, for all beam directions.

Assuming an array of 50 sensors and half-wavelength spacing, the 3-dB beamwidth is 2.02° for the $\psi_0 = 0°$ beam, 2.85° for the $\psi_0 = 45°$ beam, and 4.04° for the $\psi_0 = 60°$ beam.

2.3 ARRAY GAIN

Up to this point, we have neglected noise at the sensors in our calculations. We now wish to determine the improvement in *signal-to-noise* (S/N) *ratio* due to beamforming. That is, we form the ratio of the S/N at the output of the beamformer to the S/N at a single sensor. The output, $x_i(n)$, from the ith sensor consists of a signal sequence, $s_i(n)$, and a noise sequence, $g_i(n)$, so that $x_i(n) = s_i(n) + g_i(n)$. The signal and noise sequences are zero-mean sample sequences of uncorrelated random processes. The output of the beamformer in direction $\mathbf{u}_0$ is

$$y(n) = \sum_{i=1}^{M} a_i x_i(n - f_s \mathbf{r}_i \cdot \mathbf{u}_0/c)$$

where we have included a weighting factor of a_i for each sensor. The steering delay is expressed in numbers of samples by multiplying by the sampling frequency, f_s.

A frequency-domain description of the output of the beamformer is useful in obtaining an expression for array gain and required in Section 2.6, which deals with frequency-domain beamforming. Calculating the DFT of the output of the beamformer yields

$$Y(k) = \sum_{n=0}^{N-1} y(n)\, \mathrm{e}^{-\mathrm{j}2\pi kn/N}$$

or equivalently,

$$Y(k) = \sum_{i=1}^{M} a_i \sum_{n=0}^{N-1} x_i(n - f_s \mathbf{r}_i \cdot \mathbf{u}_0/c)\, \mathrm{e}^{-\mathrm{j}2\pi kn/N} \tag{2.7}$$

Consistent with common usage, k in (2.7) denotes the frequency index of the DFT and not the wavenumber. That is, $k = Nf/f_s$.

2.3.1 DFT of a Delayed Sequence

We assume for simplicity that the steering delay, $f_s \mathbf{r}_i \cdot \mathbf{u}_0/c$, is equal to an integer number of samples, q_i, for each sensor. That is, the beam we are forming with unit-direction vector, $\mathbf{u}_0$, is synchronous so that the delay required at each sensor for the coherent addition of the signal is expressible as an integer number of samples. We note that the second sum in (2.7) is the DFT of the delayed output of the ith sensor. After substituting $m = n - q_i$, this DFT is equal to

$$\mathrm{DFT}[x_i(n - q_i)] = \sum_{m=-q_i}^{N-1-q_i} x_i(m) \exp[-\mathrm{j}2\pi k(m + q_i)/N]$$

$$= \mathrm{e}^{-\mathrm{j}2\pi k q_i/N} \sum_{m=-q_i}^{N-1-q_i} x_i(m)\, \mathrm{e}^{-\mathrm{j}2\pi km/N}$$

For $|q_i| \ll N$, this is approximately equal to

$$\mathrm{e}^{-\mathrm{j}2\pi k q_i/N} \sum_{m=0}^{N-1} x_i(m)\, \mathrm{e}^{-\mathrm{j}2\pi km/N} = \mathrm{DFT}[\hat{x}_i(n)] \tag{2.8}$$

which is simply $\mathrm{e}^{-\mathrm{j}2\pi k q_i/N} X_i(k)$. That is, the approximation is made that a phase factor, $\mathrm{e}^{-\mathrm{j}2\pi k q_i/N}$, multiplied by the DFT of a sensor output sequence is identical to the DFT of the delayed output sequence. The implementation in (2.8) is desirable because we can operate on length-N sequences that are obtained for the same time period for each sensor.

We illustrate the difference between the delayed sequence, $x_i(n - q_i)$, and the sequence, $\hat{x}_i(n)$, that corresponds to the product of the phase factor, $e^{-j2\pi k q_i/N}$, and the DFT of the undelayed sequence $x_i(n)$. The sequence, $\hat{x}_i(n)$, is obtained by forming the inverse DFT of (2.8):

$$\hat{x}_i(n) = \frac{1}{N}\sum_{k=0}^{N-1}\left[e^{-j2\pi k q_i/N}\sum_{m=0}^{N-1} x_i(m)\, e^{-j2\pi km/N}\right]e^{j2\pi kn/N}$$

This is equivalent to

$$\hat{x}_i(n) = \sum_{m=0}^{N-1} x_i(m)\left[\frac{1}{N}\sum_{k=0}^{N-1} e^{-j2\pi k(m+q_i-n)/N}\right]$$

The factor in brackets is equal to $\delta(m + rN + q_i - n)$, where r is an integer. (The function $\delta(n)$ is the discrete delta function introduced in Section 1.1.2.) Without loss of generality, we assume that q_i is positive. Then $\delta(m + rN + q_i - n)$ equals unity for

$$m = \begin{cases} -q_i + n, & \text{if} \quad q_i \le n \le N - 1 \quad (r = 0) \\ N - q_i + n, & \text{if} \quad 0 \le n \le q_i - 1 \quad (r = -1) \end{cases}$$

The approximate sequence, $\hat{x}_i(n)$, is compared to the correct time-delayed sequence, $x_i(n - q_i)$, for $0 \le n \le N - 1$, in Table 2.2. We thus have $N - q_i$ samples that are

Table 2.2

Comparison of the Effects of a Delay in the Time Domain, $x_i(n - q_i)$, and the Inverse DFT of a Phase Shift in the Frequency Domain, $\hat{x}_i(n)$ (q_i is assumed to be positive)

n	*Time Domain* $x_i(n - q_i)$	*Frequency Domain* $\hat{x}_i(n)$	
0	$x_i(-q_i)$	$x_i(N - q_i)$	Incorrect
1	$x_i(-q_i + 1)$	$x_i(N - q_i + 1)$	
$q_i - 1$	$x_i(-1)$	$x_i(N - 1)$	
q_i	$x_i(0)$	$x_i(0)$	Correct
$q_i + 1$	$x_i(1)$	$x_i(1)$	
$N - 1$	$x_i(N - q_i - 1)$	$x_i(N - q_i - 1)$	

correct and q_i samples that are incorrect. We will discuss in Section 2.6 how this discrepancy can be accommodated.

Again, the approximation that $\hat{x}_i(n)$ equals $x_i(n - q_i)$ is reasonably accurate if N, the length of the data sequences, is large compared to $|q_i|$. This is equivalent to the condition that the observation time, $T = N/f_s$ (the time interval over which data are collected for the DFT calculation), is much longer than the time required for sound to propagate across the array. To see this, we consider the linear array of equally spaced sensors in Figure 2.5. The required delays for steering a beam in direction $\sin\psi_0$ are $|q_i| = (f_s/c)(i - 1)d\ |\sin\psi_0|$, $i = 1, \ldots, M$, so that $|q_i|$ attains a maximum value for the endfire beams for which $|\sin\psi_0| = 1$. Because the largest value of the index i is M, $q_{\max} = (M - 1)f_s d/c$. If $N \gg q_{\max}$, then $\hat{x}_i(n) \simeq x_i(n - q_i)$ for all sensors in the array. We can write this inequality in the following form:

$$N/f_s \gg q_{\max}/f_s = (M - 1)d/c = L/c$$

where L is the length of the array. Therefore, the following conditions are equivalent: (1) N is much greater than the longest delay; and (2) the observation time, N/f_s, is much larger than the time, L/c, required for sound to propagate across the array.

Example 2.3.1

Assume the following parameter values:

N = 2048
f_s = 3000 Hz
L = 100 ft
c = 5000 ft/s

The observation time is equal to 0.68 s and the propagation time across the array is 0.02 s. Therefore, for this example, the approximation that $\hat{x}_i(n)$ equals $x_i(n - q_i)$ should be quite good for all sensors.

By substituting (2.8) for the DFT of the delayed output of the ith sensor into (2.7), we can obtain an expression for the output of the beamformer in the frequency domain. This is given by

$$Y(k) = \sum_{i=1}^{M} a_i\, e^{-j2\pi k q_i/N}\, X_i(k)$$

and can be expressed as the dot product $\mathbf{w}^+\mathbf{x}$, where $^+$ denotes the complex conjugate transpose. The ith components of these vectors are $w_i = a_i\, e^{j2\pi k q_i/N}$, $x_i = X_i(k)$, and the delay for the ith sensor is $q_i = f_s \mathbf{r}_i \cdot \mathbf{u}_0/c$.

2.3.2 General Array Gain Expression for Three-Dimensional Arrays and Arbitrary Noise Fields

We are interested in the power at the output of the beamformer in direction $\mathbf{u}_0$ and for the kth frequency. This is equal to

$$\text{Power} = E\{|Y(k)|^2\} = E\{(\mathbf{w}^+\mathbf{x})(\mathbf{w}^+\mathbf{x})^+\}$$

or, equivalently,

$$\text{Power} = \mathbf{w}^+E\{\mathbf{x}\,\mathbf{x}^+\}\mathbf{w} = \mathbf{w}^+\mathbf{R}\mathbf{w} \tag{2.9}$$

where $\mathbf{R}$ is called the *spatial correlation matrix* for the kth frequency. In its most general form, the (m, n)th element of this matrix is given by

$$R_{mn} = E\{[S_m(k) + G_m(k)][S_n^*(k) + G_n^*(k)]\}$$

where $S_n(k)$ and $G_n(k)$ are the DFTs of the signal and noise sequences at the nth sensor, respectively. Because we assume that the signal and noise sequences are not correlated, this reduces to

$$R_{mn} = E\{S_m(k)S_n^*(k)\} + E\{G_m(k)G_n^*(k)\}$$

This can be expressed in terms of a normalized signal spatial correlation matrix, $\mathbf{P}$, and a normalized noise spatial correlation matrix, $\mathbf{Q}$, so that

$$R_{mn} = S_s(k)P_{mn} + S_g(k)Q_{mn}$$

The quantities, $S_s(k)$ and $S_g(k)$, are the signal and noise power spectral densities, respectively. We assumed that these densities are equal at all sensors. Substituting this expression for the spatial correlation matrix into (2.9) gives the following equation for output power in direction $\mathbf{u}_0$:

$$\text{Power} = S_s\mathbf{w}^+\mathbf{P}\mathbf{w} + S_g\mathbf{w}^+\mathbf{Q}\mathbf{w}$$

where the frequency index, k, has been omitted. By defining the array gain as the ratio of the output S/N to the S/N at one sensor, we obtain

$$\text{Array Gain} = \frac{S_s\mathbf{w}^+\mathbf{P}\mathbf{w}}{S_g\mathbf{w}^+\mathbf{Q}\mathbf{w}}\cdot\frac{1}{S_s/S_g} = \frac{\mathbf{w}^+\mathbf{P}\mathbf{w}}{\mathbf{w}^+\mathbf{Q}\mathbf{w}}$$

We next assume that the signal, $s(n)$, can be represented by a plane wave. The signal sequence received at the ith sensor therefore is $s_i(n) = s(n + f_s\mathbf{r}_i \cdot \mathbf{u}/c)$, where again $\mathbf{u}$ is the unit-direction vector to the source. Assuming that $N \gg f_s\mathbf{r}_i \cdot \mathbf{u}/c$, it follows from Section 2.3.1 that the DFT of $s_i(n)$ is approximately equal to $\exp[j2\pi(kf_s/N)(\mathbf{r}_i \cdot \mathbf{u}/c)]\, S(k)$. Then the $\mathbf{P}$ matrix is just the outer product $\mathbf{pp}^+$, where the ith component of $\mathbf{p}$ is equal to $\exp[j2\pi(kf_s/N)(\mathbf{r}_i \cdot \mathbf{u}/c)]$.

With this model for the signal, the array gain expression is

$$\text{Array Gain} = \frac{\mathbf{w}^+\mathbf{pp}^+\mathbf{w}}{\mathbf{w}^+\mathbf{Qw}} = \frac{|\mathbf{w}^+\mathbf{p}|^2}{\mathbf{w}^+\mathbf{Qw}} \tag{2.10}$$

The array gain is a function of steering direction, contained in the weight vector $\mathbf{w}$ and frequency. For a beam steered in the direction of the signal, so that $\mathbf{u} = \mathbf{u}_0$, the dot product in the numerator of (2.10) simplifies to

$$\mathbf{w}^+\mathbf{p} = \sum_{i=1}^{M} a_i$$

To this point, no assumptions have been made concerning the properties of the noise field. As we discuss in Section 3.1.2, isotropic noise is uncorrelated for a sensor spacing of d equal to $\lambda/2$. Under these conditions, the $\mathbf{Q}$ matrix in (2.10) is equal to the identity matrix $\mathbf{I}$. In addition, with the assumption that the beam is steered in the direction of the signal, the expression for array gain simplifies to

$$\text{Array Gain} = \frac{\left(\sum_{i=1}^{M} a_i\right)^2}{\sum_{i=1}^{M} a_i^2}$$

and, finally, if unity weights are used, the array gain is simply equal to M. That is, under the assumptions of uncorrelated noise, a plane wave arrival, and unity weights, the array gain in the direction of the signal is just equal to the number of sensors in the array.

2.3.3 Array Gain for a Linear Array in Isotropic Noise

We consider again the special case of a linear array consisting of M equally spaced sensors, with unity weighting. Then the ith component of the weight vector, $\mathbf{w}$, is

$$w_i = \exp[j2\pi(kf_s/N)(i-1)d\,(\sin\psi_0)/c], \quad i = 1, \ldots, M$$

or, by using the frequency relationship for the DFT where $k/N = f/f_s$,

$$w_i = \exp[j2\pi f(i-1)d(\sin\psi_0)/c]$$

As will be discussed in Section 3.1.2, the (m, n)th element of the spatial correlation matrix for isotropic noise is

$$Q_{mn} = \frac{\sin[2\pi f(m-n)d/c]}{2\pi f(m-n)d/c}$$

Substituting this expression into the denominator of (2.10) and rewriting in terms of a double sum yields

$$\sum_{m=1}^{M}\sum_{n=1}^{M} \exp[j2\pi f(n-m)d(\sin\psi_0)/c]\,\frac{\sin[2\pi f(n-m)d/c]}{2\pi f(n-m)d/c}$$

After some manipulation, the following expression for array gain can be derived:

$$\text{Array Gain} = \frac{M}{1 + \frac{2}{M}\sum_{i=1}^{M-1}(M-i)\cos[2\pi fid(\sin\psi_0)/c]\,\frac{\sin(2\pi fid/c)}{2\pi fid/c}} \tag{2.11}$$

(See, for instance, Burdic [1, eq. (11–39)].) This expression is valid for the special case of isotropic noise, a linear array of equally spaced sensors, and a steering angle equal to the signal arrival angle. For an isotropic noise field and a plane wave, the array gain in dB also is known as the *directivity index* (DI). We can see from (2.11) that the array gain for $\lambda/2$ spacing is equal to M, as we concluded earlier. For low frequencies, the noise correlation increases and consequently the array gain decreases. Array gain is plotted for an array of 10 sensors and steering angles of 0°, 45°, and 90° in Figure 2.8(a), (b), and (c), respectively. The frequency axis is normalized by $f_u = c/(2d)$, the upper design frequency of the array.

2.4 ARRAY WEIGHTING

To this point, beam patterns have been developed based on weights with unity magnitude. We compare beam patterns and DFT filter responses to determine the effects of array weighting. In Section 2.2.2, we developed the complex beam pattern for a linear array of equally spaced sensors. This was done by considering a complex sinusoid, $e^{j2\pi fn/f_s}$, propagating as a plane wave with direction $\sin\psi$. The complex

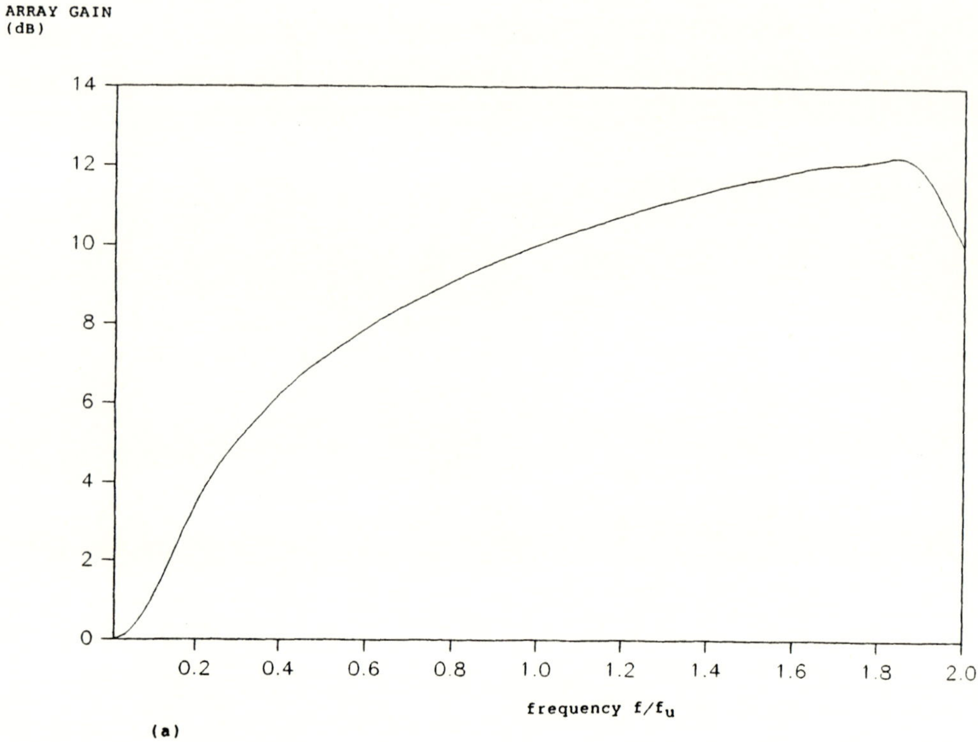

Figure 2.8 Array gain for a linear array consisting of 10 sensors and steered to (a) $\psi_0 = 0°$, (b) $\psi_0 = 45°$, and (c) $\psi_0 = 90°$.

beam pattern with a weighting function of a_i, is generated by varying ψ, and equal to

$$b(f, \psi) = \sum_{i=1}^{M} a_i \exp[j2\pi f(i-1)d(\sin\psi - \sin\psi_0)/c]$$

or, equivalently,

$$b(f, \psi) = \sum_{i=0}^{M-1} a_{i+1} \exp[j2\pi fid(\sin\psi - \sin\psi_0)/c]$$

The magnitude of this pattern is given in (2.4) for the case of unity weights.

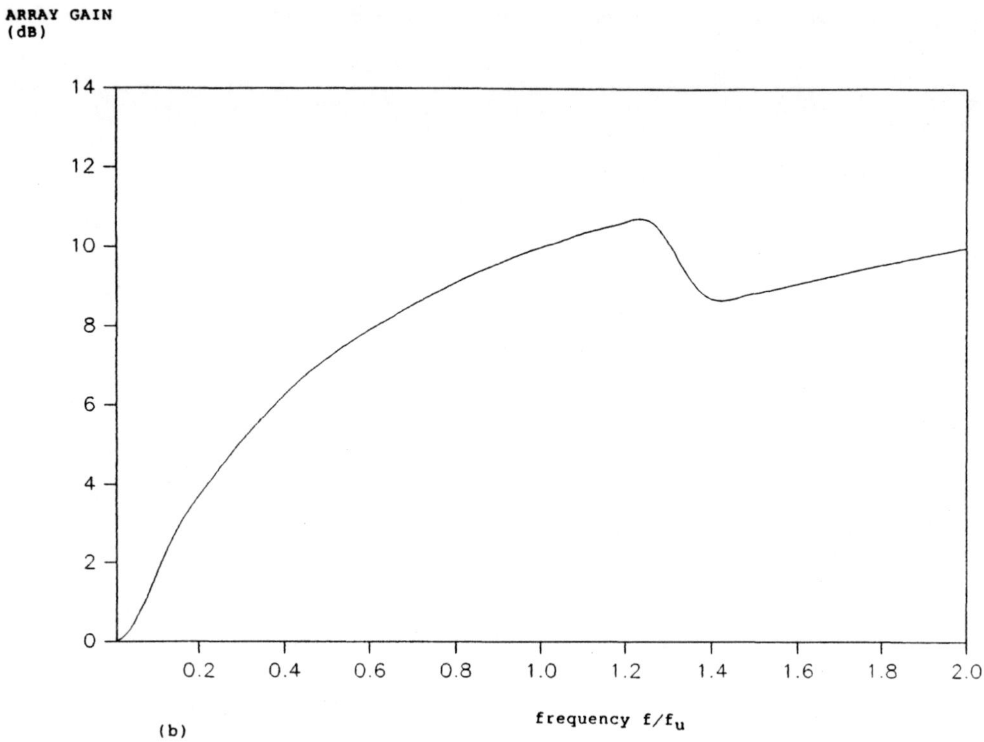

Figure 2.8(b) continued.

The response of the k_0th DFT filter was determined in Section 1.3.1 by applying the complex sinusoid, $e^{j2\pi fn/f_s}$, and varying the frequency, f. This filter response is

$$X(k_0) = \sum_{n=0}^{N-1} w(n) \exp[2\pi n(f/f_s - k_0/N)]$$

where $w(n)$ is the DFT window function. By associating the quantities:

$$f/f_s \leftrightarrow fd\,(\sin\psi)/c$$

$$k_0/N \leftrightarrow fd\,(\sin\psi_0)/c$$

it is apparent that the DFT filter response, $X(k_0)$, and the complex beam pattern,

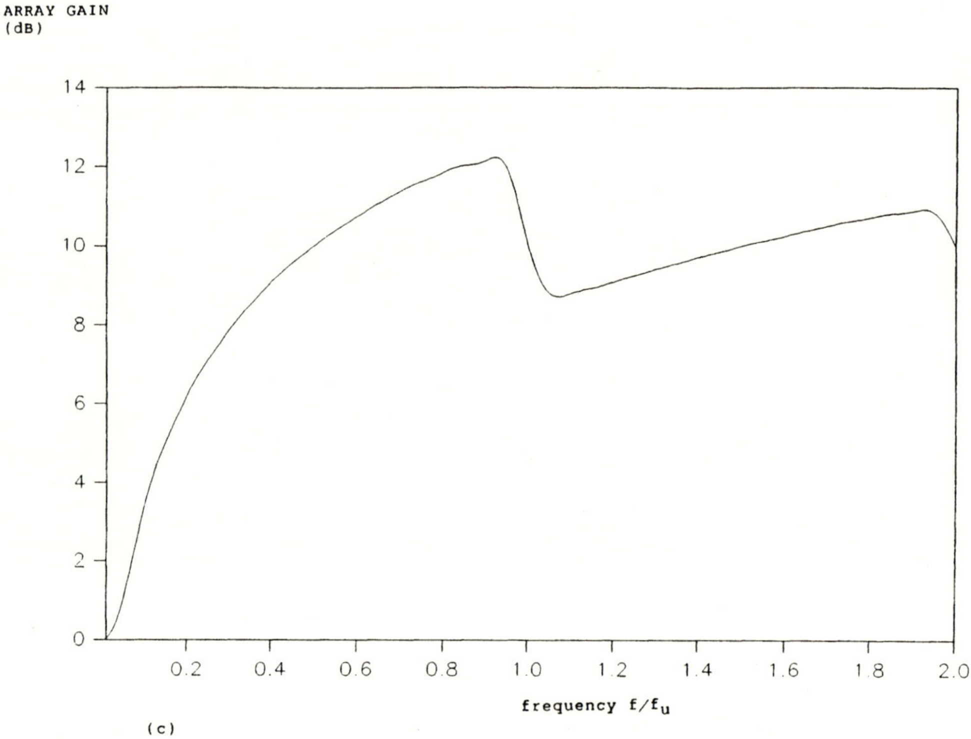

Figure 2.8(c) continued.

$b(f, \psi)$, have identical forms. There is a direct analogy, therefore, between aperture weighting for spatial analysis and the use of window functions with the DFT for spectral analysis (Section 1.3.1). A reduction in sidelobe levels in $\sin\psi$ space is accompanied by the same broadening of the mainlobe response as occurs when sidelobe levels in the frequency domain are reduced.

Common array weighting, or shading, functions include unity (uniform or rectangular), raised cosine (Hann), Taylor, and Dolph-Chebyschev. Figure 2.9 illustrates the effect of unity weighting for a linear array of 51 sensors, in a three-dimensional plot [2]. Evident in this plot is the left-right ambiguity discussed earlier. The beam is steered to $\psi_0 < 90°$, and a main lobe also is present at $\psi_0 > 90°$. The main lobe width can be seen to increase with decreasing frequency as predicted by (2.6). Also associated with unity weighting are high sidelobe levels and a narrow main lobe. The effect of Dolph-Chebyschev weighting on the same array is shown in Figure 2.10 [2]. The reduction in sidelobes is dramatic. The parameter, α, associated with this weighting function is set to 4, resulting in a uniform sidelobe level of -80 dB.

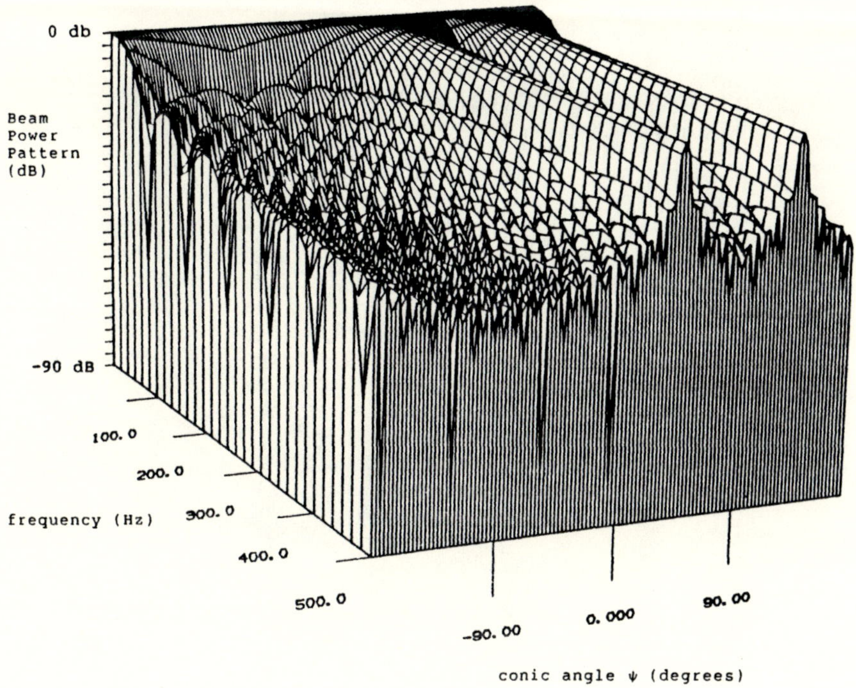

Figure 2.9 Beam power pattern for a linear array consisting of 51 sensors with unity weighting.

The main lobe, however, is considerably broader than with unity weighting. We therefore would expect to be able to detect a weak signal in the presence of a strong signal, separated by several beamwidths, with the Dolph-Chebyschev weighting. Unity weighting would be preferable if we were interested in separating, or resolving, two closely spaced signals of similar power.

2.5 TIME-DOMAIN BEAMFORMING

A simple beam-forming structure is the time-domain–delay-sum beamformer, as shown in Figure 2.11. To form a beam in a desired direction, the output of each sensor is delayed by an integer multiple of the sample spacing $\tau = 1/f_s$. The structure in this figure is repeated N_B times, where N_B is the number of beams desired. Associated with each beam direction is a set of integers, k_i, $i = 1, \ldots, M$. After the delay operation, the sensor signals are multiplied by a set of real weights, a_i, and summed. If the same set of weights is to be used for all beam directions, as often is the case, the weights are applied directly at the sensor output, before the delay-sum operation is used to form the various beams.

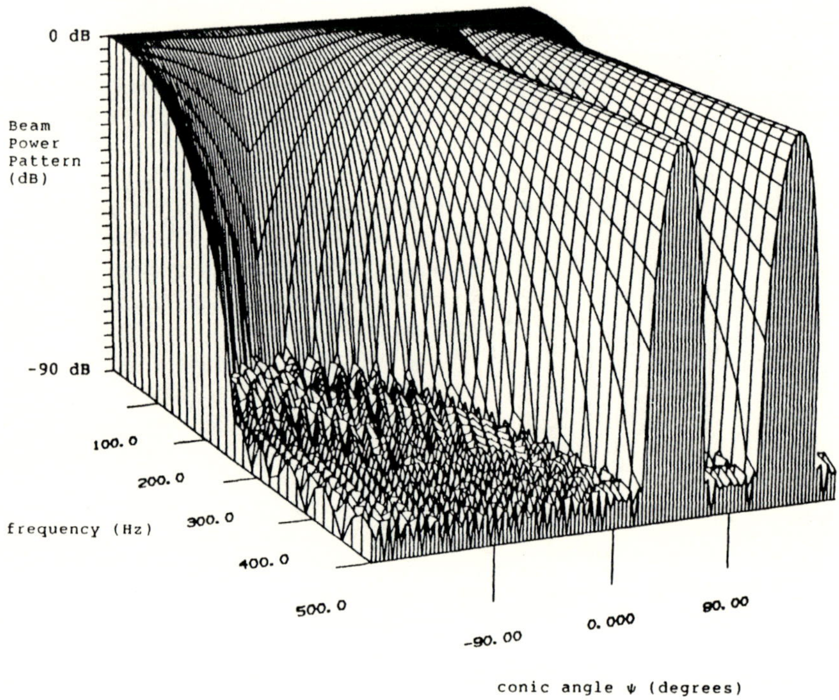

Figure 2.10 Beam power pattern for a linear array consisting of 51 sensors with Dolph-Chebyschev weighting.

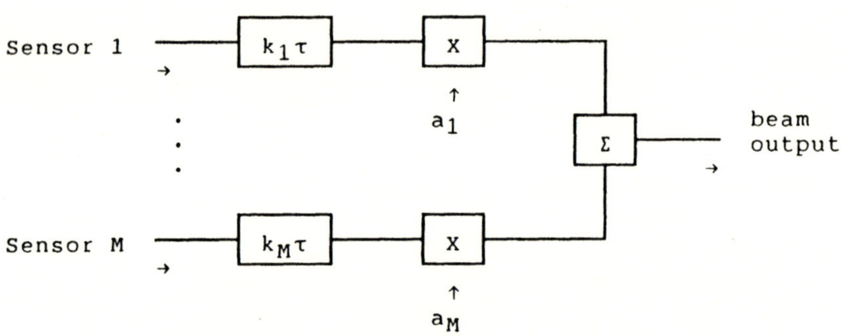

Figure 2.11 Time-domain–delay-sum beamformer.

We are interested in forming a beam in direction ψ_0 for a linear array of equally spaced sensors. The delays required to bring the output of the M sensors into time alignment are equal to $d(i-1)(\sin\psi_0)/c$, $i = 1, \ldots, M$, from Figure 2.5. Of course, we are forced to implement delays that are integer multiples of the sample spacing, τ. Therefore, the beam directions that we can realize are those for which $d(\sin\psi_0)/c = m\tau$, or

$$\sin\psi_0 = mc/(df_s), \quad m = 0, \pm 1, \pm 2, \pm 3, \ldots \tag{2.12}$$

We next determine the number of beams required to adequately cover three-dimensional space. This calculation is performed at $f = f_u = c/(2d)$, the upper design frequency of the array. Because beamwidth is inversely proportional to frequency, adequate spatial coverage at this frequency guarantees adequate coverage at lower frequencies as well.

A standard requirement is that the loss in power response must be less than 3 dB as the angle of the source changes between the *maximum response axes* (MRA) of adjacent beams. This, termed the *scalloping loss,* is analogous to the scalloping loss associated with the DFT (Section 1.3.1). From (2.6), the 3-dB width is equal to 0.88 $c/(Mfd)$, in $\sin\psi$ space. At the upper frequency of interest, this width is equal to $1.76/M$.

The number of beams required to span the region $-1 \leq \sin\psi \leq 1$ then is equal to $2/(1.76/M) + 1$, or $1.14M + 1$. A value of $M + 1$ beams is frequently specified, although the maximum scalloping loss exceeds 3 dB, because, for arbitrary values of M, the endfire beams are members of the set of equally spaced beams that are formed. The MRA of the mth beam in $\sin\psi$ space is equal to

$$\text{MRA}(m) = -1 + 2(m-1)/M, \quad m = 1, \ldots, M+1$$

These beam directions are shown in Figure 2.12 for $M = 14$. A beam is located at broadside only if the number of sensors is even. The two endfire beams (±90°) are formed for both M even and M odd.

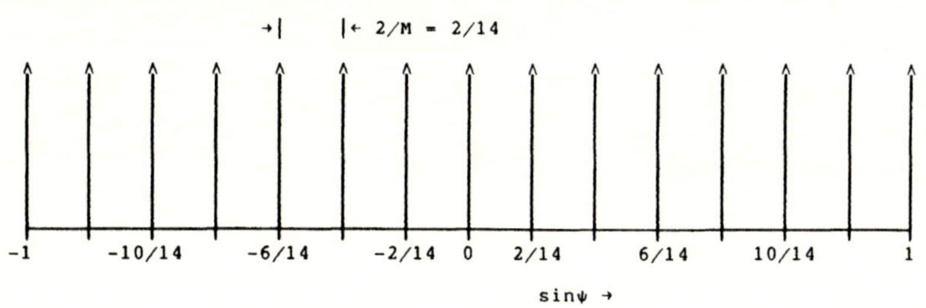

Figure 2.12 Beam directions for a linear array consisting of 14 sensors.

Now that we have an appreciation for the number of beams required, we again consider the synchronous beam directions that can be achieved for a given sensor sampling frequency, f_s. Referring to (2.12), the separation of these synchronous beams is $c/(df_s)$ in $\sin\psi$ space. At the upper design frequency of the array, the sensor spacing is equal to $c/(2f_u)$ so these beams are separated by $2f_u/f_s$.

Because the sampling frequency is typically about 2.5 times the highest frequency in the data, the separation of the beams is 0.8 in $\sin\psi$ space. That is, we are able to form only three synchronous beams for which the sensor signals can be correctly delayed and summed, as shown in Figure 2.13. Forming a beam in another direction results in a degraded beam pattern. The extent of this degradation is discussed in Section 2.5.2.

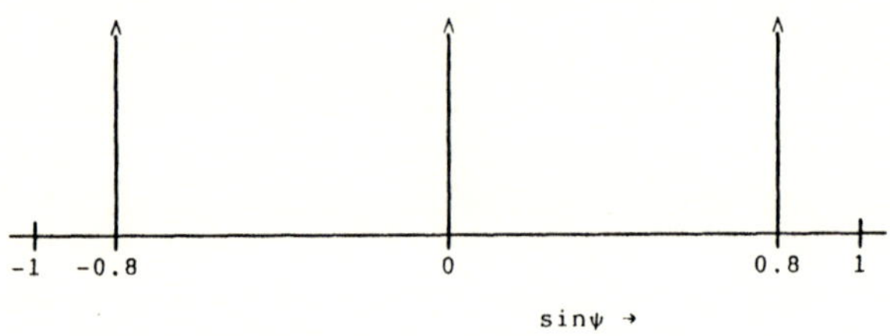

Figure 2.13 Directions of synchronous beams without interpolation.

To create more closely spaced synchronous beams, it apparently is necessary to increase the sampling frequency relative to the highest frequency in the data. That is, by oversampling the sensor outputs, closely spaced time samples then would be available, allowing more accurate realization of the required delays.

This can be achieved by sampling the analog signals from the array sensors at the required frequency for accurate beamforming. This requires the use of high-speed A/D converters, frequently located in the array itself, which may not be feasible. A second approach involves interpolating the sensor signals in the digital domain on the platform. The sampling frequency is increased, beamforming is performed, and the beam outputs are decimated to the original frequency, f_s. This approach is described next and discussed in [3 and 4].

2.5.1 Interpolation Beamformers

We denote the spacing between samples, after interpolation, by $\hat{\tau}$, and the corresponding sampling frequency by $\hat{f}_s$. Thus a sampling frequency of f_s satisfies the Nyquist criterion for the data, whereas $\hat{f}_s$ is required for reasonably accurate beamforming.

An implementation of such an interpolation beamformer is shown in Figure 2.14. Here the sensor signals are interpolated by a factor of D. The approach described in Section 1.2.3 is used, where the sequence is zero padded and low-pass filtered. A set of desired delays now is approximated more accurately by specifying integer multiples of the smaller sampling time, $\hat{\tau}$. Again, a different set of integers, k_i, is specified for each beam direction. The weighting and summation are performed at the original sampling frequency of f_s rather than at $\hat{f}_s$.

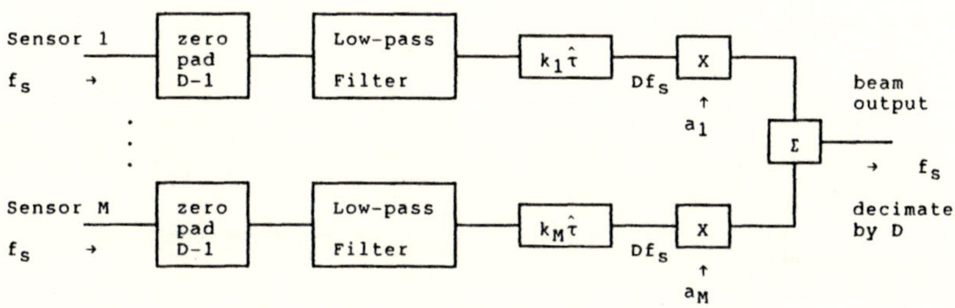

Figure 2.14 Interpolation filter before beamforming.

A variation of the structure given in Figure 2.14 that is more efficient if the number of desired beams is smaller than the number of sensors is shown in Figure 2.15. Because filtering and beamforming are linear operations, we can interchange the order in which they are executed. As shown in this figure, the sensor signals are zero padded, delayed, weighted, and then summed. The interpolation filter follows the summation and is calculated at the original sampling frequency of f_s. The equivalence of these two approaches is illustrated in Figure 2.16 and discussed next.

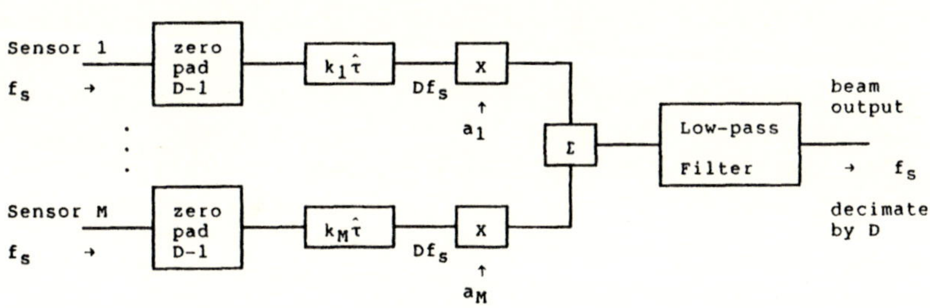

Figure 2.15 Interpolation filter after beamforming.

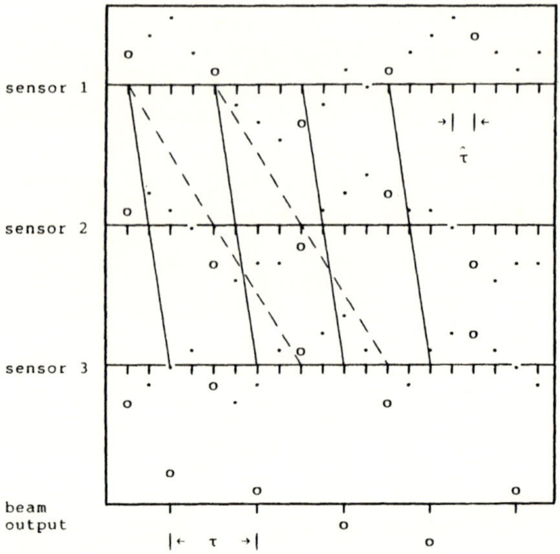

Figure 2.16 Equivalence of two interpolation beamformers (circles indicate sensor output samples at f_s, dots indicate interpolated samples for $\hat{f}_s = 4f_s$): (a) interpolation filter before beamforming; (b) interpolation filter after beamforming.

Source: After: [4] R.G. Pridham and R.A. Mucci, "A Novel Approach to Digital Beamforming," *J. Acoustical Society of America*, Vol. 63, No. 2, February 1978, pp. 425–434.

Interpolation-Filter-Before-Beamforming Structure

The output of each of the M sensors has been interpolated by D. The solid diagonal lines in Figure 2.16(a) indicate the samples that are summed to form the first synchronous beam from broadside. This beam direction is $\sin\psi = c/(d\hat{f}_s) = c/(dDf_s)$ and corresponds to $m = 1$ in (2.12). The broken diagonal lines indicate the formation of the first synchronous beam from broadside using the original (noninterpolated) sensor samples. This direction is $\sin\psi = c/(df_s)$.

Although each beam calculation utilizes the interpolated data, the beam summations are computed at the output frequency, f_s. Note that, unless a very large value of D is selected, beams still must be formed in directions that do not correspond to synchronous beams. This is required to obtain adequate spatial coverage (that is, scalloping loss that is less than 3 dB). The beam pattern degradation is reduced, of course, by using interpolated sensor data rather than the original data.

Interpolation-Filter-After-Beamforming Structure

Here the beam summations in Figure 2.15 are formed at the frequency Df_s, as indicated by the plot labeled *summer output* in Figure 2.16(b). On average, however,

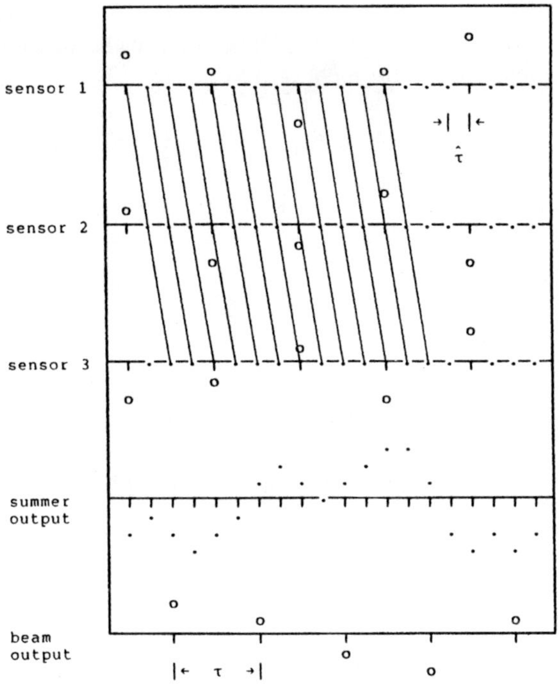

Figure 2.16(b) continued.

only $1/D$ of the sensors have nonzero values at any given time. Assuming a hardware structure that does not perform arithmetic with zero-valued inputs, the addition rate for a beam is equal to M/D additions per beam summation times Df_s beam summations per second, or Mf_s. This is the same rate as for the interpolation-filter-before-beamforming structure. The beam output is calculated at f_s, as shown in Figure 2.16(b).

Computational Requirements

Consider an array of M sensors sampled at f_s. We wish to form N_B beams, using sensor data that have been interpolated by a factor of D. A FIR filter consisting of L stages is used in the interpolation process.

The computations required to implement the interpolation-filter-before-beamforming structure (Figure 2.14) consist of L multiplications and L additions for the interpolation function for each sensor at f_s. The beamforming function requires M multiplications and M additions for each beam, also executed at f_s. The total is

$$Mf_s\,(L + N_B) \text{ multiplications/s}$$

$$Mf_s\,(L + N_B) \text{ additions/s}$$

In the event that the same set of weights is used for each beam direction, the sensor outputs are multiplied by these weights before the beamforming function is implemented. The resulting multiplication rate is

$$Mf_s\,(L + 1) \text{ multiplications/s}$$

The addition rate does not change. For the interpolation-filter-after-beamforming structure (Figure 2.15), the beamforming function requires M multiplications and M additions at f_s for each beam. The interpolation function is implemented with L multiplications and L additions at f_s, again for each beam. The requirements are

$$N_B f_s\,(L + M) \text{ multiplications/s}$$

$$N_B f_s\,(L + M) \text{ additions/s}$$

With the same set of weights for all beams, the multiplication rate becomes

$$f_s\,(M + LN_B) \text{ multiplications/s}$$

Example 2.5.1

Consider an array of 30 sensors, sampled at 2000 Hz, for which $N_B = 15$ beams are desired. An interpolation factor of $D = 4$ is assumed to be sufficient to provide beams of acceptable quality, and a FIR filter of length 100 is used. The computational requirements are shown in Table 2.3 for the two interpolation beamformers.

Table 2.3
Comparison of Computations for Two Types of Interpolation Beamformers

	Computations	
Beamformer Type	*Multiplications/s*	*Additions/s*
Interpolation-Filter-Before Beamforming		
Same Weights	$6.1 \cdot 10^6$	$6.9 \cdot 10^6$
Separate Weights	$6.9 \cdot 10^6$	$6.9 \cdot 10^6$
Interpolation-Filter-After Beamforming		
Same Weights	$3.1 \cdot 10^6$	$3.9 \cdot 10^6$
Separate Weights	$3.9 \cdot 10^6$	$3.9 \cdot 10^6$

2.5.2 Tolerance Analysis

We mentioned that differences between realizable delays and theoretical delays in the beamforming process can result in degraded beam patterns. This section considers this type of phase uncertainty, as well as uncertainties in sensor position and in sensor amplitude.

A general expression for the normalized complex beam pattern is

$$b(f, \mathbf{u}) = \frac{1}{M} \sum_{i=1}^{M} (a_i + \delta a_i) \exp\{j[2\pi f(\mathbf{r}_{ti} \cdot \mathbf{u} - \mathbf{r}_{bi} \cdot \mathbf{u}_0)/c + \delta\gamma_i]\}$$

where

δa_i = amplitude error in the *i*th sensor channel,
$\mathbf{r}_{ti}$ = true location of the *i*th sensor,
$\mathbf{r}_{bi}$ = assumed location of the *i*th sensor,
$\delta\gamma_i$ = phase error in the *i*th sensor channel.

These amplitude and phase errors may be due to variations in the characteristics of the *M* sensors, the analog amplifiers, and the antialiasing filters. We calculate the steering delays based on an assumed set of sensor locations. These assumed sensor locations can be expressed in terms of the true locations by $\mathbf{r}_{bi} = \mathbf{r}_{ti} + \boldsymbol{\alpha}_i$, where $\boldsymbol{\alpha}_i$ is a vector of position errors for the *i*th sensor. The complex beam pattern is then

$$b(f, \mathbf{u}) = \frac{1}{M} \sum_{i=1}^{M} (a_i + \delta a_i) \exp(j\{2\pi f[\mathbf{r}_{ti} \cdot (\mathbf{u} - \mathbf{u}_0) - \boldsymbol{\alpha}_i \cdot \mathbf{u}_0]/c + \delta\gamma_i\})$$

The beam power pattern, which again is equal to the squared magnitude of the complex beam pattern, is given by

$$\begin{aligned} |b(f, \mathbf{u})|^2 = \frac{1}{M^2} \sum_{m=1}^{M} \sum_{n=1}^{M} & (a_m a_n + a_m \delta a_n + a_n \delta a_m + \delta a_m \delta a_n) \\ & \cdot \exp[j(\delta\gamma_n - \delta\gamma_m)] \exp\{j[2\pi f(\boldsymbol{\alpha}_m - \boldsymbol{\alpha}_n) \cdot \mathbf{u}_0/c]\} \\ & \cdot \exp\{j[2\pi f(\mathbf{r}_{tn} - \mathbf{r}_{tm}) \cdot (\mathbf{u} - \mathbf{u}_0)/c]\} \end{aligned}$$

The amplitude errors, δa_i, phase errors, $\delta\gamma_i$, and the components of the position errors, $\boldsymbol{\alpha}_i$, are assumed to be samples of independent, zero-mean random processes. Each type of error is independent from sensor to sensor and the *i*th amplitude error is independent of the *i*th amplitude. These assumptions are used in the following derivation of (2.15), the expected value of the beam power pattern.

We are interested in the expected value of the beam power pattern as a function of these various uncertainties. We consider first the case where $m = n$, so that the phase errors cancel. Then

$$E\{|b(f, \mathbf{u})|^2\} = \frac{1}{M^2} \sum_{m=1}^{M} a_m^2 + \frac{1}{M} \sigma_a^2 \qquad (m = n) \tag{2.13}$$

where σ_a^2 is the variance of the amplitude error. Next, we consider the case where $m \neq n$. The expected value of the beam power pattern is

$$E\{|b(f, \mathbf{u})|^2\} = \frac{1}{M^2} |E\{e^{j\delta\gamma}\}|^2 \, |E\{e^{j2\pi f \boldsymbol{\alpha}\cdot\mathbf{u}_0/c}\}|^2 \cdot \sum_{m=1}^{M} \sum_{n=1}^{M} a_m a_n \exp\{j[2\pi f(\mathbf{r}_{tn} - \mathbf{r}_{tm}) \cdot (\mathbf{u} - \mathbf{u}_0)/c]\} \qquad (m \neq n) \tag{2.14}$$

The total expression for the expected value of the beam power pattern is found by adding the contributions from (2.13) and (2.14):

$$\begin{aligned} E\{|b(f, \mathbf{u})|^2\} = {} & \frac{1}{M^2} |E\{e^{j\delta\gamma}\}|^2 \, |E\{e^{j2\pi f \boldsymbol{\alpha}\cdot\mathbf{u}_0/c}\}|^2 \\ & \cdot \sum_{m=1}^{M} \sum_{n=1}^{M} a_m a_n \exp\{j[2\pi f(\mathbf{r}_{tn} - \mathbf{r}_{tm}) \cdot (\mathbf{u} - \mathbf{u}_0)/c]\} \\ & + \frac{1}{M^2} (1 - |E\{e^{j\delta\gamma}\}|^2 \, |E\{e^{j2\pi f \boldsymbol{\alpha}\cdot\mathbf{u}_0/c}\}|^2) \sum_{i=1}^{M} a_i^2 + \frac{1}{M} \sigma_a^2 \end{aligned}$$

Using the fact that the unperturbed beam power pattern is

$$|b_0\,(f, \mathbf{u})|^2 = \frac{1}{M^2} \sum_{m=1}^{M} \sum_{n=1}^{M} a_m a_n \exp\{j[2\pi f(\mathbf{r}_{tn} - \mathbf{r}_{tm}) \cdot (\mathbf{u} - \mathbf{u}_0)/c]\}$$

we can then obtain the desired result:

$$\begin{aligned} E\{|b(f, \mathbf{u})|^2\} = {} & |E\{e^{j\delta\gamma}\}|^2 \, E\{e^{j2\pi f \boldsymbol{\alpha}\cdot\mathbf{u}_0/c}\}|^2 \, |b_0(f, \mathbf{u})|^2 \\ & + \frac{1}{M^2} (1 - |E\{e^{j\delta\gamma}\}|^2 \, |E\{e^{j2\pi f \boldsymbol{\alpha}\cdot\mathbf{u}_0/c}\}|^2) \sum_{i=1}^{M} a_i^2 + \frac{1}{M} \sigma_a^2 \end{aligned} \tag{2.15}$$

If we assume no uncertainty in the sensor positions and consider a one-dimensional array, (2.15) reduces to the tolerance equation given by Steinberg [5, eq. (13.6)].

Example 2.5.2 Beam Power Pattern

A special case of (2.15) occurs when we assume no uncertainty in sensor position ($\boldsymbol{\alpha} = 0$) or in sensor weight ($\delta a = 0$) and a uniform distribution for the phase error. That is, the pdf for $\delta\gamma$ is

$$p(\delta\gamma) = \begin{cases} 1/(2\gamma_0), & |\delta\gamma| \leq \gamma_0 \\ 0, & \text{otherwise} \end{cases}$$

With this pdf, it follows that $E\{e^{j\delta\gamma}\}$ in (2.15) is

$$E\{e^{j\delta\gamma}\} = \int_{-\infty}^{\infty} e^{j\delta\gamma}\, p(\delta\gamma)\, d\delta\gamma = \frac{1}{2\gamma_0}\int_{-\gamma_0}^{\gamma_0} e^{j\delta\gamma}\, d\delta\gamma$$

Thus,

$$|E\{e^{j\delta\gamma}\}|^2 = \frac{\sin^2\gamma_0}{\gamma_0^2}$$

and (2.15) reduces to

$$E\{|b(f, \mathbf{u})|^2\} = \frac{\sin^2\gamma_0}{\gamma_0^2}|b_0(f, \mathbf{u})|^2 + \frac{1}{M^2}\left(1 - \frac{\sin^2\gamma_0}{\gamma_0^2}\right)\sum_{i=1}^{M} a_i^2$$

or, for the case of unity weights,

$$E\{|b(f, \mathbf{u})|^2\} = \frac{\sin^2\gamma_0}{\gamma_0^2}|b_0(f, \mathbf{u})|^2 + \frac{1}{M}\left(1 - \frac{\sin^2\gamma_0}{\gamma_0^2}\right) \tag{2.16}$$

Regardless of the unperturbed pattern value, the expected value of the pattern in (2.16) is bounded by a minimum value of

$$\frac{1}{M}\left(1 - \frac{\sin^2\gamma_0}{\gamma_0^2}\right)$$

We are particularly interested in phase errors that arise from quantized delays. With a sample spacing of $\tau = 1/f_s$, the magnitude of the maximum time delay error is $1/(2f_s)$. For a sinusoidal signal with a frequency of f_0, this corresponds to a maximum phase error of $\gamma_0 = \pi f_0/f_s$ rad. We can now use (2.16) to determine the expected beam power response of the array with this uniform phase error. We assume that f_0 is 725 Hz and f_s equals 3000 Hz. Then $\gamma_0 = 725\pi/3000$. Using unity weights and assuming $M = 50$ sensors, the reduction in response on the MRA (that is, $\mathbf{u} = \mathbf{u}_0$) is 0.83, or −0.83 dB, relative to the unperturbed array. (The unperturbed response of $\mathbf{u} = \mathbf{u}_0$ is equal to unity.) In addition, for unperturbed beam power pattern values, $|b_0(f, \mathbf{u})|^2$, corresponding to −20 dB, −25 dB, and −30 dB; the expected values for the perturbed pattern are −19.3 dB, −22.1 dB, and −23.6 dB, respectively. The minimum value for the expected beam power pattern is −24.5 dB and is due to delay quantization. A comparison of the perturbed and unperturbed beam power patterns is given in Figure 2.17(a) for a beam steered to 13.89°.

If we form beams using sensor data that are interpolated by a factor of 4, so that $\hat{f}_s = 4f_s$, and $f_0/\hat{f}_s = 725/12000$, the expected value of the perturbed pattern

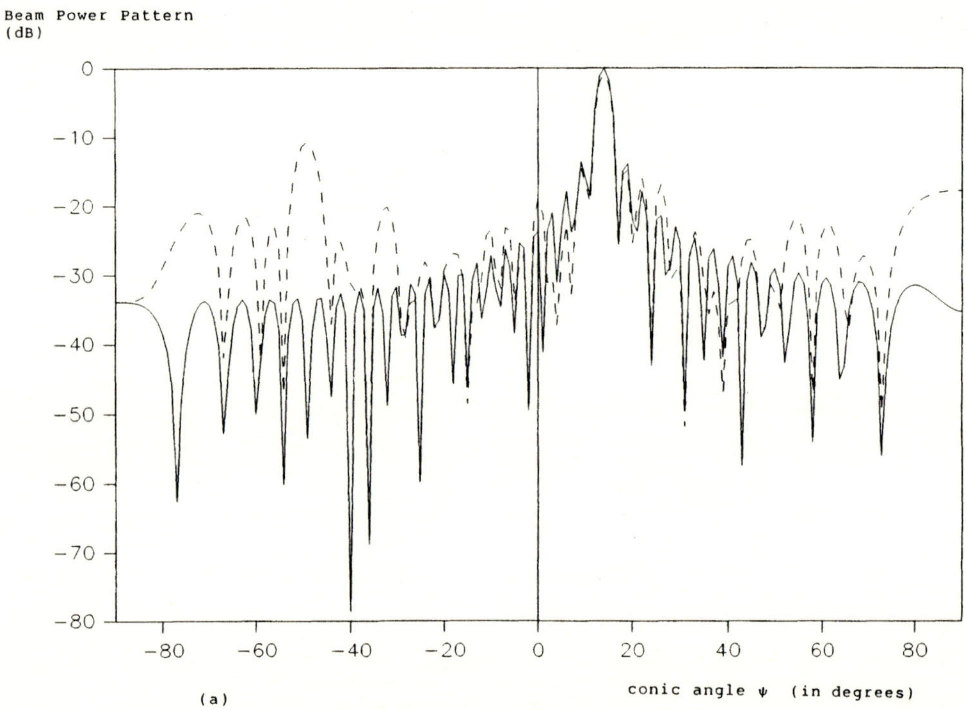

Figure 2.17 Effects of interpolation on beam power patterns using (a) delays based on f_s = 3000 Hz (dashed line), and (b) delays based on f_s = 12,000 Hz (dashed line); beam patterns using exact delays are shown with solid lines.

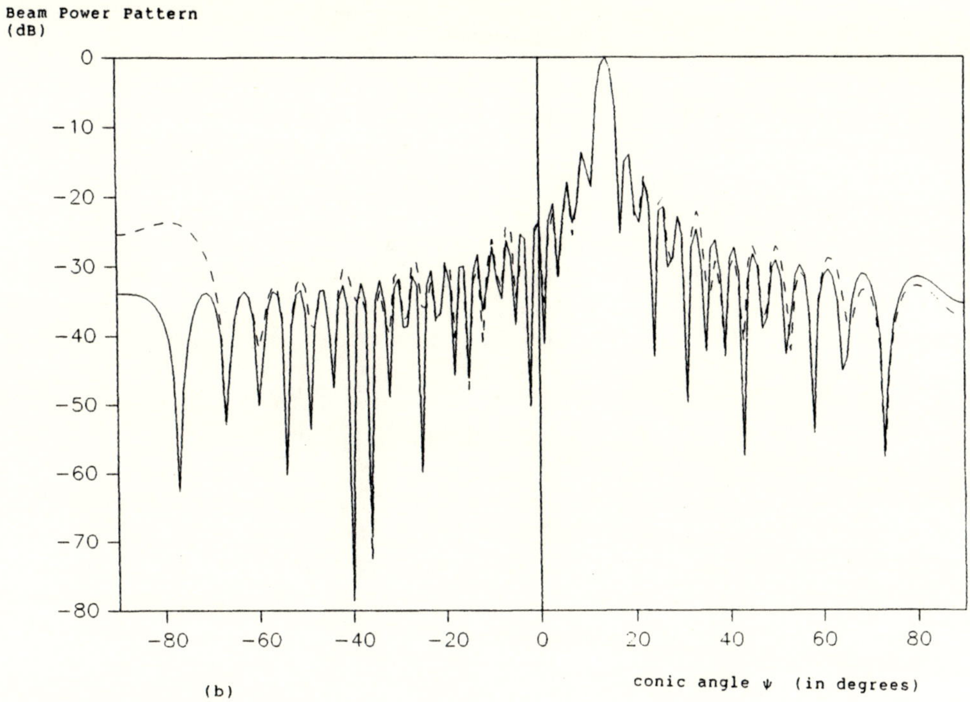

Figure 2.17(b) continued.

for the MRA is −0.05 dB, instead of −0.85 dB with no interpolation. By using this higher sampling frequency, we obtain expected values for the perturbed power pattern of −19.9 dB, −24.7 dB, and −29.1 dB, instead of the desired unperturbed power pattern values of −20 dB, −25 dB, and −30 dB, respectively. For this case, the minimum level for the expected beam power pattern is −36.2 dB. The beam power pattern using interpolated data is compared to the unperturbed pattern in Figure 2.17(b).

This example indicates the improvement in the expected value of the power pattern that results from increasing the sampling frequency of the sensor signals.

Example 2.5.3 Array Gain

This application of (2.15) involves the calculation of array gain given uncertainties in the locations of the sensors, as is the case in a sonobuoy field. An expression for array gain, from Section 2.3, is

$$\text{Array Gain} = \frac{E\{|\mathbf{w}^+ \mathbf{p}|^2\}}{\mathbf{w}^+ \mathbf{Q}\, \mathbf{w}} \tag{2.17}$$

where the expectation operator is present because of the uncertainty in sensor position. For the sonobuoy field, we assume that the noise is not correlated from sensor to sensor due to the large separation of sonobuoys relative to the wavelengths of interest. Therefore, the **Q** matrix is simply the identity matrix, **I**, and, assuming unity weighting, the denominator of (2.17) is equal to M, the number of sonobuoys. Because of the normalization of the complex beam pattern by $1/M$ in this section, the numerator is equal to $M^2\, E\{|b(f,\, \mathbf{u}_0)|^2\}$. Using (2.15) with no uncertainty in either sensor phase ($\delta\gamma = 0$) or weight ($\delta a = 0$), the numerator in (2.17) is

$$\begin{aligned} M^2\, E\{|b(f,\, \mathbf{u}_0)|^2\} &= M^2\, |E\{e^{j2\pi f \boldsymbol{\alpha}\cdot\mathbf{u}_0/c}\}|^2\, |b_0(f,\, \mathbf{u}_0)|^2 \\ &\quad + M\,(1 - |E\{e^{j2\pi f \boldsymbol{\alpha}\cdot\mathbf{u}_0/c}\}|^2) \end{aligned} \tag{2.18}$$

The peak value of the unperturbed power pattern $|b_0(f,\, \mathbf{u}_0)|^2$ is equal to unity. We assume that each component of the sensor location vector $\boldsymbol{\alpha} = \{\alpha_x,\, \alpha_y,\, \alpha_z\}$ is a Gaussian-distributed random variable with a variance of σ^2. From the assumptions at the beginning of this section, these components have zero means and are independent. Therefore,

$$E\{e^{j2\pi f \boldsymbol{\alpha}\cdot\mathbf{u}_0/c}\} = E\{e^{j2\pi f \alpha_x u_{0x}/c}\} \cdot E\{e^{j2\pi f \alpha_y u_{0y}/c}\} \cdot E\{e^{j2\pi f \alpha_z u_{0z}/c}\}$$

In particular, $E\{e^{j2\pi f \alpha_x u_{0x}/c}\}$ is found by evaluating

$$\int_{-\infty}^{\infty} e^{j2\pi f \alpha_x u_{0x}/c}\, p(\alpha_x)\, d\alpha_x = \int_{-\infty}^{\infty} e^{j2\pi f \alpha_x u_{0x}/c}\, \frac{1}{\sqrt{2\pi}\,\sigma} \exp[-\alpha_x^2/(2\sigma^2)]\, d\alpha_x$$

using a table of integrals [6, eq. (3.8964)]. The result is

$$E\{e^{j2\pi f \alpha_x u_{0x}/c}\} = \exp\left(-\frac{2\pi^2 f^2 \sigma^2 u_{0x}^2}{c^2}\right)$$

By performing the same calculations for the y and z components of $\boldsymbol{\alpha}$, we obtain

$$\begin{aligned} E\{e^{j2\pi f \boldsymbol{\alpha}\cdot\mathbf{u}_0/c}\} &= \exp\left(-\frac{2\pi^2 f^2 \sigma^2 (u_{0x}^2 + u_{0y}^2 + u_{0z}^2)}{c^2}\right) \\ &= \exp\left(-\frac{2\pi^2 f^2 \sigma^2}{c^2}\right) \end{aligned} \tag{2.19}$$

The numerator of the array gain expression (2.17) is now found by substituting (2.19) into (2.18). After dividing by the denominator of (2.17), which we determined to be equal to M, we obtain

$$\text{Array Gain} = (M - 1)\exp\left(-\frac{4\pi^2 f^2 \sigma^2}{c^2}\right) + 1 \tag{2.20}$$

This expression is plotted in Figure 2.18 for $M = 6$ sonobuoys, $c = 5000$ ft/s, and for various values of σ. This figure illustrates that, for very small uncertainties in position, beamforming performance is degraded for frequencies in excess of a few tens of Hz.

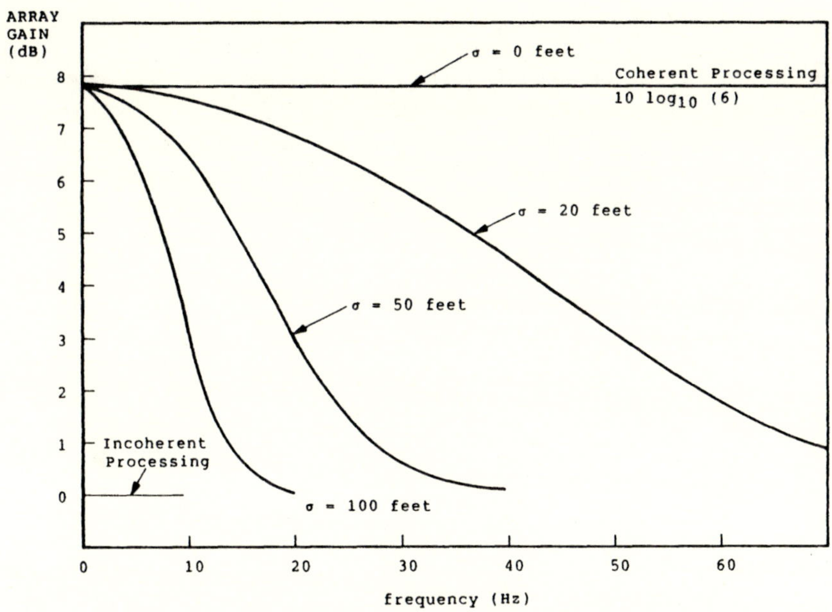

Figure 2.18 Array gain as a function of uncertainty in sensor location for a field of six sonobuoys.

In the preceding analysis, we assumed that sensor uncertainties were present in three dimensions. If uncertainties, again with standard deviations σ, exist only in the x and y directions, the expression for $E\{e^{j2\pi f\alpha\cdot\mathbf{u}_0/c}\}$ becomes

$$\exp\left(-\frac{2\pi^2 f^2 \sigma^2 (u_{0x}^2 + u_{0y}^2)}{c^2}\right) = \exp\left(-\frac{2\pi^2 f^2 \sigma^2 \cos^2\theta_0}{c^2}\right)$$

where, from Figure 2.3, $\mathbf{u}_0 = \{\cos\phi_0 \cos\theta_0, \sin\phi_0 \cos\theta_0, \sin\theta_0\}$. The array gain is then

$$(M - 1)\exp\left(-\frac{4\pi^2 f^2 \sigma^2 \cos^2\theta_0}{c^2}\right) + 1$$

which for a nonzero beam elevation of θ_0 degrees is larger (i.e., smaller loss) than the array gain (2.20) with sensor uncertainties in three dimensions. For beams that are normal to the horizontal, or x-y, plane ($|\theta_0| = 90°$); uncertainties in the x and y directions do not degrade the array gain, which then equals the unperturbed value of M.

Nonsynchronous Sampling

Another source of error in the beam pattern arises from the nonsynchronous sampling of the M sensors in the array. As is shown in Figure 2.19, one A/D converter is used to sequentially sample the sensors during an interval of time τ equal to the reciprocal of the sampling frequency. This introduces a "sampling skew" that must be corrected, to the extent possible, by adjusting the delays used in the beamformer.

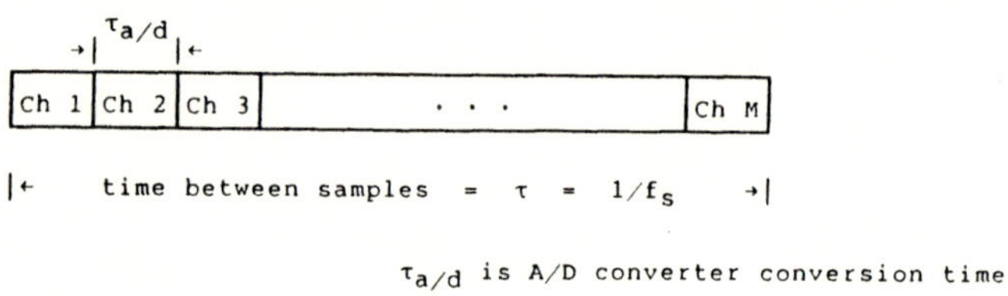

Figure 2.19 Nonsynchronous sampling of M sensors.

This sampling skew can be removed by implementing M sample-and-hold circuits. Neglecting voltage leakage, these circuits will capture the signals at each sensor and hold them until they can be processed by the A/D converter. Another approach is that of implementing M A/D converters. Data from each sensor are sampled synchronously by providing a frame pulse simultaneously to all A/D converters in the array.

2.6 FREQUENCY-DOMAIN BEAMFORMING

Up to this point, the discussion of beamforming has been restricted to a time-domain implementation. A frequency-domain structure may be more efficient in some cases.

This section derives such a structure from the time-domain beamformer and discusses a possible implementation.

We again consider a linear array of equally spaced sensors. From Figure 2.5, it follows that the output of the beamforming operation in the time domain for the mth beam is

$$y_m(n) = \sum_{i=1}^{M} a_i \, x_i[n - (i - 1)\, \Delta_m] \tag{2.21}$$

where the delay, Δ_m, is expressed in units of samples so that $\Delta_m = f_s d\,(\sin \psi_m)/c$. From Section 2.3.1, the output of the beamformer (2.21) in the frequency domain is

$$Y_m(k) = \sum_{i=1}^{M} a_i \left[\sum_{n=-(i-1)\Delta_m}^{N-1-(i-1)\Delta_m} x_i(n)\, \mathrm{e}^{-j2\pi k(i-1)\,\Delta_m/N} \right] \mathrm{e}^{-j2\pi kn/N}$$

Instead, we would like to have the frequency-domain beamformer calculate the following expression:

$$\hat{Y}_m(k) = \sum_{i=1}^{M} a_i \, \mathrm{e}^{-j2\pi k(i-1)\Delta_m/N} \sum_{n=0}^{N-1} x_i(n)\, \mathrm{e}^{-j2\pi kn/N} \tag{2.22}$$

because this is obtained simply by applying a complex weight of $a_i\, \mathrm{e}^{-j2\pi k(i-1)\Delta_m/N}$ to the DFT of the output of the ith sensor. Instead of having a different set of summation limits for each sensor, N samples are captured simultaneously for the entire array. As shown in Table 2.2, multiplying the DFT of $x_i(n)$ by $\mathrm{e}^{-j2\pi k(i-1)\Delta_m/N}$ corresponds to a time-domain sequence, $\hat{x}_i(n)$, equal to $x_i[n - (i - 1)\, \Delta_m]$ for only some values of n. The output of the frequency-domain beamformer (2.22) in the time domain is

$$\hat{y}_m(n) = \sum_{i=1}^{M} a_i \, \hat{x}_i(n)$$

where $\hat{x}_i(n)$ is the inverse DFT of (2.8) with $q_i = (i - 1)\, \Delta_m$.

As an example, we consider an array of five sensors with unity weighting and with $\Delta_m = 1$. Table 2.4(a) gives the time-domain output of the beamformer for the first two length-N data records. The sequences, $\hat{x}_i(n)$, are given in Table 2.4(b) and (c) for records 1 and 2, respectively. Because $\hat{x}_i(n) \neq x[n - (i - 1)\, \Delta_m]$ for all sensors for the first four values of n, $N - 4$ outputs of the beamformer are correct and four are not. That is, $y_m(0), \ldots, y_m(3)$ and $y_m(N), \ldots, y_m(N + 3)$ are incorrect

Table 2.4
Equivalence of Time-Domain and Frequency-Domain Beamforming Operations Using Overlap-Save Method

Record 1			*Record 2*		
(a) Time-Domain Beamforming Output for 2 Length-N Data Records					
$x_1(0)$	$x_1(4)$	$x_1(N-1)$	$x_1(N)$	$x_1(N+4)$	$x_1(2N-1)$
$x_2(-1)$	$x_2(3)$	$x_2(N-2)$	$x_2(N-1)$	$x_2(N+3)$	$x_2(2N-2)$
$x_5(-4)$	$x_5(0)$	$x_5(N-5)$	$x_5(N-4)$	$x_5(N)$	$x_5(2N-5)$
$y_m(0)$	$y_m(4)$	$y_m(N-1)$	$y_m(N)$	$y_m(N+4)$	$y_m(2N-1)$
(b) Frequency-Domain Beamforming Output for Record 1, in the Time Domain					
Incorrect Beam Output				*Correct Beam Output*	
$x_1(0)$	$x_1(1)$	$x_1(2)$	$x_1(3)$	$x_1(4)$	$x_1(N-1)$
$x_2(N-1)$	$x_2(0)$	$x_2(1)$	$x_2(2)$	$x_2(3)$	$x_2(N-2)$
$x_5(N-4)$	$x_5(N-3)$	$x_5(N-2)$	$x_5(N-1)$	$x_5(0)$	$x_5(N-5)$
				$y_m(4)$	$y_m(N-1)$
(c) Frequency-Domain Beamforming Output for Record 2, in the Time Domain					
Incorrect Beam Output				*Correct Beam Output*	
$x_1(N)$	$x_1(N+1)$	$x_1(N+2)$	$x_1(N+3)$	$x_1(N+4)$	$x_1(2N-1)$
$x_2(2N-1)$	$x_2(N)$	$x_2(N+1)$	$x_2(N+2)$	$x_2(N+3)$	$x_2(2N-2)$
$x_5(2N-4)$	$x_5(2N-3)$	$x_5(2N-2)$	$x_5(2N-1)$	$x_5(N)$	$x_5(2N-5)$
				$y_m(N+4)$	$y_m(2N-1)$
(d) Frequency-Domain Beamforming Output for Record 2, in the Time Domain, Using the Overlap-Save Method					
Incorrect Beam Output				*Correct Beam Output*	
$x_1(N-4)$	$x_1(N-3)$	$x_1(N-2)$	$x_1(N-1)$	$x_1(N)$	$x_1(2N-5)$
$x_2(2N-5)$	$x_2(N-4)$	$x_2(N-3)$	$x_2(N-2)$	$x_2(N-1)$	$x_2(2N-6)$
$x_5(2N-8)$	$x_5(2N-7)$	$x_5(2N-6)$	$x_5(2N-5)$	$x_5(N-4)$	$x_5(2N-9)$
				$y_m(N)$	$y_m(2N-5)$

and therefore should not be calculated. By using the sensor outputs for values of n such that $N - 4 \leq n \leq 2N - 5$, instead of $N \leq n \leq 2N - 1$, for the second frequency-domain beamforming calculation, the missing output for $y_m(N), \ldots, y_m(N + 3)$ can be obtained. This is the overlap-save method as discussed in [7, pp. 63–67] and illustrated in Table 2.4(d).

Beamforming therefore can be performed correctly in the frequency domain by using the overlap-save technique. Equation (2.22) can be implemented directly as a dot product of the weight vector and the vector of DFT outputs, as was done in Section 2.3.2. We also can implement (2.22) by using a DFT. An equivalence between the first sum in (2.22) and a length-M DFT can be established by selecting Δ_m such that $e^{-j2\pi k(i-1)\Delta_m/N} = e^{-j2\pi(i-1)m/M}$, or equivalently,

$$k\Delta_m/N = m/M \tag{2.23}$$

The beamforming in (2.22) then clearly is expressible as a length-M DFT:

$$\hat{Y}_m(k) = \sum_{i=0}^{M-1} a_{i+1} X_{i+1}(k)\, e^{-j2\pi im/M} \tag{2.24}$$

Again, the DFT bin number and frequency are related by $k/N = f/f_s$, where f_s is the sampling frequency of the input data to the DFT. Using (2.23), the delays associated with the mth beam are $\Delta_m = f_s\, m/(Mf)$. Finally, by using the relationship, $\Delta_m = f_s d(\sin\psi_m)/c$, the realizable beam directions in $\sin\psi$ space are

$$\sin\psi_m = c\Delta_m/(f_s d) = cm/(dMf), \quad -M/2 \le m \le M/2 \tag{2.25}$$

Therefore, the beam direction associated with each DFT bin (2.24) is a function of frequency. For example, for a frequency equal to the upper-design frequency of the array, or $f_u = c/(2d)$, the beam directions are given by $\sin\psi_m = 2m/M$. We therefore can form $M + 1$ equally spaced beams, as was the case with the time-domain beamformer.

As the frequency is reduced below the upper-design frequency of the array, however, fewer than M solutions are possible for (2.25). Therefore, fewer than M beams can be generated. For instance, for $f = f_u/2$, $\sin\psi_m = 4m/M$ and only $(M/2) + 1$ beam directions are possible because bin values $|m| > M/4$ do not correspond to realizable beam directions. This is shown in Figure 2.20 for $M = 32$ sensors. Beams are shown for $\sin\psi = \pm 1$ in this figure because at the upper design frequency, a grating lobe is present for the endfire beam (Table 2.1).

To perform beamforming over a range of frequencies, we must generate beamformer output for that range of frequencies at each of a set of fixed-beam directions. Therefore we must increase the number of beams to "fill in" the missing beams at lower frequencies in Figure 2.20. This can be done by appending the frequency-domain sensor sequence, $X_i(k)$, $i = 1, \ldots, M$, with $M(D - 1)$ zeros and performing a length-DM DFT in (2.24) for each frequency bin. As discussed in [7, pp. 384–386], this increases the number of beams generated by a factor of D. To illustrate this increase in the number of DFT outputs, we first consider the magnitude of the frequency response of a length-M DFT. This response for the k_0th bin is found from (1.13):

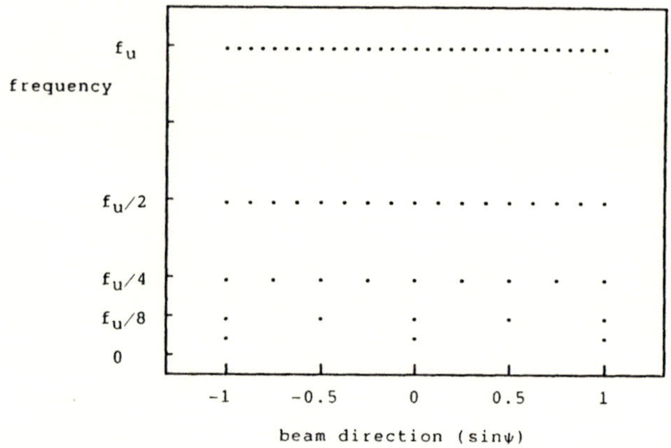

Figure 2.20 Realizable beam directions as a function of frequency for 32 sensors.

$$|H(k_0)| = \left| \frac{\sin[\pi(k_0/M - f/f_s)M]}{\sin[\pi(k_0/M - f/f_s)]} \right| \tag{2.26}$$

Of course, M such bins are located at $f = f_s k/M$, where $k = 0, \ldots, M - 1$. As a measure of bin width, the separation of first zero crossings from (2.26) is equal to $2f_s/M$.

We next consider the magnitude of the frequency response of a length-$2M$ DFT, where the input consists of M samples and M zeros (that is, $D = 2$). As in Section 1.3.1, this response is obtained by applying $e^{j2\pi fn/f_s}$ to the DFT and determining the response of the k_0th bin as the frequency is varied. Here, however, the input sequence is defined by

$$x(n) = \begin{cases} e^{j2\pi fn/f_s}, & n = 0, \ldots, M - 1 \\ 0, & n = M, \ldots, 2M - 1 \end{cases}$$

The magnitude of the resulting response is

$$|H(k_0)| = \left| \frac{\sin\{\pi[k_0/(2M) - f/f_s]M\}}{\sin\{\pi[k_0/(2M) - f/f_s]\}} \right|$$

There are now $2M$ bins located at $f = f_s k/(2M)$, $k = 0, \ldots, 2M - 1$. The bin width, however, still is $2f_s/M$. Therefore, by appending M zeros to the M input samples and performing a length-$2M$ DFT, we have generated twice as many output bins. These are separated by half the spacing but have the same width as the bins generated with a length-M DFT. The result is readily generalized for values of D greater than two.

Therefore we can more accurately obtain a desired beam direction for each frequency. This is the approach taken in [8]. In addition, a number of exact and approximate techniques for calculating the necessary beam outputs for each frequency are described in [9]. By using such techniques, the missing beam outputs at each frequency in Figure 2.20 can be generated.

A block diagram of a frequency-domain beamformer is provided in Figure 2.21. As shown, this beamformer performs the overall function of a time-domain beamformer because time-domain beam sequences are generated. The beamformer operates on blocks of length-N sequences from M sensors. The time-frequency length-N FFTs use a redundancy of $R = N/N_{\text{new}}$ (Section 1.3.1) that is greater than unity to correctly generate all output values in the beam sequences via the overlap-save approach (Table 2.4(d)). That is, the input sequences are overlapped by $N - N_{\text{new}}$ samples.

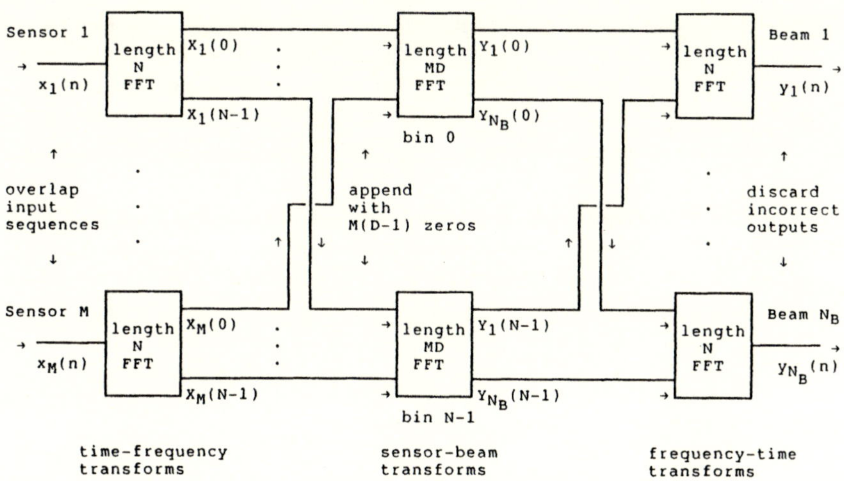

Figure 2.21 Implementation of frequency-domain beamforming.

The output of the M time-frequency FFTs are grouped in terms of frequency to generate the sequences

$$\{X_1(0), X_2(0), \ldots, X_M(0)\}, \{X_1(1), X_2(1), \ldots, X_M(1)\}, \ldots, \{X_1(N-1), X_2(N-1), \ldots, X_M(N-1)\}$$

These frequency-domain sequences are appended with $M(D - 1)$ zeros before the sensor-beam FFTs are calculated to obtain more closely spaced beams at lower frequencies. The output of the M sensor-beam FFTs are grouped in terms of beams by forming the sequences:

$$\{Y_1(0), Y_1(1), \ldots, Y_1(N-1)\}, \{Y_2(0), Y_2(1), \ldots, Y_2(N-1)\}, \ldots, \{Y_{N_B}(0), Y_{N_B}(1), \ldots, Y_{N_B}(N-1)\}$$

Length-N FFTs are used to obtain time-domain beam sequences, and $N_{\text{new}} = N/R$ valid samples are generated for each set of N input samples.

It is of interest to compare the computational requirements for the implementation of time-domain and frequency-domain beamforming. From Section 2.5.1, the number of multiplications and additions required to calculate an output for N_B beams using the interpolation-filter-before-beamforming algorithm are $M(L + 1)$ and $M(L + N_B)$, respectively. This assumes that the same set of weights is used for all beam directions. Again, L is the length of the interpolating filter. To calculate N_{new} outputs for each beam therefore requires $N_{\text{new}}M(2L + N_B + 1)$ operations.

The frequency-domain beamformer shown in Figure 2.21 requires M length-N FFTs to calculate the time-frequency transformation, N length-MD FFTs to implement the sensor-beam transformation, and N_B length-N FFTs to obtain the time-domain output sequences for N_B beams. From Section 1.3.2, a length-N radix-2 FFT requires $5N \log_2 N$ real operations. To calculate N_{new} outputs for each beam requires

$$MN + M(5N \log_2 N) + N[5MD \log_2(MD)] + N_B(5N \log_2 N)$$

real operations, where MN operations are used to implement the weighting function a_i.

Example 2.6.1

We select the following set of parameter values:

M = 64 sensors,
N = 1024,
N_{new} = 896 (the input sequences are overlapped by $N - N_{\text{new}} = 128$ samples),
L = 100-stage filter,
N_B = 64 beams,
D = 4 ($3 \times 64 = 192$ zeros are appended to the input sequence to the sensor-beam FFTs).

The time-domain implementation requires $1.52 \cdot 10^7$ real operations to generate $N_{\text{new}} = 896$ samples, and the frequency-domain beamformer requires $1.71 \cdot 10^7$ real operations.

The intent of this example is merely to illustrate the computational requirements of the two beamformers. A valid comparison must also include the capabilities of the specific hardware being used. This hardware may be optimized for performing delay-and-sum operations or it may calculate FFTs efficiently. We observe that, if only a few beams are required, then the interpolation-filter-after-beamforming structure (Figure 2.15) may prove to be computationally advantageous.

2.7 QUADRATURE BEAMFORMING

We can implement an efficient beamformer in some cases by taking advantage of the structure of the input signal. Consider a bandpass, or narrowband, signal, $x(n)$, with the spectrum shown in Figure 2.22, where the spectral width, B, is small compared to the center frequency, f_c. This problem is treated in [3]. As discussed in

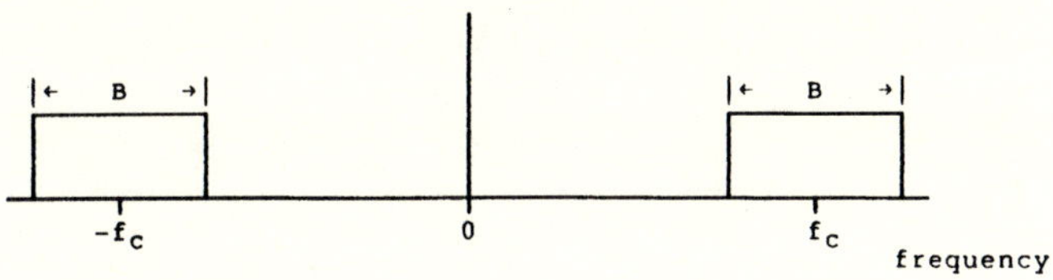

Figure 2.22 Spectrum of a bandpass signal.

Section 1.1.2, this signal can be represented by the complex envelope

$$\tilde{x}(n) = [x(n)\, e^{-j2\pi f_c n/f_s}]_{LP}$$
$$x(n) = 2\, \text{Re}[\tilde{x}(n)\, e^{j2\pi f_c n/f_s}]$$

where again LP denotes a low-pass filtering operation.

Section 1.1.2 discussed the calculation of the complex envelope in the analog domain. Figure 2.23 shows the equivalent implementation in the digital domain for

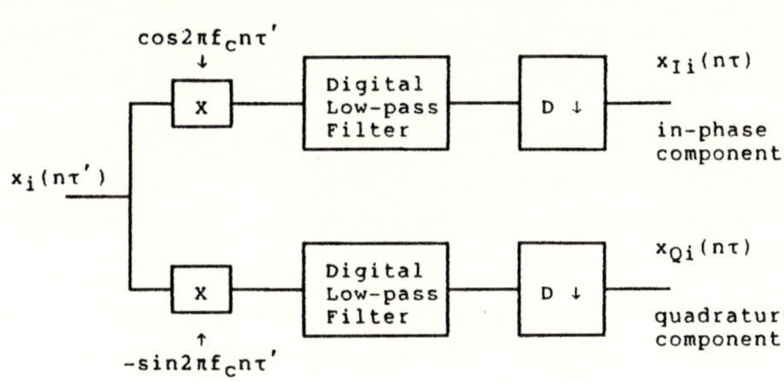

Figure 2.23 Calculation of the complex envelope in the digital domain.

the ith sensor. The output data rate can be decimated because the highest frequency in the complex envelope is $B/2$ Hz. In this figure, $f_s' = 1/\tau'$ is the sampling frequency that satisfies the Nyquist criterion for the original bandpass sequence, $x(n)$; the highest frequency in $x(n)$ is $f_c + B/2$ Hz. Again, $f_s = 1/\tau$ is the sampling frequency consistent with the complex envelope (highest frequency is $B/2$ Hz), whereas $\hat{f}_s = 1/\hat{\tau}$ is the sampling frequency required for accurate beamforming.

Because the highest frequency in the complex envelope is $B/2$ Hz, as opposed to $f_c + B/2$ Hz in the original signal, a beamforming structure based on the complex envelope can be more efficient than one operating on the original signal. To develop such a structure, we start with the conventional beamformer, steered in direction $\mathbf{u}_0$:

$$y(n\tau) = \sum_{i=1}^{M} a_i\, x_i(n\tau - \Delta_i) \tag{2.27}$$

where $\Delta_i = \mathbf{r}_i \cdot \mathbf{u}_0/c$. The output of the ith sensor can be expressed in terms of the complex envelope as

$$x_i(n\tau) = \tilde{x}_i(n\tau)\, e^{j2\pi f_c n/f_s} + \tilde{x}_i^*\,(n\tau)\, e^{-j2\pi f_c n/f_s}$$

by using the identity, Re $\{a\} = (a + a^*)/2$. Substituting this expression for $x_i(n\tau)$ into (2.27) yields

$$y(n\tau) = \sum_{i=1}^{M} a_i\, [\tilde{x}_i(n\tau - \Delta_i)\, e^{j2\pi f_c(n/f_s - \Delta_i)} + \tilde{x}_i^*\,(n\tau - \Delta_i)\, e^{-j2\pi f_c(n/f_s - \Delta_i)}] \tag{2.28}$$

The output of the beamformer also can be written in terms of its complex envelope:

$$y(n\tau) = \tilde{y}(n\tau)\, e^{j2\pi f_c n/f_s} + \tilde{y}^*(n\tau)\, e^{-j2\pi f_c n/f_s} \tag{2.29}$$

The following expression for a quadrature beamformer is obtained from (2.28) and (2.29):

$$\tilde{y}(n\tau) = \sum_{i=1}^{M} a_i\, \tilde{x}_i\,(n\tau - \Delta_i)\, e^{-j2\pi f_c \Delta_i} \tag{2.30}$$

The delays Δ_i correspond to a desired direction of $\mathbf{u}_0$. A set of beams is formed in parallel by implementing (2.30) with different sets of Δ_i. As discussed in Section 2.5.1, it often is necessary to interpolate the sensor signals in order to form beams of adequate quality. Here the complex envelope is interpolated to a sampling frequency of $\hat{f}_s$. The implementation is seen more readily by writing (2.30) in terms of the in-phase and quadrature components of the outputs of the beamformer, $y_I(n\tau)$ and $y_Q(n\tau)$, respectively. These components are

$$y_I(n\tau) = \sum_{i=1}^{M} a_i \, [x_{Ii}(n\tau - k_i\hat{\tau}) \cos(2\pi f_c \Delta_i) + x_{Qi}(n\tau - k_i\hat{\tau}) \sin(2\pi f_c \Delta_i)]$$

$$y_Q(n\tau) = \sum_{i=1}^{M} a_i \, [x_{Qi}(n\tau - k_i\hat{\tau}) \cos(2\pi f_c \Delta_i) - x_{Ii}(n\tau - k_i\hat{\tau}) \sin(2\pi f_c \Delta_i)] \quad (2.31)$$

In (2.31), $k_i\hat{\tau}$ is the delay for the ith sensor that most closely approximates the desired delay Δ_i. The beam output is formed at f_s, the sampling frequency that satisfies the Nyquist criterion for the complex envelope. Additional beams are formed by implementing additional sets of integers k_i. This quadrature beamformer is shown in Figure 2.24.

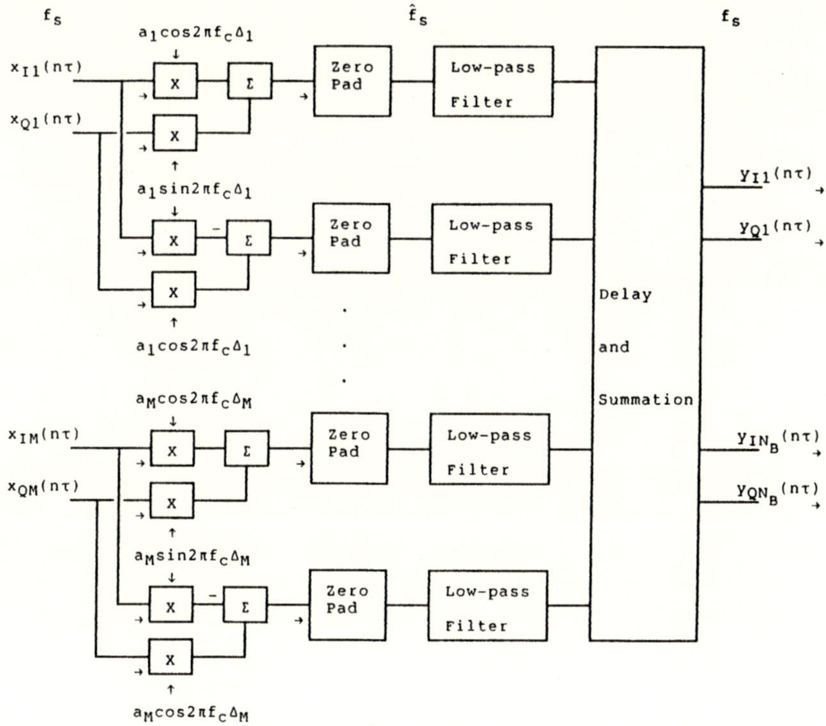

Figure 2.24 Quadrature beamformer.

REFERENCES

1. Burdic, W.S., *Underwater Acoustic System Analysis*, Prentice-Hall, Englewood Cliffs, NJ, 1984.
2. Hallenborg, P.A., private communication.
3. Pridham, R.G., and R.A. Mucci, "Digital Interpolation Beamforming for Low-Pass and Bandpass Signals," *Proc. IEEE*, Vol. 67, No. 6, June 1979, pp. 904–919.
4. Pridham, R.G., and R.A. Mucci, "A Novel Approach to Digital Beamforming," *J. Acoustical Society of America*, Vol. 63, No. 2, February 1978, pp. 425–434.
5. Steinberg, B.D., *Principles of Aperture and Array System Design*, John Wiley & Son, New York, 1976.
6. Gradshteyn, I.S., and I.M. Ryzhik, *Tables of Integrals, Series, and Products*, Academic Press, New York, 1980.
7. Rabiner, L.R., and B. Gold, *Theory and Application of Digital Signal Processing*, Prentice-Hall, Englewood Cliffs, NJ, 1975.
8. Weber, M.E., and R. Heisler, "A Frequency-Domain Beamforming Algorithm for Wideband, Coherent Signal Processing," *J. Acoustical Society of America*, Vol. 76, No. 4, October 1984, pp. 1132–1144.
9. Maranda, B., "Efficient Digital Beamforming in the Frequency Domain," *J. Acoustical Society of America*, Vol. 86, No. 5, November 1989, pp. 1813–1819.

Chapter 3
Optimum Detection and Estimation in Passive Sonar

Here we briefly examine typical signal and noise environments encountered in passive sonar. Using a criterion (Neyman-Pearson) for optimizing performance, we derive optimum passive broadband, narrowband, and modulation detectors. Next, maximum likelihood estimators are obtained for signal parameters such as arrival angle, power, frequency, and delay. In Chapter 4, the performance of practical, realizable detectors and estimators is compared to the performance of these optimum structures.

3.1 SIGNAL AND NOISE ENVIRONMENTS

In this section, we present a short overview of signal and noise fields encountered in sonar. The characteristics of such fields are described in detail in books such as those by Urick [1] and Burdic [2].

3.1.1 Signal Models

Temporally, the spectrum of the signal of interest consists of a broadband component and narrowband, or discrete frequency, lines. Propeller noise is amplitude modulated. Discrete lines occur at the rotational frequency of the propulsion shaft and at the blade frequency [1, p. 308]. (The blade frequency is equal to the shaft frequency times the number of blades.) Spatially, the signal is regarded as a plane wave relative to the receiving array so that the wavefronts are parallel. That is, the source is located in the far field. Wavefront curvature is treated briefly in Section 9.2.

In this chapter, the effects of the channel on the signal are limited to absorption and multipath generation. Absorption in the medium can be modeled by the following equation for a frequency-dependent attenuation coefficient, $\alpha(f)$ [1, p. 102]:

$$\alpha(f) = \frac{0.1f^2}{1+f^2} + \frac{40f^2}{4100+f^2} + 2.75 \cdot 10^{-4} f^2 \tag{3.1}$$

where $\alpha(f)$ is in dB per kiloyard, and f is in kHz.

As a result of multipaths, a signal emanating from a source arrives at a receiving array of sensors via two or more distinct propagation paths. Associated with each path is a weight and a delay. Therefore, in a two-path environment, the sequence

$$x(n) = a_1\, s(n - n_1) + a_2\, s(n - n_2)$$

would be received. (For simplicity, we assume that the delays are equal to whole numbers of samples.) Signal and noise acoustic pressure levels often are measured in terms of a basic unit of pressure, the micropascal (μPa), which is defined as

$$1\ \mu\text{Pa} = 10^{-5}\ \text{dyne/cm}^2 = 0.67 \cdot 10^{-22}\ \text{W/cm}^2$$

The utility of this convention is that typical signal and noise levels, in dB relative to 1 μPa, are positive quantities.

Table 3.1 gives typical source levels as a function of frequency. These source levels are in units of dB relative to a μPa in a 1-Hz band and a distance of 1 yard. This is denoted by dB//μPa/Hz @ 1 yd.

Table 3.1
Typical Source Levels (in dB//μPa/Hz @ 1 yd)

Frequency (Hz)	*Source Level* (dB)
100	150
300	142
500	136
1000	130
3000	118

3.1.2 Noise Models

Ambient ocean noise can be modeled as a superposition of an anisotropic noise field from the ocean surface and an isotropic noise field [3]. A space-time correlation function is obtained by observing the noise pressure at two points in the ocean, spatially separated by a distance, d, and temporally separated by a time interval, τ. The orientation of this pair of points relative to the surface is specified by an angle, γ. The geometry is shown in the insets in Figure 3.1. The correlation function due to the anisotropic surface component is

$$\rho(d, \gamma, \tau) = \rho(d, \gamma) \cos 2\pi f \tau$$

where $\rho(d, \gamma)$ is plotted in Figure 3.1 for $\gamma = 0°$ and $\gamma = 90°$. The directionality of the surface noise sources is $g(\alpha) = \cos\alpha$ in Figure 3.1(a) and $g(\alpha) = \cos^2\alpha$ in Figure 3.1(b), where α is measured from vertical. The function in Figure 3.1(b) fits observed data well for the frequency range of 400–1000 Hz [1, p. 206] and [2, p. 312].

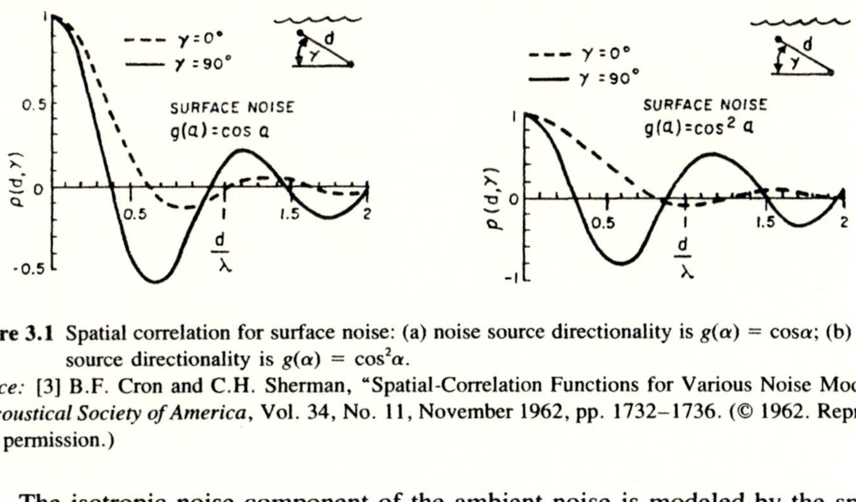

Figure 3.1 Spatial correlation for surface noise: (a) noise source directionality is $g(\alpha) = \cos\alpha$; (b) noise source directionality is $g(\alpha) = \cos^2\alpha$.
Source: [3] B.F. Cron and C.H. Sherman, "Spatial-Correlation Functions for Various Noise Models," *J. Acoustical Society of America*, Vol. 34, No. 11, November 1962, pp. 1732–1736. (© 1962. Reprinted with permission.)

The isotropic noise component of the ambient noise is modeled by the space-time correlation function [3, eq. (9)]:

$$\rho(d, \tau) = [\sin(kd)/(kd)] \cos 2\pi f \tau$$

where

$$k = 2\pi f/c$$

This function, plotted in Figure 3.2 for $\tau = 0$, corresponds fairly closely to the observed ambient noise behavior for the case of low sea state. As was observed in Section 2.3.3, the isotropic component of noise is not correlated from sensor to sensor when the sensor spacing is equal to multiples of half the wavelength. A plot of average deep-water ambient noise is given in Figure 3.3 [1]. The low-frequency

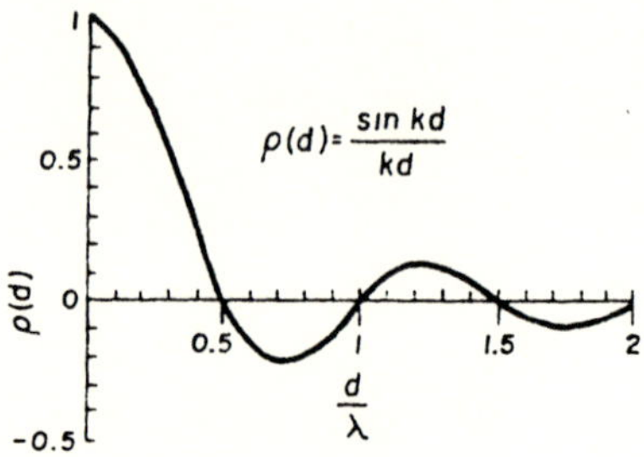

Figure 3.2 Spatial correlation for volume or isotropic noise.
Source: [3] B.F. Cron and C.H. Sherman, "Spatial-Correlation Functions for Various Noise Models," *J. Acoustical Society of America*, Vol. 34, No. 11, November 1962, pp. 1732–1736. (© 1962. Reprinted with permission.)

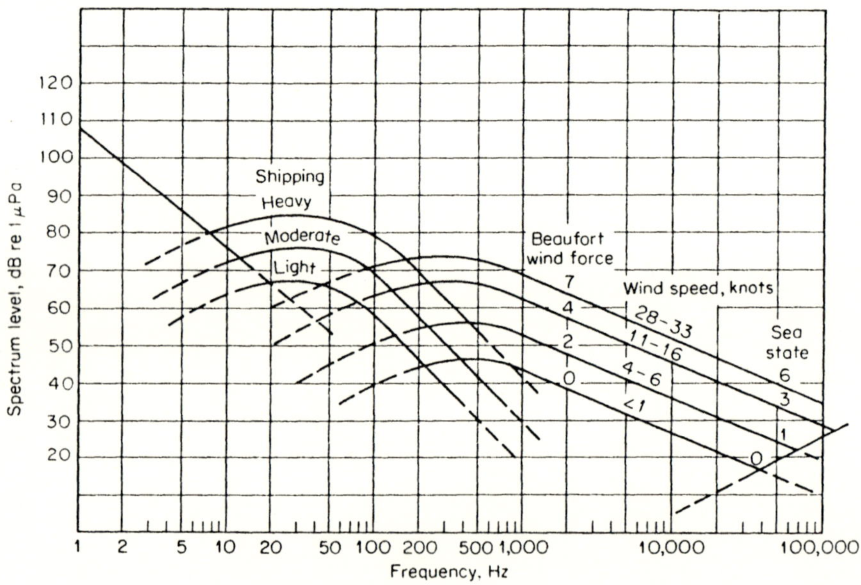

Figure 3.3 Average deep-water ambient noise.
Source: [1] R.J. Urick, *Principles of Underwater Sound*, McGraw-Hill, New York. (© 1975. Reprinted with permission.)

component is seismic in nature whereas the high-frequency contribution arises from molecular noise. An estimate of the noise spectrum is found by combining the appropriately chosen components due to distant shipping and sea state in Figure 3.3.

Flow noise often is considered to be spatially uncorrelated from sensor to sensor. It is treated as a zero-mean stationary Gaussian random process. Flow noise exhibits an f^{-3} dependency at high frequencies [1, Section 11.3] as sketched in Figure 3.4.

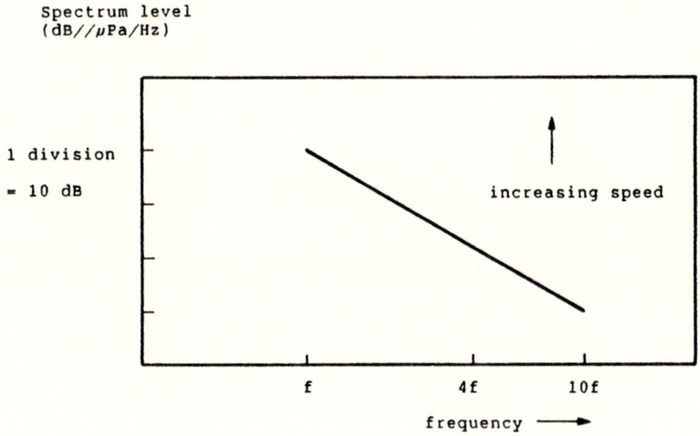

Figure 3.4 Hypothetical flow noise levels.

Beamformer Output S/N

The S/N in dB in a 1-Hz band at the output of the beamformer is given by

$$S/N = \mathrm{SL} - \mathrm{TL} - (\mathrm{NL} - \mathrm{DI}) \tag{3.2}$$

where

- SL = source level (dB//μPa/Hz @ 1 yd),
- TL = transmission loss; for spherical spreading, it is equal to the sum of a spreading component, $20 \log_{10} R$, and an absorption component, $\alpha(f) R \cdot 10^{-3}$
- R = range, in yards,
- $\alpha(f)$ = frequency-dependent absorption coefficient (3.1),
- NL = noise level (dB//μPa/Hz); this consists of both ambient and flow noise,
- DI = directivity index of the receiving array; this is the array gain in dB for isotropic noise and a plane wave signal (Section 2.3.3).

As an example, we calculate the S/N at the beamformer output in a 1-Hz band centered on 300 Hz. From Table 3.1, the source level is 142 dB//μPa/Hz @ 1 yd. At a range of 5000 yards, the loss due to spherical spreading is 74 dB, and that due to absorption is only 0.05 dB. A value of 80 dB//μPa/Hz is assumed for flow noise and the ambient noise level is about 67 dB//μPa for sea state 3 from Figure 3.3. Therefore, the sonar is operating in a flow noise-limited environment. As a result, the noise is spatially uncorrelated, and the array gain is equal to the number of sensors (Section 2.3.3). For an array of 50 sensors, the DI is 17 dB. From (3.2), the S/N at the output of the beamformer is

$$S/N = 142 - 74 - (80 - 17) = 5 \text{ dB/Hz at } 300 \text{ Hz}$$

We observe that the S/N at the input to the array is only -12 dB in a 1-Hz band at 300 Hz.

3.2 OPTIMUM DETECTION

We assume that the various parameters associated with the signal and noise fields are known. The question addressed in this section is whether the signal in fact is present. Detectors are obtained that are optimum for a certain detection criterion and for a given set of assumptions. Of course, any departure from these assumptions violates the conclusion of optimality. Two excellent references for the material in Section 3.2 are Whalen [4] and Van Trees [5].

3.2.1 Hypothesis Testing

We assume that there are two hypotheses concerning a received sequence, $x(n)$. The first of these, the null or H_0 hypothesis, assumes that the received sequence consists only of noise so that $x(n) = g(n)$. In the second, the H_1 hypothesis, the signal is assumed to be present so that $x(n) = s(n) + g(n)$. We let D_0 denote the choice of H_0 hypothesis and D_1 denote the choice of the H_1 hypothesis. Four outcomes based on a set of samples of $x(n)$ are possible:

1. We correctly identify the presence of a signal. The probability of this outcome is called the *probability of detection* (P_D), denoted by $P(D_1 \mid H_1)$.
2. We declare that a signal is present when in fact there is no signal. This is the *probability of false alarm* (P_{fa}), denoted by $P(D_1 \mid H_0)$.
3. We incorrectly declare that no signal is present. This is denoted by $P(D_0 \mid H_1)$.
4. We correctly declare that the received sequence consists only of noise. This final outcome is denoted by $P(D_0 \mid H_0)$.

Depending on the particular application, we might know the a priori probability that a signal is present, $P(H_1)$, or absent, $P(H_0)$. This situation might arise in active sonar systems and communication systems. In addition, we might be interested in assigning costs, or penalties, to each of the four outcomes.

In passive sonar, in general, we do not have the *a priori* probabilities $P(H_0)$ and $P(H_1)$ and we do not assign costs to the four possible outcomes of the experiment. The *Neyman-Pearson* (NP) criterion can be applied in this case because only a knowledge of the signal and noise pdfs is required. The criterion maximizes P_D for a specified P_{fa}.

The NP detector uses a likelihood ratio test of the following form:

$$\lambda_T = \frac{p_1(y)}{p_0(y)} \geqslant \beta$$

where

λ_T = test statistic,
y = single observation of $x(n)$,
$p_1(y)$ = pdf of y given that H_1 is true,
$p_0(y)$ = pdf of y given that H_0 is true.

The decision rule for the NP detector is

$$\text{Choose } H_1\text{:} \quad \text{when } \lambda_T \geqslant \beta$$

$$\text{Choose } H_0\text{:} \quad \text{otherwise}$$

Example 3.2.1

$$\text{Let } H_1\text{:} \quad y = 2 + g$$

$$H_0\text{:} \quad y = g$$

where g is a sample of a zero-mean Gaussian random process with unity variance. An optimum (in the NP sense) detection rule is desired for $P_{fa} = 0.1$. That is, we wish to obtain a decision rule to differentiate between the hypotheses of only noise and a value of 2 plus noise. The functions $p_0(y)$ and $p_1(y)$ are sketched in Figure 3.5. We form the likelihood ratio and obtain the test statistic:

$$\lambda_T = \frac{p_1(y)}{p_0(y)} = \frac{(1/\sqrt{2\pi}) \exp[-(y-2)^2/2]}{(1/\sqrt{2\pi}) \exp[-y^2/2]}$$

$$= \exp[2(y-1)] \tag{3.3}$$

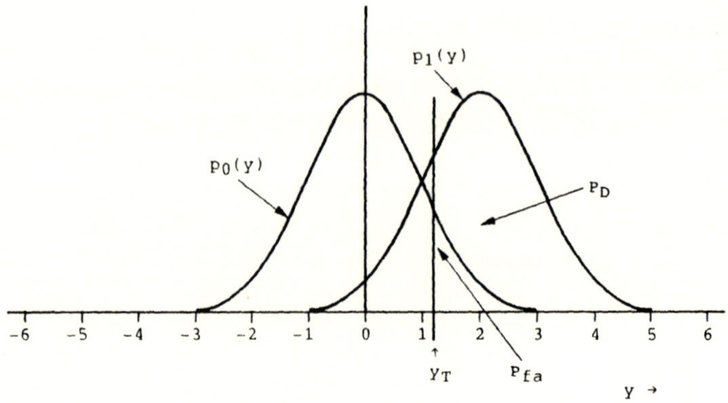

Figure 3.5 Probability density functions for a known signal in noise.

Next, an integration limit, y_T, is calculated by equating the area under $p_0(y)$ from y_T to infinity and the specified value of P_{fa}. That is,

$$P_{fa} = 0.1 = \int_{y_T}^{\infty} p_0(y)\, dy = \frac{1}{\sqrt{2\pi}} \int_{y_T}^{\infty} \exp(-y^2/2)\, dy$$

The resulting integration limit is $y_T = 1.28$. The value for P_D is found by integrating from y_T to infinity under $p_1(y)$:

$$P_D = \int_{y_T}^{\infty} p_1(y)\, dy = \frac{1}{\sqrt{2\pi}} \int_{y_T}^{\infty} \exp[-(y-2)^2/2]\, dy$$

or $P_D = 0.77$. The areas corresponding to P_{fa} and P_D are shown in Figure 3.5. The threshold, $\beta = 1.75$, then is obtained by evaluating (3.3) with $y = y_T$. The decision rule is

Choose H_1: when $\lambda_T \geqslant \beta = 1.75$

Choose H_0: otherwise

An equivalent decision rule is

Choose H_1: when $y \geqslant 1.28$

Choose H_0: otherwise

Example 3.2.2

We expand the previous example by considering N time measurements. That is, for a known signal, $s(n)$, in zero mean Gaussian noise, $g(n)$, $n = 0, \ldots, N-1$, the hypotheses are

$$\mathrm{H}_1\colon \mathbf{x} = \mathbf{s} + \mathbf{g}$$

$$\mathrm{H}_0\colon \mathbf{x} = \mathbf{g}$$

Using the multivariate Gaussian distribution, the pdf for H_0 is

$$p_0(\mathbf{x}) = \frac{\exp[-(1/2)\,\mathbf{x}^T\,\mathbf{R}_g^{-1}\,\mathbf{x}]}{(2\pi)^{N/2}\,|\mathbf{R}_g|^{1/2}}$$

where $\mathbf{R}_g$ is the noise correlation matrix. Similarly, the H_1 pdf is

$$p_1(\mathbf{x}) = \frac{\exp[-(1/2)(\mathbf{x}-\mathbf{s})^T\,\mathbf{R}_g^{-1}\,(\mathbf{x}-\mathbf{s})]}{(2\pi)^{N/2}\,|\mathbf{R}_g|^{1/2}} \tag{3.4}$$

We form the likelihood ratio

$$\begin{aligned}\lambda'_T &= \frac{p_1(\mathbf{x})}{p_0(\mathbf{x})}\\ &= \exp[(1/2)(\mathbf{s}^T\,\mathbf{R}_g^{-1}\,\mathbf{x} + \mathbf{x}^T\,\mathbf{R}_g^{-1}\,\mathbf{s} - \mathbf{s}^T\,\mathbf{R}_g^{-1}\,\mathbf{s})]\end{aligned} \tag{3.5}$$

Because $\exp(z)$ is a monotonically increasing function of its argument, we can form a test statistic based on the argument in (3.5). By noting that the last term in this argument does not depend on the observed vector $\mathbf{x}$ and the first term is merely the transpose of the second, a simpler expression for the test statistic is

$$\lambda_T = \mathbf{s}^T\,\mathbf{R}_g^{-1}\,\mathbf{x}$$

The test statistic, λ_T, is a Gaussian random variable because the samples of $\mathbf{x}$ are jointly Gaussian random variables.

Receiver-Operating Characteristic Curves for Gaussian Variables

We are interested in obtaining an equation relating P_D, P_{fa}, and S/N at the output of the detector. The area under the Gaussian pdf from a threshold, y, to infinity is required to obtain P_D and P_{fa}. This area is

$$\frac{1}{\sqrt{2\pi}\,\sigma}\int_y^{\infty} \exp[-(t-\mu)^2/(2\sigma^2)]\,dt$$

By letting $w = (t - \mu)/\sigma$, this area is equal to

$$\frac{1}{\sqrt{2\pi}} \int_{(y-\mu)/\sigma}^{\infty} \exp(-w^2/2)\, \mathrm{d}w = \operatorname{erfc}[(y - \mu)/\sigma]$$

where the complementary error function is defined by

$$\operatorname{erfc}(y) = \frac{1}{\sqrt{2\pi}} \int_{y}^{\infty} \exp(-t^2/2)\, \mathrm{d}t$$

The P_{fa} can be expressed in terms of erfc(z) by

$$P_{fa} = P(\lambda_T > \beta\,|\mathrm{H}_0) = \operatorname{erfc}\left(\frac{\beta - E\{\lambda_T|\mathrm{H}_0\}}{\sigma\{\lambda_T|\mathrm{H}_0\}}\right)$$

Solving for the threshold, β, in terms of the specified value of P_{fa} gives

$$\beta = \sigma\{\lambda_T|\mathrm{H}_0\} \operatorname{erfc}^{-1}(P_{fa}) + E\{\lambda_T|\mathrm{H}_0\} \tag{3.6}$$

In a similar fashion, P_D can be expressed in terms of erfc(z):

$$P_D = P(\lambda_T > \beta|\mathrm{H}_1) = \operatorname{erfc}\left(\frac{\beta - E\{\lambda_T|\mathrm{H}_1\}}{\sigma\{\lambda_T|\mathrm{H}_1\}}\right)$$

The output of the detector is the test statistic, λ_T. We are interested in the change in the expected value of this quantity when a signal is present, as opposed to the noise-only case. The deflection

$$\sqrt{d} = \frac{E\{\lambda_T|\mathrm{H}_1\} - E\{\lambda_T|\mathrm{H}_0\}}{\sigma\{\lambda_T|\mathrm{H}_0\}} \tag{3.7}$$

is the normalized change in λ_T in the presence of the signal. The square of this quantity, or d, is the S/N at the output of the detector and 10 log$d$ is termed the *detection index*. The desired expression relating d, P_D, and P_{fa} can now be obtained:

$$P_D = \operatorname{erfc}\left[\frac{\operatorname{erfc}^{-1}(P_{fa}) - \sqrt{d}}{v}\right] \tag{3.8}$$

where

$$v = \frac{\sigma\{\lambda_T|\mathrm{H}_1\}}{\sigma\{\lambda_T|\mathrm{H}_0\}}$$

This expression defines the *receiver-operating characteristic* (ROC) curves. We can plot, for example, P_D as a function of P_{fa} and parameterized on d, as shown in Figure 3.6.

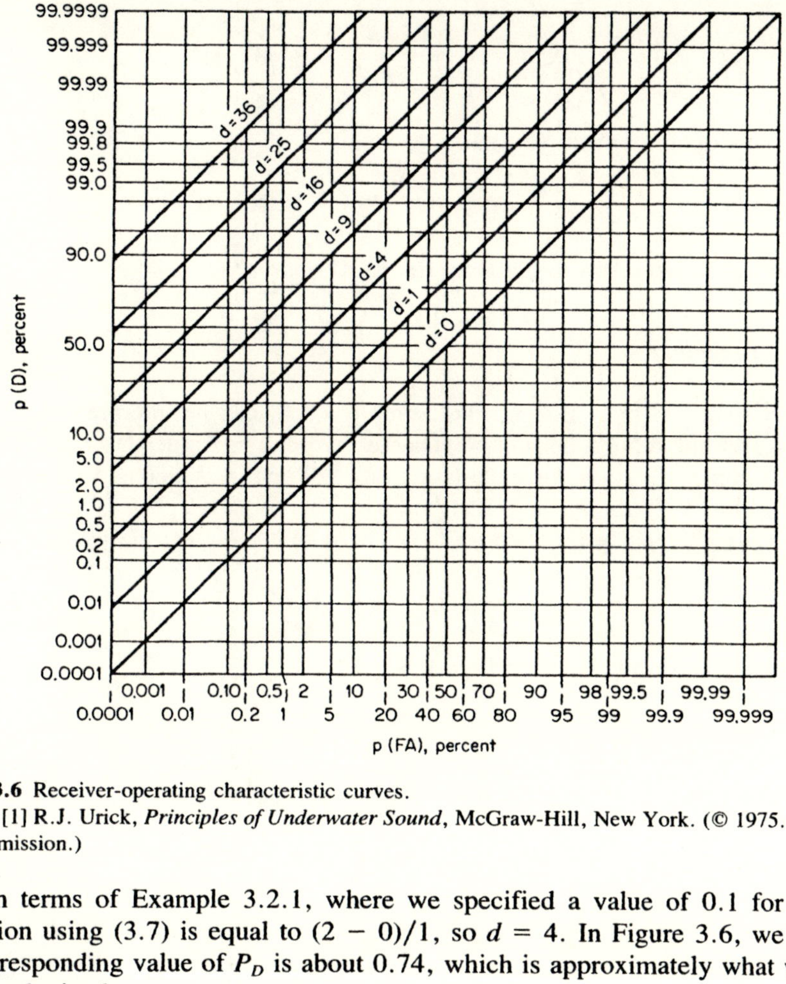

Figure 3.6 Receiver-operating characteristic curves.
Source: [1] R.J. Urick, *Principles of Underwater Sound*, McGraw-Hill, New York. (© 1975. Reprinted with permission.)

In terms of Example 3.2.1, where we specified a value of 0.1 for P_{fa}, the deflection using (3.7) is equal to $(2 - 0)/1$, so $d = 4$. In Figure 3.6, we see that the corresponding value of P_D is about 0.74, which is approximately what was previously obtained.

For a known waveform in Gaussian noise (Example 3.2.2) the ROC curves are generated as follows: The expectations and variances for the test statistic, λ_T, under the H_0 and H_1 hypotheses are

$$E\{\lambda_T|H_0\} = E\{\mathbf{s}^T \mathbf{R}_g^{-1} \mathbf{g}\} = 0$$

$$E\{\lambda_T|H_1\} = E\{\mathbf{s}^T \mathbf{R}_g^{-1} (\mathbf{s} + \mathbf{g})\} = \mathbf{s}^T \mathbf{R}_g^{-1} \mathbf{s}$$

$$\sigma^2\{\lambda_T|H_0\} = \sigma^2\{\lambda_T|H_1\} = \mathbf{s}^T \mathbf{R}_g^{-1} \mathbf{s}$$

Using (3.6) and (3.7), the threshold, β, and the deflection, $\sqrt{d}$, are given by

$$\beta = (\mathbf{s}^T \mathbf{R}_g^{-1} \mathbf{s})^{1/2} \operatorname{erfc}^{-1}(P_{fa})$$

$$\sqrt{d} = \frac{\mathbf{s}^T \mathbf{R}_g^{-1} \mathbf{s}}{(\mathbf{s}^T \mathbf{R}_g^{-1} \mathbf{s})^{1/2}} = (\mathbf{s}^T \mathbf{R}_g^{-1} \mathbf{s})^{1/2}$$

Finally, P_D is obtained from (3.8):

$$P_D = \operatorname{erfc}[\operatorname{erfc}^{-1}(P_{fa}) - (\mathbf{s}^T \mathbf{R}_g^{-1} \mathbf{s})^{1/2}]$$

This equation is then used to generate the ROC curves.

Correlation receiver

The test statistic, $\lambda_T = \mathbf{s}^T \mathbf{R}_g^{-1} \mathbf{x}$, for Example 3.2.2, can be interpreted as a correlation receiver, or replica correlator. That is, we can examine this expression and deduce an implementation for calculating λ_T. Following Whalen [4, Section 6.6], we can perform a Cholesky factorization of $\mathbf{R}_g^{-1}$ and obtain $\mathbf{R}_g^{-1} = \mathbf{L}^T \mathbf{L}$, where $\mathbf{L}$ is a lower triangular matrix (Section 7.2.4). The test statistic can be written as

$$\lambda_T = \mathbf{s}^T \mathbf{L}^T \mathbf{L} \mathbf{x} = (\mathbf{L} \mathbf{s})^T (\mathbf{L} \mathbf{x})$$

This implementation is shown in Figure 3.7.

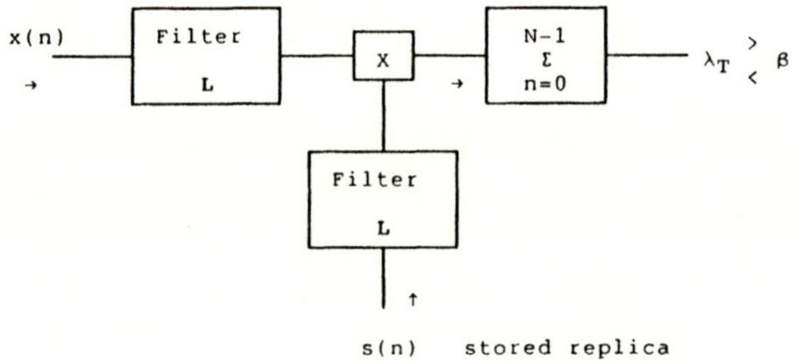

Figure 3.7 Replica correlator.

3.2.2 Broadband Detector for a Linear Array of Sensors

This section treats a generalization of Example 3.2.2 in which we consider N outputs from each of M sensors, rather than from a single sensor. The goal is to obtain the

NP detector for a linear array of M equally spaced sensors. We observe the sequences

$$x_i(n) = v_i(n) + g_i(n), \quad i = 1, \ldots, M$$
$$n = 0, \ldots, N-1$$

The signal, $v_i(n)$, and noise, $g_i(n)$, are sample sequences of uncorrelated stationary Gaussian random processes.

A standard approach in the analysis of this type of problem is to use the DFT of these time-domain sequences, where from (1.12)

$$X_i(k) = \sum_{n=0}^{N-1} x_i(n)\, e^{-j2\pi nk/N}, \quad k = 0, \ldots, N-1$$

The utility of this approach results from the fact that, for a sufficiently long observation time, $T = N/f_s$, where f_s is the sampling frequency, the DFT outputs are not correlated for different frequencies. (These outputs are approximately uncorrelated if T is large compared to the inverse bandwidth of $x_i(n)$.) For real sequences, frequency indices greater than $N/2$ need not be used because the DFT outputs for $0 \leqslant k \leqslant N/2$ completely describe $x_i(n)$. This follows from the complex conjugate symmetry of the DFT, where $X(N-k) = X^*(k)$ for $0 \leqslant k \leqslant N/2$. It is convenient to organize the DFTs of the sensors by frequency index to form the vector

$$\mathbf{x} = \{X_1(0), \ldots, X_M(0), X_1(1), \ldots, X_M(1), \ldots, X_1(N/2-1), \ldots, X_M(N/2-1)\}^T$$

A signal vector, $\mathbf{v}$, and a noise vector, $\mathbf{g}$, are formed in the same manner. The corresponding *cross-power spectral density* (cpsd) matrices are defined by $\mathbf{S}_x = E\{\mathbf{x}\ \mathbf{x}^+\}$, $\mathbf{S}_v = E\{\mathbf{v}\ \mathbf{v}^+\}$, and $\mathbf{S}_g = E\{\mathbf{g}\ \mathbf{g}^+\}$, respectively. These $MN/2 \times MN/2$ matrices are block-diagonal in structure, because of the assumption of uncorrelated DFT outputs for different frequencies. Each block is $M \times M$ and corresponds to one frequency. As an example, the cpsd matrix for $\mathbf{x}$ is provided in Table 3.2.

Table 3.2

Structure of the Cross-Power Spectral Density Matrix (each block is $M \times M$)

$$\mathbf{S}_x = \begin{bmatrix} [\mathbf{S}_x(0)] & & \\ & [\mathbf{S}_x(1)] & \\ & & [\mathbf{S}_x(N/2-1)] \end{bmatrix}$$

The Broadband Detection Problem

We now express this problem in terms of the hypotheses:

$$\begin{aligned} H_1: &\quad \mathbf{x} = \mathbf{v} + \mathbf{g} \\ H_0: &\quad \mathbf{x} = \mathbf{g} \end{aligned} \tag{3.9}$$

Using the multivariate complex Gaussian distribution, we obtain

$$p_0(\mathbf{x}) = \frac{\exp(-\mathbf{x}^+ \mathbf{S}_g^{-1} \mathbf{x})}{\pi^{MN/2} |\mathbf{S}_g|}$$

$$\tag{3.10}$$

$$p_1(\mathbf{x}) = \frac{\exp[-\mathbf{x}^+ (\mathbf{S}_v + \mathbf{S}_g)^{-1} \mathbf{x}]}{\pi^{MN/2} |\mathbf{S}_v + \mathbf{S}_g|}$$

The likelihood ratio corresponding to (3.9) is then equal to

$$\lambda_T' = \frac{|\mathbf{S}_g|}{|\mathbf{S}_v + \mathbf{S}_g|} \exp\{\mathbf{x}^+ [\mathbf{S}_g^{-1} - (\mathbf{S}_v + \mathbf{S}_g)^{-1}] \mathbf{x}\}$$

Using the matrix identity,

$$\mathbf{A}^{-1} - (\mathbf{B} + \mathbf{A})^{-1} = \mathbf{A}^{-1} \mathbf{B} (\mathbf{B} + \mathbf{A})^{-1} \tag{3.11}$$

the likelihood ratio can be written as

$$\lambda_T' = \frac{|\mathbf{S}_g|}{|\mathbf{S}_v + \mathbf{S}_g|} \exp[\mathbf{x}^+ \mathbf{S}_g^{-1} \mathbf{S}_v (\mathbf{S}_v + \mathbf{S}_g)^{-1} \mathbf{x}]$$

We assume that the matrices in (3.11), as well as in other matrix identities in this section, have consistent dimensions and that the indicated inverses exist. Because $\mathbf{S}_v$ and $\mathbf{S}_g$ are block diagonal as shown in Table 3.2, it is convenient to use the log-likelihood ratio:

$$\ln \lambda_T' = \ln \frac{|\mathbf{S}_g|}{|\mathbf{S}_v + \mathbf{S}_g|} + \sum_{k=0}^{N/2-1} \mathbf{x}^+(k)\, \mathbf{S}_g^{-1}(k) \mathbf{S}_v(k) [\mathbf{S}_v(k) + \mathbf{S}_g(k)]^{-1} \mathbf{x}(k) \tag{3.12}$$

where $\mathbf{x}(k)$ is a length-M vector. The $M \times M$ matrices, $\mathbf{S}_v(k)$ and $\mathbf{S}_g(k)$, are the signal and noise spatial correlation matrices, respectively, for the kth frequency. The first term in (3.12) consists of known quantities, so we can use as a test statistic

$$\lambda_T = \sum_{k=0}^{N/2-1} \mathbf{x}^+(k)\, \mathbf{S}_g^{-1}(k)\mathbf{S}_v(k)[\mathbf{S}_v(k) + \mathbf{S}_g(k)]^{-1}\, \mathbf{x}(k) \tag{3.13}$$

Plane Wave Arrival

Equation (3.13) is quite general. For instance, we can let $v_i(n)$ be a broadband signal propagating as a plane wave with a conic angle, ψ, relative to the linear array of sensors, so that

$$v_i(n) = s[n + (i-1)\,\Delta] \quad \text{where } \Delta = f_s d(\sin\psi)/c$$

Assuming that $N \gg (i-1)\Delta$ and that Δ is an integer, we have from Section 2.3.1 that the DFT of $v_i(n)$ is approximately equal to

$$V_i(k) = \exp[\mathrm{j}2\pi k(i-1)\Delta/N]\, S(k)$$

or, in vector form:

$$\mathbf{v}(k) = S(k)\, \mathbf{p}(k)$$

The vector of phase terms, $\mathbf{p}(k)$, defined in Section 2.3.2, is equal to

$$\mathbf{p}(k) = \{1, \exp(\mathrm{j}2\pi k\Delta/N), \ldots, \exp[\mathrm{j}2\pi k(M-1)\Delta/N]\}^T$$

With this definition for $V_i(k)$, we can calculate the signal spatial correlation matrix for the kth frequency:

$$\begin{aligned}\mathbf{S}_v(k) &= E\{\mathbf{v}(k)\, \mathbf{v}^+(k)\} = E\{S(k)\mathbf{p}(k)\, \mathbf{p}^+(k)S^*(k)\}\\ &= S_s(k)\, \mathbf{p}(k)\mathbf{p}^+(k)\end{aligned}$$

where $S_s(k)$ is the power spectral density of $s(n)$ at the kth frequency. The test statistic (3.13) is then given by

$$\lambda_T = \sum_{k=0}^{N/2-1} \mathbf{x}^+(k)\, \mathbf{S}_g^{-1}(k)S_s(k)\mathbf{p}(k)\mathbf{p}^+(k) \cdot [S_s(k)\mathbf{p}(k)\mathbf{p}^+(k) + \mathbf{S}_g(k)]^{-1}\, \mathbf{x}(k) \tag{3.14}$$

Using the matrix identity

$$\mathbf{AB}^+(\mathbf{C} + \mathbf{BAB}^+)^{-1} = (\mathbf{A}^{-1} + \mathbf{B}^+\mathbf{C}^{-1}\mathbf{B})^{-1}\mathbf{B}^+\mathbf{C}^{-1}$$

and associating **B** with $\sqrt{S_s(k)}\mathbf{p}(k)$, **C** with $\mathbf{S}_g(k)$, and **A** with the scalar of 1, (3.14) becomes

$$\lambda_T = \sum_{k=0}^{N/2-1} \frac{\mathbf{x}^+(k)\,\mathbf{S}_g^{-1}(k)\sqrt{S_s(k)}\mathbf{p}(k)\sqrt{S_s(k)}\mathbf{p}^+(k)\mathbf{S}_g^{-1}(k)\,\mathbf{x}(k)}{1 + \sqrt{S_s(k)}\mathbf{p}^+(k)\mathbf{S}_g^{-1}\sqrt{S_s(k)}\mathbf{p}(k)}$$

which is equivalent to

$$\lambda_T = \sum_{k=0}^{N/2-1} \frac{S_s(k)\,|\mathbf{p}^+(k)\,\mathbf{S}_g^{-1}(k)\,\mathbf{x}(k)|^2}{1 + S_s(k)\mathbf{p}^+(k)\mathbf{S}_g^{-1}(k)\mathbf{p}(k)} \tag{3.15}$$

To visualize the form of the NP detector, we define the filter

$$F^2(k) = \frac{S_s(k)/S_g^2(k)}{1 + S_s(k)\,\mathbf{p}^+(k)\,\mathbf{Q}^{-1}(k)\,\mathbf{p}(k)/S_g(k)} \tag{3.16}$$

where the noise spatial correlation matrix has been replaced with $\mathbf{S}_g(k) = S_g(k)\,\mathbf{Q}(k)$. The quantity, $S_g(k)$, is the noise power spectral density at the kth frequency (assumed to be the same for all sensors). The following expression for the test statistic is obtained by substituting (3.16) into (3.15):

$$\lambda_T = \sum_{k=0}^{N/2-1} |F(k)\,\mathbf{p}^+(k)\,\mathbf{Q}^{-1}(k)\,\mathbf{x}(k)|^2 \tag{3.17}$$

We can consider the product, $\mathbf{p}^+(k)\,\mathbf{Q}^{-1}(k) = \mathbf{h}^+(k)$, as a vector of filters, one for each sensor. The product

$$\mathbf{h}^+(k)\,\mathbf{x}(k) = \sum_{i=1}^{M} H_i^*(k)\,X_i(k)$$

is the sum of filtered sensor outputs. The test statistic (3.17) then becomes

$$\lambda_T = \sum_{k=0}^{N/2-1} \left| F(k) \left[\sum_{i=1}^{M} H_i^*(k)\,X_i(k) \right] \right|^2$$

Time-Domain Representation

Using a general form of Parseval's identity,

$$\sum_{n=0}^{N-1} x_1(n)x_2^*(n) = \frac{1}{N}\sum_{k=0}^{N-1} X_1(k)X_2^*(k) \tag{3.18}$$

the test statistic can be expressed in terms of time-domain sequences. Because we are dealing with real input sequences so that $X(N-k) = X^*(k)$, $0 \leq k \leq N/2$, it follows from (3.18) that

$$\sum_{n=0}^{N-1} x(n)x^*(n) = \frac{2}{N}\sum_{k=0}^{N/2-1} X(k)X^*(k)$$

where it is assumed that $X(0)$ and $X(N/2)$ are zero. The sensor signals, $x_i(n)$, are convolved with filter impulse responses, $h_i(n)$, and summed. The output of this summation is convolved with the filter impulse response, $f(n)$. Finally the test statistic, λ_T, is obtained by summing the squared magnitude of the output of filter, $f(n)$, over the N samples available. The expression is

$$\lambda_T = \frac{N}{2}\sum_{n=0}^{N-1}\left| f(n) \otimes \left[\sum_{i=1}^{M} h_i^*(n) \otimes x_i(n)\right]\right|^2$$

where $\otimes$ denotes the convolution operator. The set of operations involved in calculating the test statistic is implemented by the detector in Figure 3.8.

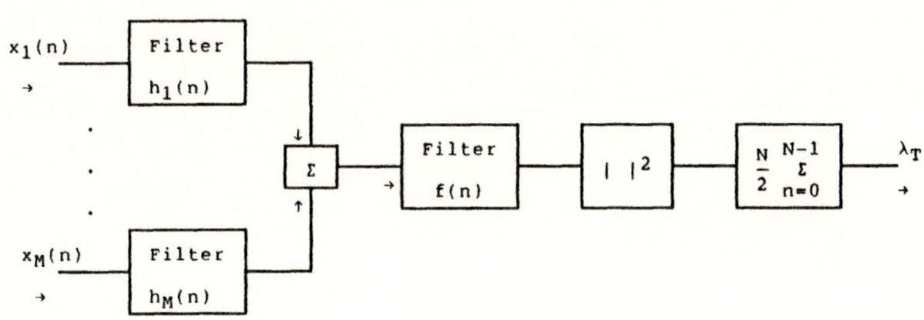

Figure 3.8 Neyman-Pearson broadband detector for an array of sensors.

For a long observation time, λ_T can be assumed to be Gaussian distributed from the central limit theorem. We then can derive the Gaussian ROC curves (3.8). The required quantities are

$$\sqrt{d} = \frac{\sum_{k=0}^{N/2-1} F^2(k)S_s(k)|\mathbf{p}^+(k)\mathbf{Q}^{-1}(k)\mathbf{p}(k)|^2}{\left[\sum_{k=0}^{N/2-1} F^4(k)S_g^2(k)|\mathbf{p}^+(k)\mathbf{Q}^{-1}(k)\mathbf{p}(k)|^2\right]^{1/2}}$$

$$\sigma^2\{\lambda_T|\mathrm{H}_1\} = \sum_{k=0}^{N/2-1} F^4(k)|\mathbf{p}^+(k)\mathbf{Q}^{-1}(k)\mathbf{p}(k)|^2[S_s(k)\mathbf{p}^+(k)\mathbf{Q}^{-1}(k)\mathbf{p}(k) + S_g(k)]^2$$

$$\sigma^2\{\lambda_T|\mathrm{H}_0\} = \sum_{k=0}^{N/2-1} F^4(k)S_g^2(k)|\mathbf{p}^+(k)\mathbf{Q}^{-1}(k)\mathbf{p}(k)|^2 \tag{3.19}$$

$$E\{\lambda_T|\mathrm{H}_0\} = \sum_{k=0}^{N/2-1} F^2(k)S_g(k)|\mathbf{p}^+(k)\mathbf{Q}^{-1}(k)\mathbf{p}(k)|$$

With these quantities, the threshold (3.6) and the ROC curves (3.8) can be determined.

For the kth frequency, the array gain from Section 2.3.2 is

$$\text{Array gain} = \frac{|\mathbf{w}^+(k)\mathbf{p}(k)|^2}{\mathbf{w}^+(k)\mathbf{Q}(k)\mathbf{w}(k)}$$

In the NP detector, $\mathbf{w}(k)$ corresponds to the vector of filters, $\mathbf{h}(k) = \mathbf{Q}^{-1}(k)\mathbf{p}(k)$. Then the array gain is

$$\text{Array Gain} = \frac{|\mathbf{p}^+(k)\mathbf{Q}^{-1}(k)\mathbf{p}(k)|^2}{\mathbf{p}^+(k)\mathbf{Q}^{-1}(k)\mathbf{Q}(k)\mathbf{Q}^{-1}(k)\mathbf{p}(k)} = \mathbf{p}^+(k)\mathbf{Q}^{-1}(k)\mathbf{p}(k)$$

It is instructive to consider some simple cases.

Spatially Uncorrelated Noise

If we assume that the noise is spatially uncorrelated, the normalized noise spatial correlation matrix, $\mathbf{Q}(k)$, is equal to the $M \times M$ identity matrix. Then the filters,

$H_i(k)$, are just equal to the phase terms in the vector, $\mathbf{p}(k)$. In the time domain, this corresponds to a shift by $(i - 1)\Delta$. Thus the first portion of the detector in Figure 3.8 reduces to the conventional time-domain beamformer in Figure 2.11. The required quantities in (3.19) for calculating the ROC curves (3.8) are

$$\sqrt{d} = \frac{M \sum_{k=0}^{N/2-1} F^2(k) S_s(k)}{\left[\sum_{k=0}^{N/2-1} F^4(k) S_g^2(k) \right]^{1/2}}$$

$$\sigma^2\{\lambda_T | \mathrm{H}_1\} = M^2 \sum_{k=0}^{N/2-1} F^4(k) [M S_s(k) + S_g(k)]^2$$

$$\sigma^2\{\lambda_T | \mathrm{H}_0\} = M^2 \sum_{k=0}^{N/2-1} F^4(k) S_g^2(k)$$

$$E\{\lambda_T | \mathrm{H}_0\} = M \sum_{k=0}^{N/2-1} F^2(k) S_g(k)$$

Defining an effective noise bandwidth, β_{eff},

$$\beta_{\mathrm{eff}} = \frac{\left[\sum_{k=0}^{N/2-1} F^2(k) S_g(k) \right]^2}{\sum_{k=0}^{N/2-1} F^4(k) S_g^2(k)}$$

the expression for deflection is

$$\sqrt{d} = \sqrt{\beta_{\mathrm{eff}}}\, M \frac{\sum_{k=0}^{N/2-1} F^2(k) S_s(k)}{\sum_{k=0}^{N/2-1} F^2(k) S_g(k)}$$

The effective noise bandwidth is an indication of the number of independent samples available for integration. Using the Schwartz inequality, it follows that $\beta_{\mathrm{eff}} \leq N/2$, or half the number of data samples available for processing. This corresponds to the case of white signal and noise spectra, which is treated next.

White Signal and Noise Spectra

The assumption of white signal and noise spectra means that $S_s(k) = S_s$ and $S_g(k) = S_g$ for $0 \leq k \leq N/2$. The quantities required to obtain the ROC curves are

$$\sqrt{d} = M\sqrt{N/2}\, S_s/S_g$$
$$\sigma^2\{\lambda_T|H_1\} = M^2(N/2)F^4(MS_s + S_g)^2$$
$$\sigma^2\{\lambda_T|H_0\} = M^2(N/2)F^4S_g^2$$
$$E\{\lambda_T|H_0\} = M(N/2)F^2S_g$$

so that (3.8) becomes

$$P_D = \text{erfc}\left[\frac{\text{erfc}^{-1}(P_{fa}) - M\sqrt{N/2}S_s/S_g}{1 + MS_s/S_g}\right]$$

The expression for deflection points out the effects of the coherent gain, M, resulting from the beamforming, and the incoherent gain, $\sqrt{N/2}$, due to postdetection integration. The coherent gain, which involves a knowledge of phase, is proportional to the number of sensors. The incoherent gain, which does not involve a knowledge of phase, is proportional to the square root of the observation time, or equivalently, the square root of the number of independent samples integrated.

3.2.3 Narrowband Detector

A random process is a narrowband random process if the spectrum has the form shown in Figure 3.9, where the bandwidth is much smaller than the center frequency. A sample sequence, $g(n)$, of a zero-mean narrowband random process can be described by its complex envelope (Section 1.1.2):

$$g(n) = 2\,\text{Re}[\tilde{g}(n)\exp(j2\pi f_c n/f_s)]$$

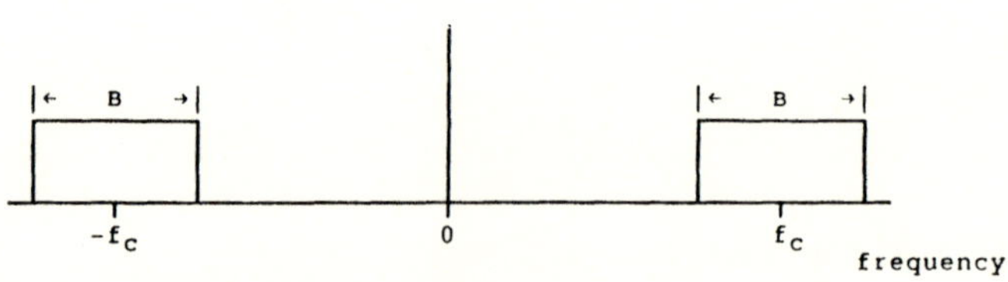

Figure 3.9 Spectrum of a narrowband random process.

Expanding the complex factors yields

$$g(n) = 2g_I(n) \cos 2\pi f_c n/f_s - 2g_Q(n) \sin 2\pi f_c n/f_s$$

where $g_I(n)$ and $g_Q(n)$ denote the real (in-phase) and imaginary (quadrature) components of $\tilde{g}(n)$, respectively. By defining $x(n) \equiv 2g_I(n)$ and $y(n) \equiv 2g_Q(n)$, we obtain a standard representation for a narrowband random process:

$$g(n) = x(n) \cos 2\pi f_c n/f_s - y(n) \sin 2\pi f_c n/f_s$$

Assuming that we are dealing with a Gaussian narrowband random process, then $x(n)$ and $y(n)$ are Gaussian distributed because they can be derived from $g(n)$ via linear operations. Furthermore, we can show that $x(n)$ and $y(n)$ have identical power spectral densities and, therefore, identical autocorrelation functions. In particular, the variances of $x(n)$ and $y(n)$ are equal and equal to the variance of $g(n)$.

We can define the envelope, $z(n)$, and phase, $\theta(n)$, of the narrowband process by

$$\begin{aligned} z(n) &= [x^2(n) + y^2(n)]^{1/2} \\ \theta(n) &= \tan^{-1}[y(n)/x(n)] \end{aligned} \tag{3.20}$$

The pdf of the envelope is the Rayleigh function

$$p(z) = \frac{z}{\sigma_g^2} \exp\left(-\frac{z^2}{2\sigma_g^2}\right), \quad z \geqslant 0 \tag{3.21}$$

where σ_g^2 is the variance of $g(n)$. For a sinusoidal signal and a sample sequence of a narrowband Gaussian process, the received sequence is

$$x(n) = A \cos(2\pi f_c n/f_s + \theta) + g(n)$$

and the pdf of the envelope is the Rician function

$$p(z) = \frac{z}{\sigma_g^2} \exp\left(-\frac{z^2 + A^2}{2\sigma_g^2}\right) I_0\left(\frac{Az}{\sigma_g^2}\right), \quad z \geqslant 0 \tag{3.22}$$

where I_0 is the modified Bessel function of zero order defined by [4, p. 104]

$$I_0(x) = \sum_{n=0}^{\infty} \frac{x^{2n}}{2^{2n}(n!)^2}$$

With $A = 0$, the Rician pdf (3.22) reduces to the Rayleigh pdf (3.21), as we would expect. These pdfs are plotted in Figure 3.10 in terms of the normalized variables, $v = z/\sigma_g$ and $\alpha = A/\sigma_g$. The Rayleigh and Rician pdfs are of interest because they will serve as the pdfs for the H_0 and H_1 hypotheses, respectively, in the narrowband detection problem. Consequently, we will equate a desired value of P_{fa} to an area under the Rayleigh pdf to determine a threshold. Then the area under the Rician pdf from this threshold to infinity will be calculated to determine the P_D.

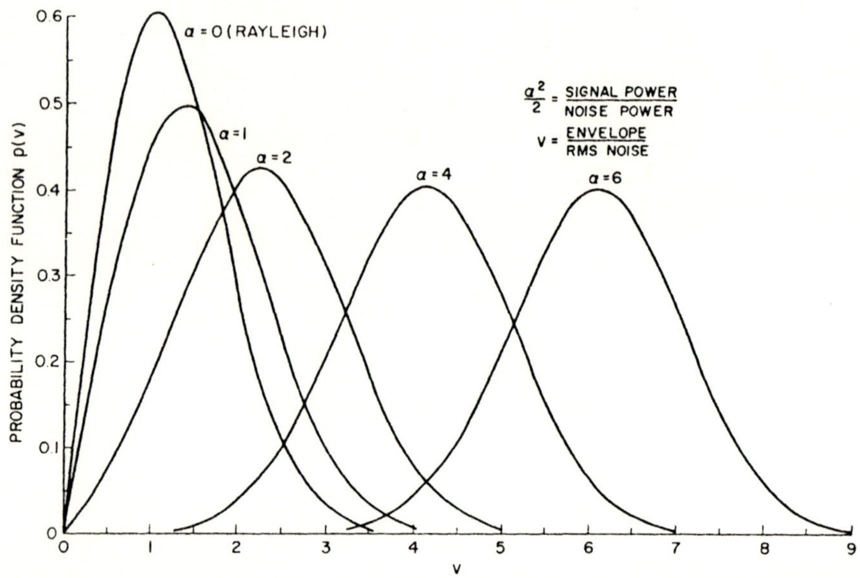

Figure 3.10 Rayleigh and Rician probability density functions; normalized parameters are $v = z/\sigma_g$, $\alpha = A/\sigma_g$).

Source: [4] A.D. Whalen, *Detection of Signals in Noise*, Academic Press, New York. (© 1971. Reprinted with permission.)

Because this is a detection problem, we assume that the signal parameters (with the exception of phase) are known and wish to determine whether the signal is present. This situation arises in active sonar, for instance, where a continuous-wave pulse of known frequency is transmitted. We in general would not expect to know the phase angle, θ, however. This parameter is assumed to be uniformly distributed between zero and 2π rad. The detection problem then is one of choosing between the following hypotheses, based on N measurements.

The Narrowband Detection Problem

$$\mathrm{H}_1\text{: } x(n) = A \sin(2\pi f_c n/f_s + \theta) + g(n), \quad n = 0, \ldots, N-1$$
$$\mathrm{H}_0\text{: } x(n) = g(n)$$

As in the broadband detection problem (Section 3.2.2), we wish to develop the NP detector. This derivation can be found in many sources. In particular, Whalen [4, Chapter 7] develops the NP detector for continuous, rather than sampled, waveforms. The noise process is assumed to be uncorrelated so that the noise correlation matrix is $\mathbf{R}_g = \sigma_g^2 \mathbf{I}$. The required likelihood functions are found by using the multivariate Gaussian distribution (3.4). For H_0 this function is

$$p_0(\mathbf{x}) = \frac{1}{(2\pi)^{N/2}\sigma_g^N} \exp\left[-\frac{1}{2\sigma_g^2}\sum_{n=0}^{N-1} x^2(n)\right]$$

The likelihood function for H_1, conditioned on θ, is

$$p_1(\mathbf{x}|\theta) = \frac{1}{(2\pi)^{N/2}\sigma_g^N} \exp\left\{-\frac{1}{2\sigma_g^2}\sum_{n=0}^{N-1} [x(n) - A\sin(2\pi f_c n/f_s + \theta)]^2\right\}$$

The desired likelihood function, $p_1(\mathbf{x})$, is found by integrating $p_1(\mathbf{x}|\theta)$ with respect to phase. The result is

$$p_1(\mathbf{x}) = \frac{1}{2\pi}\int_0^{2\pi} p_1(\mathbf{x}|\theta)\, d\theta$$

The likelihood ratio is

$$\lambda_T(\theta) = \frac{\dfrac{1}{2\pi}\displaystyle\int_0^{2\pi} \exp\left\{-\frac{1}{2\sigma_g^2}\sum_{n=0}^{N-1} [x(n) - A\sin(2\pi f_c n/f_s + \theta)]^2\right\} d\theta}{\exp\left[-\dfrac{1}{2\sigma_g^2}\displaystyle\sum_{n=0}^{N-1} x^2(n)\right]} \tag{3.23}$$

The sum in the numerator of (3.23) contains the term

$$\sum_{n=0}^{N-1} A^2 \sin^2(2\pi f_c n/f_s + \theta) \tag{3.24}$$

If the input data sequence contains an integer number of half cycles of $\sin(2\pi f_c n/f_s)$, then (3.24) equals $A^2N/2$. This condition is equivalent to $2Nf_c/f_s = m$, where m is an integer. If this condition is not satisfied, a second term is present in (3.24). This term can be neglected, however, if the observation time of N/f_s is much larger than one period of the sinusoidal signal $(1/f_c)$. With this assumption, (3.23) becomes

$$\lambda_T(\theta) = \exp\left(-\frac{A^2N}{4\sigma_g^2}\right)\frac{1}{2\pi}\int_0^{2\pi}\exp\left[\frac{A}{\sigma_g^2}\sum_{n=0}^{N-1}x(n)\sin(2\pi f_c n/f_s + \theta)\right]d\theta \quad (3.25)$$

Expanding the sum in (3.25) gives

$$\cos\theta\sum_{n=0}^{N-1}x(n)\sin(2\pi f_c n/f_s) + \sin\theta\sum_{n=0}^{N-1}x(n)\cos(2\pi f_c n/f_s)$$

By defining the quantities q and θ_0 so that

$$\begin{aligned} q\cos\theta_0 &\equiv \sum_{n=0}^{N-1}x(n)\sin(2\pi f_c n/f_s) \\ q\sin\theta_0 &\equiv \sum_{n=0}^{N-1}x(n)\cos(2\pi f_c n/f_s) \end{aligned} \quad (3.26)$$

the likelihood ratio (3.25) becomes

$$\lambda_T(\theta) = \exp\left(-\frac{A^2N}{4\sigma_g^2}\right)\frac{1}{2\pi}\int_0^{2\pi}\exp\left[\frac{Aq}{\sigma_g^2}\cos(\theta-\theta_0)\right]d\theta \quad (3.27)$$

Using the definition for the modified Bessel function [4, eq. (4-50)],

$$I_0(y) = \frac{1}{2\pi}\int_0^{2\pi}\exp[y\cos(\theta-\theta_0)]\,d\theta$$

the dependence on θ is removed and (3.27) simplifies to

$$\lambda_T = \exp\left(-\frac{A^2N}{4\sigma_g^2}\right)I_0\left(\frac{Aq}{\sigma_g^2}\right)$$

Because the modified Bessel function is a monotonically increasing function with respect to its argument, we can simply specify a decision rule for the optimum detector that is based on q, where, from (3.26),

$$q^2 = \left[\sum_{n=0}^{N-1} x(n)\sin(2\pi f_c n/f_s)\right]^2 + \left[\sum_{n=0}^{N-1} x(n)\cos(2\pi f_c n/f_s)\right]^2$$

H_0 and H_1 pdfs for q

The implementation, shown in Figure 3.11, is called the *quadrature receiver*. In this figure, the output of the summing operations are identified as a and b so that $q^2 = a^2 + b^2$. Such output is Gaussian distributed because the input noise is Gaussian distributed. We therefore can identify q as the envelope of a narrowband Gaussian process (3.20), so that the pdf of q is Rayleigh under the H_0 hypothesis and Rician under the H_1 hypotheses. It remains to develop the likelihood functions for q. We start by determining the first and second moments of a and b. For the H_1 hypothesis,

$$E\{a|\theta\} = \sum_{n=0}^{N-1} A\sin(2\pi f_c n/f_s + \theta)\sin(2\pi f_c n/f_s)$$

$$\approx (AN/2)\cos\theta$$

$$\sigma^2\{a|\theta\} \approx \sigma_g^2 N/2 \equiv \sigma^2$$

Similarly,

$$E\{b|\theta\} = \sum_{n=0}^{N-1} A\sin(2\pi f_c n/f_s + \theta)\cos(2\pi f_c n/f_s)$$

$$\approx (AN/2)\sin\theta$$

$$\sigma^2\{b|\theta\} \approx \sigma_g^2 N/2$$

$$E\{ab|\theta\} \approx E\{a|\theta\}E\{b|\theta\}$$

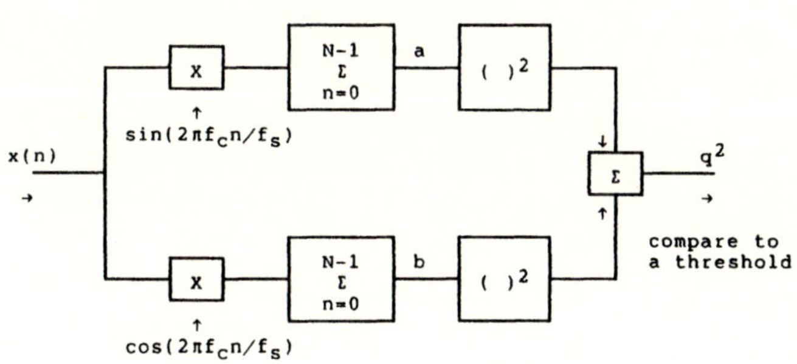

Figure 3.11 Quadrature receiver.

Therefore, a and b are not correlated. Because a and b are Gaussian distributed, they also are independent ($p(a, b) = p(a)p(b)$). These approximations are valid for $N/f_s \gg 1/f_c$. Equality occurs if $2Nf_c/f_s$ is equal to an integer. We now are in a position to write the joint pdf for a and b, conditioned on θ. This has the following form:

$$p_1(a, b|\theta) = \frac{1}{2\pi\sigma^2} \exp\left(-\frac{1}{2\sigma^2}\{[a - (AN/2)\cos\theta]^2 + [b - (AN/2)\sin\theta]^2\}\right)$$

Applying the transformation $a = q\cos\theta_0$, $b = q\sin\theta_0$, $\theta_0 = \tan^{-1}(b/a)$ to this pdf and simplifying yields

$$p_1(q, \theta_0|\theta) = \frac{q}{2\pi\sigma^2} \exp\left[\frac{q^2 + (AN/2)^2}{-2\sigma^2}\right] \exp\left[\frac{qAN}{2\sigma^2}\cos(\theta - \theta_0)\right]$$

As was done with (3.27), we integrate with respect to $\theta_0 (0 \leqslant \theta_0 \leqslant 2\pi)$ and obtain

$$p_1(q|\theta) = \frac{q}{\sigma^2} \exp\left[\frac{q^2 + (AN/2)^2}{-2\sigma^2}\right] I_0\left(\frac{qAN}{2\sigma^2}\right)$$

Finally, $p_1(q)$ is found by integrating with respect to θ, where θ again is assumed to be uniformly distributed over the interval zero to 2π rad. This pdf is

$$p_1(q) = \frac{q}{\sigma^2} \exp\left[\frac{q^2 + (AN/2)^2}{-2\sigma^2}\right] I_0\left(\frac{qAN}{2\sigma^2}\right) \tag{3.28}$$

which is the Rician pdf (3.22) with A replaced by $AN/2$ and σ_g^2 replaced by σ^2. The pdf for q under the H_0 hypothesis is just (3.28) with $A = 0$:

$$p_0(q) = \frac{q}{\sigma^2} \exp\left(\frac{q^2}{-2\sigma^2}\right) \tag{3.29}$$

Receiver-Operating Characteristic Curves for the Narrowband Detector

We again are interested in generating a set of ROC curves. A threshold is defined by using (3.29) for a specified value of P_{fa} with the following equation:

$$P_{fa} = \int_{\kappa}^{\infty} p_0(q)\,dq = \int_{\kappa}^{\infty} \frac{q}{\sigma^2} \exp\left(\frac{q^2}{-2\sigma^2}\right) dq \tag{3.30}$$

which simplifies to $P_{fa} = \exp[-\kappa^2/(2\sigma^2)]$. Therefore, the threshold, κ, is equal to $(-2\sigma^2 \ln P_{fa})^{1/2}$. Next, P_D is calculated by using (3.28):

$$P_D = \int_{\kappa}^{\infty} p_1(q)\, \mathrm{d}q = \int_{\kappa}^{\infty} \frac{q}{\sigma^2} \exp\left[\frac{q^2 + (AN/2)^2}{-2\sigma^2}\right] I_0\left(\frac{qAN}{2\sigma^2}\right) \mathrm{d}q$$

With the change of variable, $\nu = q/\sigma$, and setting $\alpha = AN/(2\sigma)$, we have

$$P_D = \int_{\beta}^{\infty} \nu \exp\left(-\frac{\nu^2 + \alpha^2}{2}\right) I_0(\nu\alpha)\, \mathrm{d}\nu = Q(\alpha, \beta) \tag{3.31}$$

where $\beta = \kappa/\sigma$. The function, $Q(\alpha, \beta)$, shown in Figure 3.12, is the Marcum Q-function, or one minus the cumulative Rician pdf. Expressions (3.30) and (3.31) specify the operation of the narrowband detector. Figure 3.13 shows a family of ROC curves with S/N (in dB) as a function of P_D and parameterized on P_{fa}. The

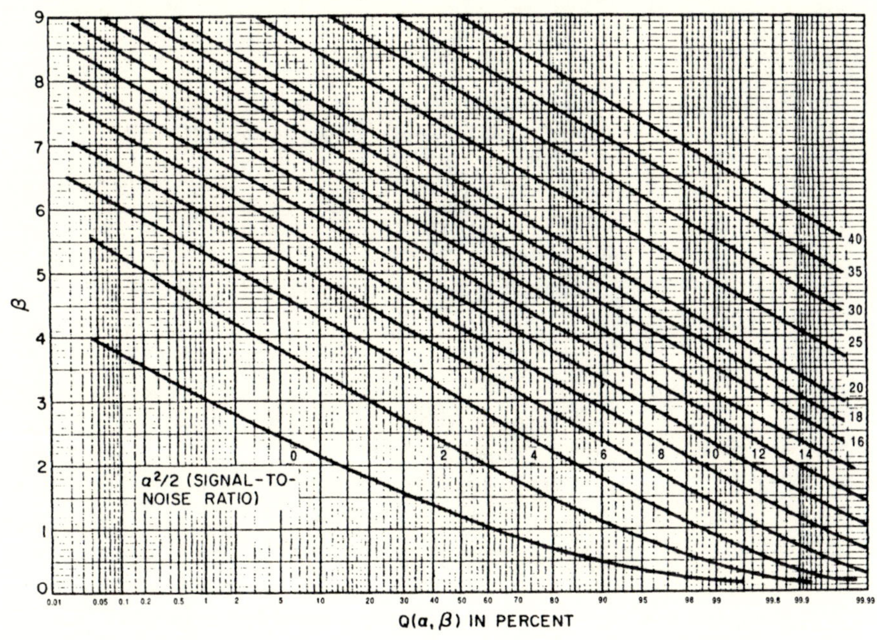

Figure 3.12 Marcum Q-function.
Source: [4] A.D. Whalen, *Detection of Signals in Noise*, Academic Press, New York. (© 1971. Reprinted with permission.)

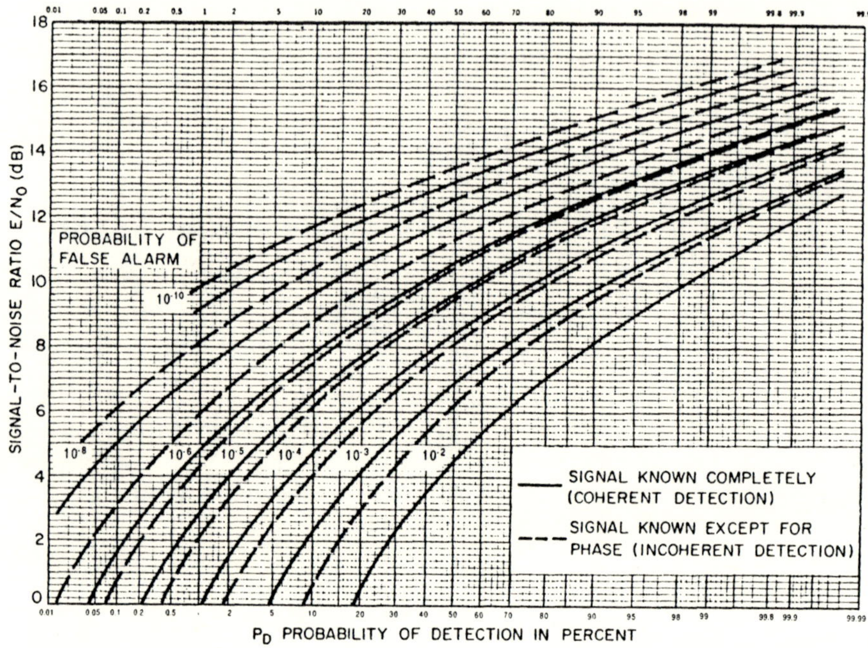

Figure 3.13 Receiver-operating characteristic curves for the narrowband detector.
Source: [4] A.D. Whalen, *Detection of Signals in Noise*, Academic Press, New York. (© 1971. Reprinted with permission.)

dashed lines are applicable because we assume no *a priori* knowledge of phase. The S/N used in Figure 3.13 is

$$S/N = \frac{1}{2}\left(\frac{AN}{2}\right)^2 \frac{1}{\sigma^2} = \alpha^2/2$$

Example 3.2.3

We wish to determine P_D for $P_{fa} = 10^{-4}$ and $S/N = 8$ (9 dB). The normalized threshold, β, is equal to $(-2 \ln P_{fa})^{1/2}$ or 4.3. For an S/N value of 8, the corresponding value of α is $(2\ S/N)^{1/2}$, or 4. Then the Marcum-Q function is equal to about 0.44, from Figure 3.12. That is, for a value of P_{fa} of 10^{-4} and a S/N of 8, the P_D equals 0.44. The same result is obtained from Figure 3.13 where $10 \log_{10} S/N = 9$ dB is used.

3.2.4 Modulation Detector

We consider a broadband sequence, $s(n)$, from a single sensor or beam, that may be amplitude modulated by a function of the form, $1 + m\cos(2\pi f_m n/f_s + \theta)$, where m is the modulation index and f_m is the modulation frequency. The hypotheses for this case are

$$H_1: x(n) = [1 + m\cos(2\pi f_m n/f_s + \theta)]s(n) + g(n)$$

$$H_0: x(n) = s(n) + g(n), \quad n = 0, \ldots, N-1$$

We wish to obtain the optimum detector in the NP sense to detect the presence of this modulation component. The sequences, $s(n)$ and $g(n)$, are sample sequences of zero-mean uncorrelated stationary Gaussian random processes. We proceed as with the broadband detector in Section 3.2.2 and calculate the DFT of the sensor outputs. Letting $v(n) = [1 + m\cos(2\pi f_m n/f_s + \theta)]s(n)$, the output is

$$V(k) = S(k) + (m/2)\,e^{j\theta}S(k - k_m) + (m/2)\,e^{-j\theta}S(k + k_m)$$

where $S(k)$ is the DFT of $s(n)$. We assume that the modulation index is small compared to unity and that the modulation frequency satisfies $f_m = k_m f_s/N$. That is, the modulation line is centered in the k_mth bin. The pdfs for the two hypotheses are

$$p_1(\mathbf{x}|\theta) = \frac{1}{\pi^{N/2}|\mathbf{S}_v + \mathbf{S}_g|}\exp[-\mathbf{x}^+(\mathbf{S}_v + \mathbf{S}_g)^{-1}\mathbf{x}]$$

$$p_0(\mathbf{x}) = \frac{1}{\pi^{N/2}|\mathbf{S}_s + \mathbf{S}_g|}\exp[-\mathbf{x}^+(\mathbf{S}_s + \mathbf{S}_g)^{-1}\mathbf{x}]$$

The modulated signal, unmodulated signal, and noise cpsd matrices are defined by $\mathbf{S}_v = E\{\mathbf{v}\mathbf{v}^+\}$, $\mathbf{S}_s = E\{\mathbf{s}\mathbf{s}^+\}$, and $\mathbf{S}_g = E\{\mathbf{g}\mathbf{g}^+\}$, respectively. Invoking the assumption that the observation time is sufficiently long so that the DFT output is uncorrelated for different frequencies, $\mathbf{S}_s$ and $\mathbf{S}_g$ are diagonal matrices with elements $S_s(k)$ and $S_g(k)$, the signal and noise power spectral densities for the kth frequency, respectively. Retaining only terms that are linear in m results in a Hermitian matrix, $\mathbf{S}_v$, that is tridiagonal in form as shown in Table 3.3. (A square matrix, $\mathbf{A}$, with complex elements is Hermitian if $\mathbf{A} = \mathbf{A}^+$.) The likelihood ratio is

$$\lambda_T(\theta) = \frac{|\mathbf{S}_s + \mathbf{S}_g|}{|\mathbf{S}_v + \mathbf{S}_g|}\exp\{-\mathbf{x}^+[(\mathbf{S}_v + \mathbf{S}_g)^{-1} - (\mathbf{S}_s + \mathbf{S}_g)^{-1}]\mathbf{x}\} \tag{3.32}$$

Table 3.3
Tridiagonal Form of $\mathbf{S}_v$ for Small Values of the Modulation Index (m)

		Column 1 ↓		*Column $k_m + 1$* ↓			*Column N/2* ↓
row 1	→	$x(1)$		$y(1)$			
row $k_m + 1$	→	$y^*(1)$	$x(k_m + 1)$			$y(k_m + 1)$	
row $N/2 - k_m$	→		$y^*(N/2 - 2k_m)$		$x(N/2 - k_m)$		$y(N/2 - k_m)$
row $N/2$	→				$y^*(N/2 - k_m)$		$x(N/2)$

For the ith row:
The (i, i)th element of $\mathbf{S}_v$ is $x(i) = S_s(i - 1)$.
The $(i, i + k_m)$th element of $\mathbf{S}_v$ is

$$y(i) = (m/2)\, e^{-j\theta}\, [S_s(i - 1) + S_s(k_m + i - 1)], \quad i \leq N/2 - k_m.$$

$S_v(i + k_m, i) = S_v^*(i, i + k_m)$.

An example of an $\mathbf{S}_v$ matrix from simulated data is found in Table 3.4. This was generated using a length-16 DFT, $k_m = 3$, $m = 0.1$, $\theta = 30°$, and a uniformly distributed zero-mean sequence. The expected value of $S_s(k)$ is 1.0 for all k. Therefore, from Table 3.3, the expected values of the upper diagonal elements are $0.1\, e^{-j30}\, (1.0 + 1.0)/2 = 0.1 \cos 30 - j\, 0.1 \sin 30 = 0.0866 - j\, 0.05$.

The difference of the two inverse matrices in (3.32) is

$$(\mathbf{S}_v + \mathbf{S}_g)^{-1} - (\mathbf{S}_s + \mathbf{S}_g)^{-1} = -(\mathbf{S}_s + \mathbf{S}_g)^{-1}(\mathbf{S}_v - \mathbf{S}_s)(\mathbf{S}_v + \mathbf{S}_g)^{-1} \tag{3.33}$$

where (3.33) is obtained by setting $\mathbf{A} = \mathbf{S}_s + \mathbf{S}_g$ and $\mathbf{B} = \mathbf{S}_v - \mathbf{S}_s$ in (3.11). Because m is assumed to be small compared to unity, $(\mathbf{S}_v + \mathbf{S}_g)^{-1} \approx (\mathbf{S}_s + \mathbf{S}_g)^{-1}$, and (3.33) is approximately equal to

$$-(\mathbf{S}_s + \mathbf{S}_g)^{-1}(\mathbf{S}_v - \mathbf{S}_s)(\mathbf{S}_s + \mathbf{S}_g)^{-1} \tag{3.34}$$

The form of $(\mathbf{S}_v - \mathbf{S}_s)$ is that given in Table 3.3 without the main diagonal. Recalling that $\mathbf{S}_s$ and $\mathbf{S}_g$ are diagonal matrices and by using (3.34), the exponent in (3.32) is

$$\begin{aligned} &\frac{m}{2} \sum_{k=k_m}^{N/2-k_m-1} \frac{X^*(k)X(k - k_m)\, e^{j\theta}[S_s(k) + S_s(k - k_m)]}{[S_s(k) + S_g(k)][S_s(k - k_m) + S_g(k - k_m)]} \\ &+ \frac{m}{2} \sum_{k=k_m}^{N/2-k_m-1} \frac{X^*(k)X(k + k_m)\, e^{-j\theta}[S_s(k) + S_s(k + k_m)]}{[S_s(k) + S_g(k)][S_s(k + k_m) + S_g(k + k_m)]} \end{aligned} \tag{3.35}$$

Table 3.4
An Example of an $\mathbf{S}_v$ Matrix Obtained with Simulated Data

(a) Real Part

1.011	0.000	−0.002	0.089	0.002	−0.004	−0.002	0.000
0.000	1.003	−0.001	0.003	0.091	−0.004	−0.002	0.003
−0.002	−0.001	1.008	0.006	−0.001	0.084	0.003	−0.008
0.089	0.003	0.006	1.001	0.001	0.002	0.084	−0.001
0.002	0.091	−0.001	0.001	1.006	−0.001	0.000	0.084
−0.004	−0.004	0.084	0.002	−0.001	1.014	−0.006	0.003
−0.002	−0.002	0.003	0.084	0.000	−0.006	1.000	0.000
0.000	0.003	−0.008	−0.001	0.084	0.003	0.000	0.996

(b) Imaginary Part

0.000	0.002	0.001	−0.045	−0.003	0.001	0.003	0.002
−0.002	0.000	0.001	0.007	−0.055	−0.004	0.002	−0.006
−0.001	−0.001	0.000	0.007	−0.002	−0.055	0.001	0.002
0.045	−0.007	−0.007	0.000	−0.003	−0.004	−0.050	0.001
0.003	0.055	0.002	0.003	0.000	−0.001	−0.002	−0.044
−0.001	0.004	0.055	0.004	0.001	0.000	−0.002	−0.001
−0.003	−0.002	−0.001	0.050	0.002	0.002	0.000	−0.005
−0.002	0.006	−0.002	−0.001	0.044	0.001	0.005	0.000

where the summation limits are selected so that the frequency indices for $S_s(k)$ and $S_g(k)$ in (3.35) are in the range of $0 \leqslant k \leqslant N/2$. To make the problem tractable, we assume that the signal power spectral density, $S_s(k)$, is varying slowly over a frequency interval equal to f_m, which often is a fairly good assumption. We can then use the following approximations:

$$S_s(k) + S_s(k - k_m) \approx [2S_s(k)]^{1/2}[2S_s(k - k_m)]^{1/2}$$

$$S_s(k) + S_s(k + k_m) \approx [2S_s(k)]^{1/2}[2S_s(k + k_m)]^{1/2}$$

By defining the filter,

$$F(k) = \frac{[2S_s(k)]^{1/2}}{[S_s(k) + S_g(k)]} \tag{3.36}$$

it follows that (3.35) is approximately equal to

$$\frac{m}{2}\sum_{k=k_m}^{N/2-k_m-1} X^*(k)F^*(k)\,X(k-k_m)F(k-k_m)\,e^{j\theta} + \frac{m}{2}\sum_{k=k_m}^{N/2-k_m-1} X^*(k)F^*(k)\,X(k+k_m)F(k+k_m)\,e^{-j\theta} \tag{3.37}$$

Time-Domain Implementation of the Modulation Detector

Parseval's identity (3.18) is used to obtain the equivalent time-domain detector. We can show that the sums in (3.37) are equal to half of the result obtained by summing over $0 \leqslant k \leqslant N-1$. Therefore, the time-domain summation is scaled by 1/2. The exponent in (3.32) is

$$\begin{aligned}&\frac{mN}{4}\sum_{n=0}^{N-1}[x(n)\otimes f(n)]^*[x(n)\otimes f(n)]\,e^{j2\pi k_m n/N}\,e^{j\theta}\\&+\frac{mN}{4}\sum_{n=0}^{N-1}[x(n)\otimes f(n)]^*[x(n)\otimes f(n)]\,e^{-j2\pi k_m n/N}\,e^{-j\theta}\\&=\frac{mN}{2}\sum_{n=0}^{N-1}|x(n)\otimes f(n)|^2\cos(2\pi k_m n/N+\theta)\end{aligned} \tag{3.38}$$

We are now in a position to calculate the likelihood ratio by substituting (3.38) into (3.32):

$$\lambda_T(\theta) = \frac{|\mathbf{S}_s+\mathbf{S}_g|}{|\mathbf{S}_v+\mathbf{S}_g|}\exp\left[\frac{mN}{2}\sum_{n=0}^{N-1}|x(n)\otimes f(n)|^2\cos(2\pi k_m n/N+\theta)\right] \tag{3.39}$$

We proceed as with the narrowband detector (Section 3.2.3) and remove the dependency on θ. By defining the quantities q and θ_0 so that

$$q\cos\theta_0 \equiv \sum_{n=0}^{N-1}|x(n)\otimes f(n)|^2\cos(2\pi k_m n/N)$$

$$q\sin\theta_0 \equiv \sum_{n=0}^{N-1}|x(n)\otimes f(n)|^2\sin(2\pi k_m n/N)$$

the likelihood ratio (3.39) becomes

$$\lambda_T(\theta) = A\exp[(mN/2)q\cos(\theta+\theta_0)], \quad \text{where } A = \frac{|\mathbf{S}_s+\mathbf{S}_g|}{|\mathbf{S}_v+\mathbf{S}_g|}$$

We can show that the determinant, $|\mathbf{S}_v + \mathbf{S}_g|$, is not a function of θ. Then λ_T is obtained by calculating

$$\lambda_T = \frac{A}{2\pi}\int_0^{2\pi} \exp[(mN/2)q\cos(\theta + \theta_0)]\, d\theta$$
$$= AI_0(mNq/2)$$

Because the Bessel function is a monotonically increasing function of its argument, we can calculate q and compare this to a threshold, where q^2 is defined by

$$q^2 = \left[\sum_{n=0}^{N-1} |x(n) \otimes f(n)|^2 \cos(2\pi k_m n/N)\right]^2 + \left[\sum_{n=0}^{N-1} |x(n) \otimes f(n)|^2 \sin(2\pi k_m n/N)\right]^2 \quad (3.40)$$

This detector is shown in Figure 3.14. Again, the assumptions that led to this structure are that the modulation index is small and the signal power spectral density is varying slowly over a frequency interval equal to the modulation frequency. The calculation of q^2 (3.40) is similar to that in the narrowband detector (Figure 3.11) except that the input sequence has been squared. As shown in Figure 3.14, $x(n)$ is convolved with the filter impulse response, $f(n)$, and the square of the magnitude of the filter output is formed. This is similar to the filtering and squaring operations in the broadband detector in Figure 3.8. At this point, the signal is applied to a quadrature receiver. This is reasonable because the spectrum of the signal contains a narrowband component at the modulation frequency, f_m. The structure in Figure 3.14 therefore is an interesting combination of the matched filter and squaring operations of the broadband detector and the quadrature receiver that appears in the narrowband detector.

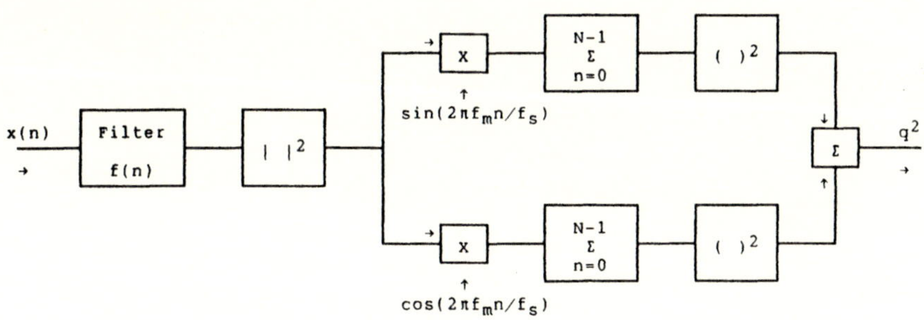

Figure 3.14 Neyman-Pearson modulation detector.

We close this section by noting that the filters in the modulation detector (3.36) and the broadband detector (3.16) both reduce to the form $F^2(k) = S_s(k)/[S_g(k)]^2$ for low S/N.

3.3 PARAMETER ESTIMATION

In this section, we assume that the signal is present but the value of one or more of its parameters is unknown yet nonrandom. These parameters may include arrival angle, narrowband signal frequency or amplitude, and delay between a pair of sensors or beams. The problem is to estimate a vector of such signal parameters, $\boldsymbol{\alpha}$, based on a set of measurements $x(n) = s(n, \boldsymbol{\alpha}) + g(n)$. The texts by Whalen [4] and Van Trees [5] are excellent references for the subject of parameter estimation.

As in Section 3.2, we assume little a priori knowledge. We assume that neither cost functions (that penalize errors made in estimating the values of the parameters) nor *a priori* knowledge of the parameters (such as parameter pdfs) is available. A commonly used estimator in this case is the *maximum likelihood estimator* (MLE). This estimator maximizes the likelihood function $p_1(\mathbf{r}|\boldsymbol{\alpha})$. (The likelihood function for the H_1 hypothesis is used here because we are assuming that the signal is present. The subscript of 1 is omitted in the remainder of this section.) The MLE of the *i*th parameter is the solution to the equation:

$$\frac{\partial p(\mathbf{x}|\boldsymbol{\alpha})}{\partial \alpha_i} = 0$$

and this is denoted by $\hat{\alpha}_i$. The MLE is that value of a parameter for which the probability of having obtained the observed inputs is a maximum.

3.3.1 Cramer-Rao Lower Bound

An estimator, $\hat{\alpha}$, is said to be biased if the expected value of the estimator differs from the true value of α. A consistent estimator is one for which both the bias and the variance tend to zero as the number of observations increase. We can derive an expression for the minimum variance that an unbiased estimator can attain. This, called the *Cramer-Rao lower bound* (CRLB), is equal to

$$E\left\{\left[\frac{\partial \ln p(\mathbf{x}|\alpha)}{\partial \alpha}\right]^2\right\}^{-1} = -E\left\{\frac{\partial^2 \ln p(\mathbf{x}|\alpha)}{\partial \alpha^2}\right\}^{-1} \tag{3.41}$$

The unbiased estimator with minimum variance (that is, an estimator that attains the CRLB) is unique; it is called an *efficient estimator*. If an efficient estimator exists, it is the MLE. Although the MLE is not an efficient estimator in all cases, it is asymptotically efficient as the number of independent observations approaches infinity [5, p. 71].

Example 3.3.1

Consider a sample sequence, $x(n)$, $n = 1, \ldots, N$, of a stationary uncorrelated Gaussian random process. Estimate the mean given that the variance is known.

The likelihood function is the multivariate Gaussian pdf (3.4):

$$p(\mathbf{x}|\mu) = \frac{1}{(2\pi)^{N/2}\sigma^N} \exp\left\{-\sum_{n=1}^{N} \frac{[x(n) - \mu]^2}{2\sigma^2}\right\}$$

It is more convenient to use the log-likelihood function:

$$\ln p(\mathbf{x}|\mu) = -\frac{N}{2}\ln(2\pi\sigma^2) - \sum_{n=1}^{N} \frac{[x(n) - \mu]^2}{2\sigma^2}$$

Differentiating this expression with respect to μ and equating to zero yields

$$\hat{\mu} = \frac{1}{N}\sum_{n=1}^{N} x(n)$$

which is just the sample mean, as expected. By calculating the statistics of $\hat{\mu}$, we can show that this estimator is unbiased. In addition, the variance of $\hat{\mu}$ is equal to σ^2/N, which is the same as the CRLB obtained from (3.41). Therefore, this estimator is efficient.

Simultaneous Estimation of Parameters

When more than one parameter is to be estimated, CRLBs for the variances of the estimators can be obtained by using the *Fisher information matrix* (FIM), where the (i, j)th element of this matrix is given by

$$J_{ij} = -E\left\{\frac{\partial^2 \ln p(\mathbf{x}|\boldsymbol{\alpha})}{\partial\alpha_i\partial\alpha_j}\right\}$$

For two parameters, the FIM is

$$\mathbf{J} = \begin{bmatrix} -E\left\{\dfrac{\partial^2 \ln p(\mathbf{x}|\boldsymbol{\alpha})}{\partial\alpha_1^2}\right\} & -E\left\{\dfrac{\partial^2 \ln p(\mathbf{x}|\boldsymbol{\alpha})}{\partial\alpha_1\partial\alpha_2}\right\} \\ -E\left\{\dfrac{\partial^2 \ln p(\mathbf{x}|\boldsymbol{\alpha})}{\partial\alpha_2\partial\alpha_1}\right\} & -E\left\{\dfrac{\partial^2 \ln p(\mathbf{x}|\boldsymbol{\alpha})}{\partial\alpha_2^2}\right\} \end{bmatrix}$$

The CRLB for α_i is equal to Ψ_{ii}, where $\mathbf{\Psi}$ is $\mathbf{J}^{-1}$, the inverse of the FIM. A pair of estimates, α_1 and α_2, that attains these bounds is jointly efficient. Returning to the two-parameter case, the CRLB for α_1 is

$$\text{CRLB}(\alpha_1) = \Psi_{11} = J_{22}/(J_{11}J_{22} - J_{12}^2)$$

$$= -\frac{1}{E\left\{\dfrac{\partial^2 \ln p(\mathbf{x}|\boldsymbol{\alpha})}{\partial \alpha_1^2}\right\}} \frac{1}{\left(1 - \dfrac{J_{12}^2}{J_{11}J_{22}}\right)}$$

The first factor in this expression is the CRLB when only the first parameter is unknown (3.41). The second factor is greater than or equal to unity. It equals unity when the off-diagonal elements, J_{12}, of the FIM are zero. Therefore, if $J_{12} = 0$, the estimates are uncoupled, and uncertainty in one parameter does not affect the minimum variance that can be achieved for the other parameter. If $J_{12} \neq 0$, the estimates are coupled, and uncertainty in one parameter increases the minimum variance (the CRLB) for the other parameter.

In summary, the MLE approach is implemented by obtaining the likelihood, or log-likelihood function for hypothesis H_1. That is, the signal is known to be present. The likelihood function is maximized with respect to each parameter.

3.3.2 Broadband Signal Parameter Estimation

We are interested in obtaining MLE estimates of conic angle, signal power, and noise power using the linear array of M equally spaced sensors considered in Section 3.2.2. We assume for simplicity that the broadband signal and noise power spectral densities, $S_s(k)$ and $S_g(k)$, respectively, are constant over the frequency range of $0 \leq f \leq f_s/2$. Therefore, $S_s(k) = S_s$ and $S_g(k) = S_g$ for $k = 0, \ldots, N/2$. Furthermore, we assume that the noise is not correlated from sensor to sensor or, equivalently, that the noise spatial correlation matrix, $\mathbf{Q}(k)$, is the identity matrix. Spatially, the signal is a plane wave with a conic angle of ψ relative to the array. Starting with the likelihood function under the H_1 hypothesis (3.10), we apply the matrix identity

$$(\mathbf{A} + \mathbf{BCD})^{-1} = \mathbf{A}^{-1} - \mathbf{A}^{-1}\mathbf{B}(\mathbf{C}^{-1} + \mathbf{DA}^{-1}\mathbf{B})^{-1}\mathbf{DA}^{-1}$$

to $(\mathbf{S}_v + \mathbf{S}_g) = (S_s\mathbf{pp}^+ + S_g\mathbf{I})$ by defining $\mathbf{A} = S_g\mathbf{I}$, $\mathbf{B} = \mathbf{p}$, $\mathbf{C} = S_s$ (a scalar), and $\mathbf{D} = \mathbf{p}^+$. Then,

$$\mathbf{x}^+(\mathbf{S}_v + \mathbf{S}_g)^{-1}\mathbf{x} = (\mathbf{x}^+\mathbf{x})/S_g - \frac{S_s/S_g^2}{1 + MS_s/S_g}\mathbf{x}^+\mathbf{p}\,\mathbf{p}^+\mathbf{x}$$

Using the definition for F^2 in (3.16), (3.10) becomes

$$p(\mathbf{x}|\boldsymbol{\alpha}) = \frac{\exp[-(\mathbf{x}^+\mathbf{x})/S_g + F^2\,\mathbf{x}^+\mathbf{p}\,\mathbf{p}^+\mathbf{x}]}{\pi^{MN/2}\,|S_s\mathbf{pp}^+ + S_g\mathbf{I}|} \tag{3.42}$$

The kth block of $S_s\mathbf{pp}^+ + S_g\mathbf{I}_{MN/2}$ is $S_s\mathbf{p}(k)\mathbf{p}^+(k) + S_g\mathbf{I}_M$, where the dimension of the identity matrix is indicated by the subscript. As mentioned in Section 7.1.1, $S_s\mathbf{p}(k)\mathbf{p}^+(k) + S_g\mathbf{I}_M$ has one eigenvalue equal to $(S_g + MS_s)$ and $M - 1$ eigenvalues that all equal S_g. Because the determinant of a matrix is equal to the product of its eigenvalues, $|S_s\mathbf{pp}^+ + S_g\mathbf{I}_M| = S_g^M\,(1 + MS_s/S_g)$. By forming the log-likelihood function, (3.42) becomes

$$\ln p(\mathbf{x}|\boldsymbol{\alpha}) = \ln \prod_{k=0}^{N/2-1} \frac{1}{(\pi S_g)^M\,(1 + M\,S_s/S_g)} - \sum_{k=0}^{N/2-1} [\mathbf{x}^+(k)\,\mathbf{x}(k)]/S_g + \sum_{k=0}^{N/2-1} |F\,\mathbf{x}^+(k)\,\mathbf{p}(k)|^2 \qquad (3.43)$$

where

$$\boldsymbol{\alpha} = \{\psi, S_s, S_g\}$$

$$F^2 = \frac{S_s/(S_g)^2}{1 + M\,S_s/S_g}$$

$$p_i(k) = \exp[j2\pi k(i-1)\,\Delta/N]$$

$$\Delta = f_s d(\sin\psi)/c$$

Estimate Arrival Angle

The arrival angle, ψ, occurs only in the third term of (3.43). The MLE for ψ is

$$\hat{\psi} = \max_{\psi} \sum_{k=0}^{N/2-1} |F\,\mathbf{x}^+(k)\,\mathbf{p}(k)|^2 \qquad (3.44)$$

Using Parseval's identity, the realization of this equation in the time domain is shown in Figure 3.15.

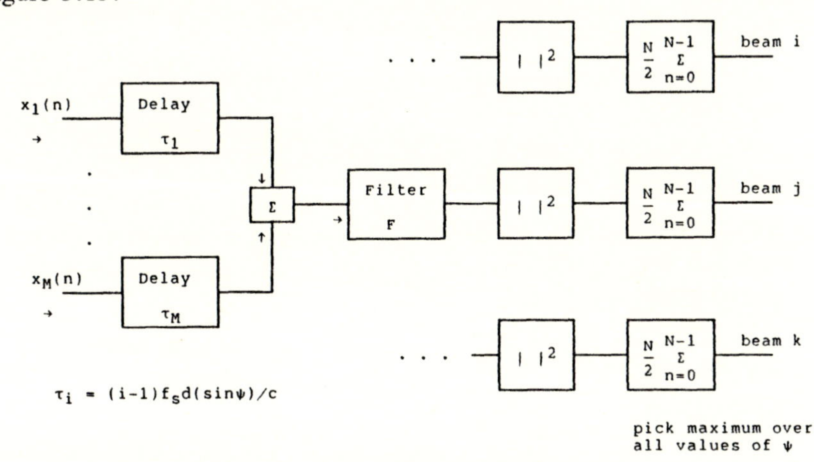

Figure 3.15 Maximum likelihood estimator for arrival angle (broadband).

Estimate Signal Power Spectral Density

The signal power spectral density appears in the first and third terms of (3.43). We differentiate (3.43) with respect to S_s and obtain

$$\frac{\partial \ln p(\mathbf{x}|\boldsymbol{\alpha})}{\partial S_s} = \frac{M^2N}{2\,S_g^2(1 + MS_s/S_g)^2} \cdot \left[-S_s + \frac{2}{M^2N} \sum_{k=0}^{N/2-1} \mathbf{x}^+(k)\mathbf{p}(k)\mathbf{p}^+(k)\mathbf{x}(k) - \frac{S_g}{M} \right] \tag{3.45}$$

By setting (3.45) equal to zero, the MLE of signal power spectral density is

$$\hat{S}_s = \frac{2}{M^2N} \sum_{k=0}^{N/2-1} |\mathbf{x}^+(k)\mathbf{p}(k)|^2 - \frac{S_g}{M} \tag{3.46}$$

Here we are assuming that S_g is known. The steering vector, $\mathbf{p}(k)$, is evaluated at $\hat{\psi}$, the MLE for angle (3.44). That is, we are estimating S_s in the direction corresponding to $\hat{\psi}$. Using Parseval's identity, (3.46) can be realized in the time domain as shown in Figure 3.16.

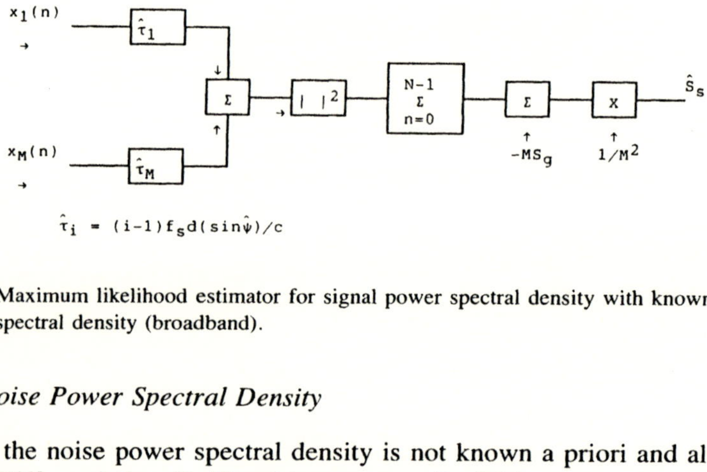

Figure 3.16 Maximum likelihood estimator for signal power spectral density with known noise power spectral density (broadband).

Estimate Noise Power Spectral Density

In general, the noise power spectral density is not known a priori and also must be estimated. Differentiating (3.43) with respect to S_g yields

$$\frac{\partial \ln p(\mathbf{x}|\boldsymbol{\alpha})}{\partial S_g} = \frac{-MN[S_g + (M-1)\,S_s]}{2S_g^2(1 + MS_s/S_g)} + \frac{1}{S_g^2} \sum_{k=0}^{N/2-1} \mathbf{x}^+(k)\mathbf{x}(k) - \frac{S_s(2S_g + MS_s)}{S_g^2\,(S_g + MS_s)^2} \sum_{k=0}^{N/2-1} \mathbf{x}^+(k)\mathbf{p}(k)\mathbf{p}^+(k)\mathbf{x}(k) \tag{3.47}$$

To find the MLE for S_g, (3.47) is set equal to zero. We use $\hat{S}_s$ from (3.46) and evaluate $\mathbf{p}(k)$ at $\hat{\psi}$. After some manipulation, the MLE for S_g is

$$\hat{S}_g = \frac{2}{MN} \sum_{k=0}^{N/2-1} \mathbf{x}^+(k)\mathbf{x}(k) - \hat{S}_s$$

This estimate is used to replace S_g in Figure 3.16. The resulting estimator for signal power spectral density for the case of unknown noise power spectral density is shown in Figure 3.17.

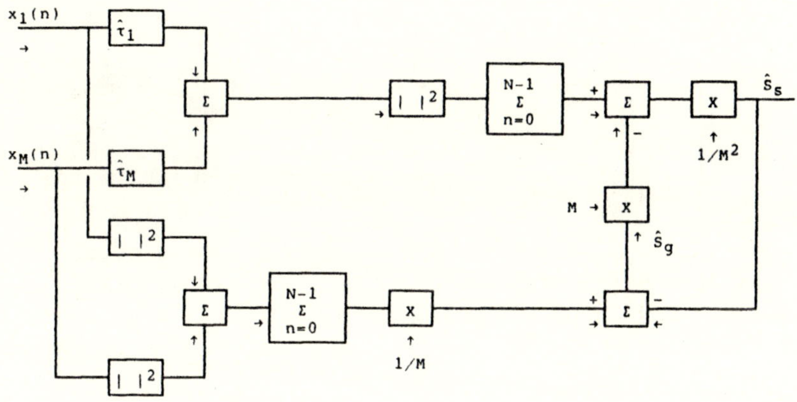

Figure 3.17 Maximum likelihood estimator for signal power spectral density with unknown noise power spectral density (broadband).

It is of interest to examine the variances of the three parameters in this example. We can show that the FIM has the following structure:

$$\begin{bmatrix} -E\left\{\dfrac{\partial^2 \ln p(\mathbf{x}|\boldsymbol{\alpha})}{\partial \psi^2}\right\} & 0 & 0 \\ 0 & -E\left\{\dfrac{\partial^2 \ln p(\mathbf{x}|\boldsymbol{\alpha})}{\partial S_s^2}\right\} & -E\left\{\dfrac{\partial^2 \ln p(\mathbf{x}|\boldsymbol{\alpha})}{\partial S_s\, \partial S_g}\right\} \\ 0 & -E\left\{\dfrac{\partial^2 \ln p(\mathbf{x}|\boldsymbol{\alpha})}{\partial S_g\, \partial S_s}\right\} & -E\left\{\dfrac{\partial^2 \ln p(\mathbf{x}|\boldsymbol{\alpha})}{\partial S_g^2}\right\} \end{bmatrix}$$

A knowledge of arrival angle has no impact on the variance of the estimate of either S_s or S_g because of the zeros in the (1, 2) and (1, 3) locations of the FIM. Similarly, a knowledge of either S_s or S_g does not improve the quality of the angle estimate. Because of the nonzero (2, 3) element, the estimates of S_s and S_g are coupled. Therefore, a knowledge of S_g results in a reduction in the variance of the estimate of S_s. The CRLBs for $\hat{S}_s$ and $\hat{S}_g$ are equal to Ψ_{22} and Ψ_{33}, respectively, where $\mathbf{\Psi}$ is the inverse of the FIM. These bounds are

$$\hat{S}_s\text{:} \begin{cases} \dfrac{2[M^2S_s^2 + 2MS_sS_g + MS_g^2/(M-1)]}{NM^2} & \text{if } S_g \text{ is unknown} \\[2ex] \dfrac{2(S_g + MS_s)^2}{NM^2} & \text{if } S_g \text{ is known} \end{cases}$$

$$\hat{S}_g\text{:} \begin{cases} \dfrac{2\,S_g^2}{N(M-1)} & \text{if } S_s \text{ is unknown} \\[2ex] \dfrac{2\,S_g^2\,(S_g + MS_s)^2}{NM[S_g^2 + 2(M-1)S_sS_g + (M-1)MS_s^2]} & \text{if } S_s \text{ is known} \end{cases} \tag{3.48}$$

We can show that both $\hat{S}_s$ and $\hat{S}_g$ are unbiased and the variance of each estimate is equal to the CRLB for that estimator. Thus $\hat{S}_s$ and $\hat{S}_g$ are jointly efficient. It follows from (3.48) that, as the number of sensors is increased, the CRLBs for $\hat{S}_s$ with and without knowledge of S_g approach the same value. The same is true for the corresponding CRLBs for $\hat{S}_g$.

We consider again the MLE for arrival angle but now include nonwhite signal and noise power spectral densities, $S_s(k)$ and $S_g(k)$, respectively. The CRLB for the variance of $\hat{\psi}$ is found from (3.41):

$$\frac{6\,c^2}{(2\pi d\cos\psi)^2M^2(M^2-1)}\left\{\sum_{k=0}^{N/2-1}\frac{S_s^2(k)\,f_k^2}{S_g^2(k)[1 + MS_s(k)/S_g(k)]}\right\}^{-1}$$

where $f_k = kf_s/N$, the frequency of the kth DFT bin. This CRLB is equal to that obtained by MacDonald and Schultheiss [6, eq. (16)]. The split-beam tracker described in [6] achieves an estimate for ψ that has a variance equal to $4[1 - (1/M)^2]/3$ times the CRLB for long observation times. Therefore, for $M = 2$ sensors, $\hat{\psi}$ is asymptotically efficient, and for large M the variance of the estimate provided by the split-beam tracker is only 4/3 times the CRLB.

3.3.3 Narrowband Signal Parameter Estimation

We consider the following input consisting of a sinusoidal signal and a sample sequence of a zero-mean white Gaussian noise process with variance σ_g^2:

$$x(n) = A\cos(2\pi fn/f_s + \theta) + g(n), \quad n = 1, \ldots, N$$

We are interested in calculating the MLEs of frequency, amplitude, and phase so that the parameter vector is $\boldsymbol{\alpha} = \{f, A, \theta\}$. From Section 3.2.3, the likelihood function is

$$p(\mathbf{x}|\boldsymbol{\alpha}) = \frac{1}{(2\pi)^{N/2}\,\sigma_g^N} \exp\left\{-\frac{1}{2\sigma_g^2}\sum_{n=1}^{N}[x(n) - A\cos(2\pi fn/f_s + \theta)]^2\right\}$$

This problem has been analyzed by Kenefic and Nuttall [7]. They define the following quantities:

$$a_{11}(f) = \frac{2}{N}\sum_{n=1}^{N}\cos^2(2\pi fn/f_s)$$

$$a_{12}(f) = \frac{1}{N}\sum_{n=1}^{N}\sin(4\pi fn/f_s)$$

$$a_{22}(f) = \frac{2}{N}\sum_{n=1}^{N}\sin^2(2\pi fn/f_s)$$

and the data-dependent quantities

$$I(f) = \sum_{n=1}^{N} x(n)\cos(2\pi fn/f_s)$$

$$Q(f) = \sum_{n=1}^{N} x(n)\sin(2\pi fn/f_s)$$

The MLEs for amplitude and phase are

$$\hat{A} = [C^2(\hat{f}) + S^2(\hat{f})]^{1/2}$$

$$\hat{\theta} = \tan^{-1}[S(\hat{f})/C(\hat{f})]$$

where

$$C(\hat{f}) = \frac{2}{N}\frac{a_{22}(\hat{f})I(\hat{f}) - a_{12}(\hat{f})Q(\hat{f})}{a_{11}(\hat{f})a_{22}(\hat{f}) - a_{12}^2(\hat{f})}$$

$$S(\hat{f}) = \frac{2}{N}\frac{a_{12}(\hat{f})I(\hat{f}) - a_{11}(\hat{f})Q(\hat{f})}{a_{11}(\hat{f})a_{22}(\hat{f}) - a_{12}^2(\hat{f})}$$

Here the MLE, $\hat{f}$, is the value of frequency that maximizes

$$\frac{1}{N}\frac{a_{22}(f)I^2(f) - 2a_{12}(f)I(f)Q(f) + a_{11}(f)Q^2(f)}{a_{11}(f)a_{22}(f) - a_{12}^2(f)} \tag{3.49}$$

We can show that $a_{11}(f) \approx 1$, $a_{12}(f) \approx 0$, and $a_{22}(f) \approx 1$ when the frequency is in the following range:

$$f_s/(2N) < f < f_s(N-1)/(2N)$$

When this inequality is satisfied, which occurs for most frequencies of interest, (3.49) reduces to

$$\frac{1}{N}\left[I^2(f) + Q^2(f)\right] = \frac{1}{N}\left|\sum_{n=1}^{N} x(n)\, e^{-j2\pi fn/f_s}\right|^2$$

But this is basically a quadrature receiver (Figure 3.18), where f is varied for maximum output. Rife and Boorstyn [8] develop a similar result for the case of a complex sinusoidal signal in white Gaussian noise. Here the input is

$$x(n) = A\cos(2\pi fn/f_s + \theta) + g(n), \quad n = 0, \ldots, N-1$$

$$y(n) = A\sin(2\pi fn/f_s + \theta) + \hat{g}(n)$$

where $y(n)$ is the Hilbert transform of $x(n)$, and the authors treat zero-mean uncorrelated narrowband Gaussian noise with variance of σ_g^2.

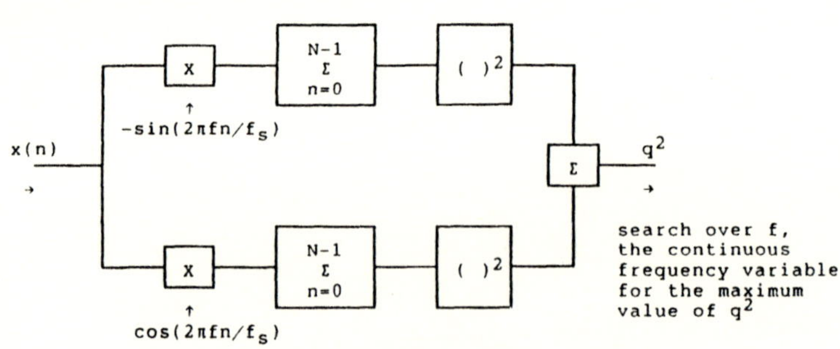

Figure 3.18 Maximum likelihood estimator for frequency (narrowband).

The FIM for this case can be shown to have the following form:

$$\begin{bmatrix} -E\left\{\dfrac{\partial^2 \ln p(\mathbf{x}|\boldsymbol{\alpha})}{\partial f^2}\right\} & 0 & -E\left\{\dfrac{\partial^2 \ln p(\mathbf{x}|\boldsymbol{\alpha})}{\partial f\,\partial \theta}\right\} \\ 0 & -E\left\{\dfrac{\partial^2 \ln p(\mathbf{x}|\boldsymbol{\alpha})}{\partial A^2}\right\} & 0 \\ -E\left\{\dfrac{\partial^2 \ln p(\mathbf{x}|\boldsymbol{\alpha})}{\partial \theta\,\partial f}\right\} & 0 & -E\left\{\dfrac{\partial^2 \ln p(\mathbf{x}|\boldsymbol{\alpha})}{\partial \theta^2}\right\} \end{bmatrix}$$

It is apparent from the structure of the FIM that a knowledge of amplitude has no effect on the quality of either the frequency or the phase estimates. The CRLB for each of the parameters is [8]

$$\hat{f}: \begin{cases} \dfrac{3\,\sigma_g^2 f_s^2}{2\pi^2 A^2 N(N-1)(2N-1)}, & \text{if } \theta \text{ is known} \\ \dfrac{3\,\sigma_g^2 f_s^2}{\pi^2 A^2 N(N^2-1)}, & \text{if } \theta \text{ is unknown} \end{cases}$$

$$\hat{A}: \frac{\sigma_g^2}{N}$$

$$\hat{\theta}: \begin{cases} \dfrac{\sigma_g^2}{A^2 N}, & \text{if } f \text{ is known} \\ \dfrac{2\,\sigma_g^2 (2N-1)}{A^2 N(N+1)}, & \text{if } f \text{ is unknown} \end{cases}$$

3.3.4 Delay Estimation

We consider two spatially separated sensors receiving the sequences, $x_1(n)$ and $x_2(n)$, $n = 0, \ldots, N-1$, where

$$x_1(n) = s(n) + g_1(n)$$

$$x_2(n) = \alpha s(n + D) + g_2(n)$$

and where $s(n)$, $g_1(n)$, and $g_2(n)$ are sample sequences of uncorrelated zero-mean Gaussian random processes. We are interested in obtaining the MLE for the relative delay, D. This problem can be viewed as a special case of the broadband angle estimation problem where the number of sensors is equal to two. We again form a vector of DFT output:

$$\mathbf{x} = \{X_1(0), X_2(0), X_1(1), X_2(1), \ldots, X_1(N/2-1), X_2(N/2-1)\}^T$$

The likelihood function is (3.10)

$$p(\mathbf{x}|D) = \frac{\exp[-\mathbf{x}^+ (\mathbf{S}_v + \mathbf{S}_g)^{-1}\mathbf{x}]}{\pi^N |\mathbf{S}_v + \mathbf{S}_g|} \tag{3.50}$$

Here $v_1(n) = s(n)$ and $v_2(n) = \alpha s(n + D)$. Following the development in Section 3.2.2, the $\mathbf{S}_v$ and $\mathbf{S}_n$ matrices are block diagonal (Table 3.2), where each block is 2 × 2. The kth block (corresponding to the kth frequency) of the spatial correlation matrix for $\{V_1(k), V_2(k)\}^T$ is given by

$$\mathbf{S}_v(k) = S_s(k)\begin{bmatrix} 1 & \alpha e^{-j2\pi kD/N} \\ \alpha e^{j2\pi kD/N} & 1 \end{bmatrix}$$

We have assumed that $N \gg D$ so that $V_2(k) \approx \alpha e^{j2\pi kD/N}\, S(k)$. The noise spatial correlation matrix is

$$\mathbf{S}_g(k) = \begin{bmatrix} S_{g1}(k) & 0 \\ 0 & S_{g2}(k) \end{bmatrix}$$

and the kth 2 × 2 block of $(\mathbf{S}_v + \mathbf{S}_g)$ has the following form:

$$\mathbf{S}_v(k) + \mathbf{S}_g(k) = \begin{bmatrix} S_s(k) + S_{g1}(k) & \alpha S_s(k)\, e^{-j2\pi kD/N} \\ \alpha S_s(k)\, e^{j2\pi kD/N} & S_s(k) + S_{g2}(k) \end{bmatrix} \tag{3.51}$$

It is straightforward to show that $|\mathbf{S}_v + \mathbf{S}_g|$ is not a function of D because $g_1(n)$ and $g_2(n)$ are assumed to be uncorrelated. After inverting (3.51) and substituting into (3.50), the log-likelihood function can be written as

$$\begin{aligned} \ln p(\mathbf{x}|D) = &- \ln \pi^N - \ln|\mathbf{S}_v + \mathbf{S}_g| \\ &- \sum_{k=0}^{N/2-1} \frac{|X_1(k)|^2[S_s(k) + S_{g2}(k)] + |X_2(k)|^2[S_s(k) + S_{g1}(k)]}{[S_s(k) + S_{g1}(k)][S_s(k) + S_{g2}(k)] - \alpha^2 S_s^2(k)} \\ &+ \sum_{k=0}^{N/2-1} \frac{\alpha S_s(k)[X_1(k)X_2^*(k)\, e^{j2\pi kD/N} + X_1^*(k)X_2(k)\, e^{-j2\pi kD/N}]}{[S_s(k) + S_{g1}(k)][S_s(k) + S_{g2}(k)] - \alpha^2 S_s^2(k)} \end{aligned} \tag{3.52}$$

Only the last summation in (3.52) is a function of D. Because the sensor sequences, $x_1(n)$ and $x_2(n)$, are assumed to be real, the second part of this summation is just the continuation of the first for values of k in the range $N/2 \leq k < N$. This follows from the complex conjugate symmetry of the DFT of a real sequence. The MLE for D therefore can be found by maximizing the following expression:

$$\sum_{k=0}^{N-1} \frac{\alpha S_s(k) X_1(k) X_2^*(k)\, e^{j2\pi kD/N}}{[S_s(k) + S_{g1}(k)][S_s(k) + S_{g2}(k)] - \alpha^2 S_s^2(k)} \tag{3.53}$$

We can identify the function

$$F^2(k) = \frac{\alpha S_s(k)}{[S_s(k) + S_{g1}(k)][S_s(k) + S_{g2}(k)] - \alpha^2 S_s^2(k)}$$

as a filter, so that (3.53) has the form

$$\sum_{k=0}^{N-1} [X_1(k)F(k)]\,[X_2(k)F(k)]^*\, e^{j2\pi kD/N} \tag{3.54}$$

Therefore, (3.54) can be viewed as an inverse DFT where the time index is given by D. The implementation of (3.54) is shown in Figure 3.19. This structure calculates an estimate of the cross-correlation function, $\hat{R}_{12}(D)$. The MLE of D is the delay corresponding to the largest value of $\hat{R}_{12}(D)$. The implementation in (3.54) is a discrete-frequency equivalent of the MLE for delay discussed by Knapp and Carter [9] and Carter [10]. Using the general formulation of Parseval's identity (3.18) in which we define $Y_1(k) = X_1(k)\, F(k)$ and $Y_2(k) = X_2(k)\, F(k)$, (3.54) is equivalent to

$$N \sum_{n=0}^{N-1} y_1(n) y_2^*(n - D)$$

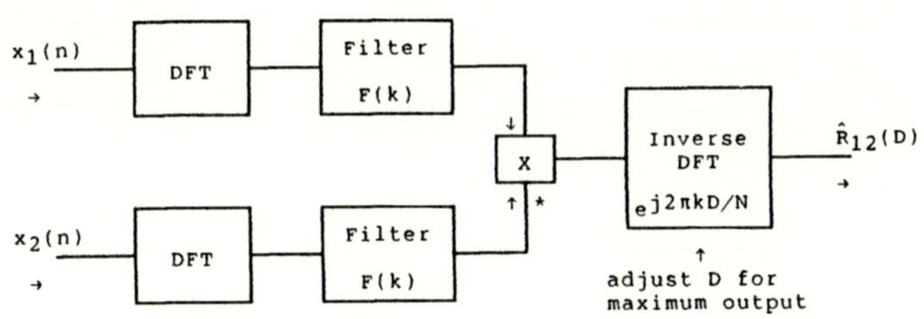

Figure 3.19 Maximum likelihood estimator for delay (frequency domain).

This time-domain implementation is shown in Figure 3.20. In [9] and [10] the MLE for delay is shown to be asymptotically efficient for sufficiently large values of N. Using (3.41), the CRLB for the variance of $\hat{D}$ is

$$\left\{ \sum_{k=0}^{N-1} \frac{(2\pi f_k)^2 \, \alpha^2 S_s^2(k)}{[S_s(k) + S_{g1}(k)][S_s(k) + S_{g2}(k)] - \alpha^2 S_s^2(k)} \right\}^{-1}$$

which is the discrete-frequency equivalent of [9, eq. (A4)]. Friedlander [11] calculates the CRLB for the case of a single sensor in a multipath environment where the received sequence is $x(n) = s(n) + \alpha s(n - D) + g(n)$.

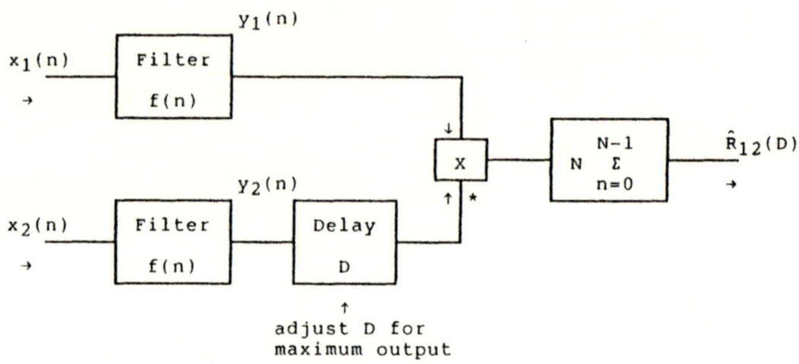

Figure 3.20 Maximum likelihood estimator for delay (time domain).

3.3.5 Conclusion

We have derived a number of estimators in Section 3.3. We can express the MLEs of signal and noise power spectral densities in Section 3.3.2 in terms of the data. Therefore, we can calculate the variances of these estimators directly and compare them to the corresponding CRLBs. In contrast, the MLEs of arrival angle (Section 3.3.2), frequency (Section 3.3.3), and delay (Section 3.3.4) cannot be expressed in terms of the data. These MLEs are found by determining the parameter value for which a certain function attains a maximum value. We cannot directly determine the variance of these estimators, although they are expected to be asymptotically efficient for long observation times.

REFERENCES

1. Urick, R.J., *Principles of Underwater Sound*, McGraw-Hill, New York, 1975.
2. Burdic, W.S., *Underwater Acoustic System Analysis*, Prentice-Hall, Englewood Cliffs, NJ, 1984.

3. Cron, B.F., and C.H. Sherman, "Spatial-Correlation Functions for Various Noise Models," *J. Acoustical Society of America*, Vol. 34, No. 11, November 1962, pp. 1732–1736.
4. Whalen, A.D., *Detection of Signals in Noise*, Academic Press, New York, 1971.
5. Van Trees, H.L., *Detection, Estimation, and Modulation Theory, Part I*, John Wiley & Sons, New York, 1968.
6. MacDonald, V.H., and P.M. Schultheiss, "Optimum Passive Bearing Estimation in a Spatially Incoherent Noise Environment," *J. Acoustical Society of America*, Vol. 46, No. 1 (Part 1), 1969, pp. 37–43.
7. Kenefic, R.J., and A.H. Nuttall, "Maximum Likelihood Estimation of the Parameters of a Tone Using Real Discrete Data," *IEEE J. Oceanic Engineering*, Vol. OE-12, No. 1, January 1987, pp. 279–280.
8. Rife, D.C., and R.R. Boorstyn, "Single-Tone Parameter Estimation from Discrete-Time Observations," *IEEE Trans. on Information Theory*, Vol. IT-20, No. 5, September 1974, pp. 591–598.
9. Knapp, C.H., and G.C. Carter, "The Generalized Correlation Method for Estimation of Time Delay," *IEEE Trans. on Acoustics, Speech, and Signal Processing*, Vol. ASSP-24, No. 4, August 1976, pp. 320–327.
10. Carter, G.C., "Coherence and Time Delay Estimation," *Proc. IEEE*, Vol. 75, No. 2, February 1987, pp. 236–255.
11. Friedlander, B., "On the Cramer-Rao Bound for Time Delay and Doppler Estimation," *IEEE Trans. on Information Theory*, Vol. IT-30, No. 3, May 1984, pp. 575–580.

Chapter 4
Practical Implementations of Optimum Sonar Detectors and Estimators

In this chapter we consider practical realizations of the optimum detectors and estimators developed in Chapter 3. A combination of inadequate processor power and lack of *a priori* knowledge of the signal and noise environments leads to suboptimum implementations.

4.1 ADAPTING OPTIMIZED RECEIVERS TO REAL-WORLD CONDITIONS

Operational sonar systems typically do not perform adaptive space-time processing. They do not sense the spatial correlation of the noise field, as represented by the spatial correlation matrix, **Q**, that appears in the Neyman-Pearson detector. The processing resources required to estimate **Q** have been prohibitive for fielded systems. As far as the optimum filter in the NP detector, $F(k)$, is concerned, a simple bandpass filter often is implemented because the required signal and noise spectra in general are not known. This approach offers some degree of robustness to an unknown, varying environment; however, it is not matched to any specific situation.

Because of limited processing resources, systems are matched only to discrete values of the parameter to be estimated. For the estimation of arrival angle, a multiple-beam system is implemented. For frequency estimation, a system providing a bank of narrowband filters is provided. For delay estimation, multiple delay bins are used. As a result, processing losses are incurred when the actual parameter value differs from the values to which the system is matched.

Neither the stability nor the duration of the signal is generally known in the passive case. To compensate, multiple frequency resolutions and multiple integration times are provided.

4.2 BACKGROUND NOISE ESTIMATION AND REMOVAL

For each type of processing, such as broadband, narrowband, and delay estimation, data for display often are generated with a large dynamic range due to slowly varying backgrounds. This can create a problem when the data are to be displayed on a printer or *cathode ray tube* (CRT) with only a limited number of bits available (perhaps three or four); or we want to implement a detection function that automatically determines when a threshold is exceeded. In these cases, we must estimate and remove the variation in noise background.

A signal of interest in a nonuniform background may appear in the output of a processor as in Figure 4.1. For a narrowband processor, the horizontal axis represents frequency. The background variation may be due to the frequency behavior of the ambient noise or flow noise spectra given in Figures 3.3 and 3.4, respectively. For a broadband processor, where the horizontal axis represents arrival angle, a nonuniform background can be attributed to a nonisotropic noise field.

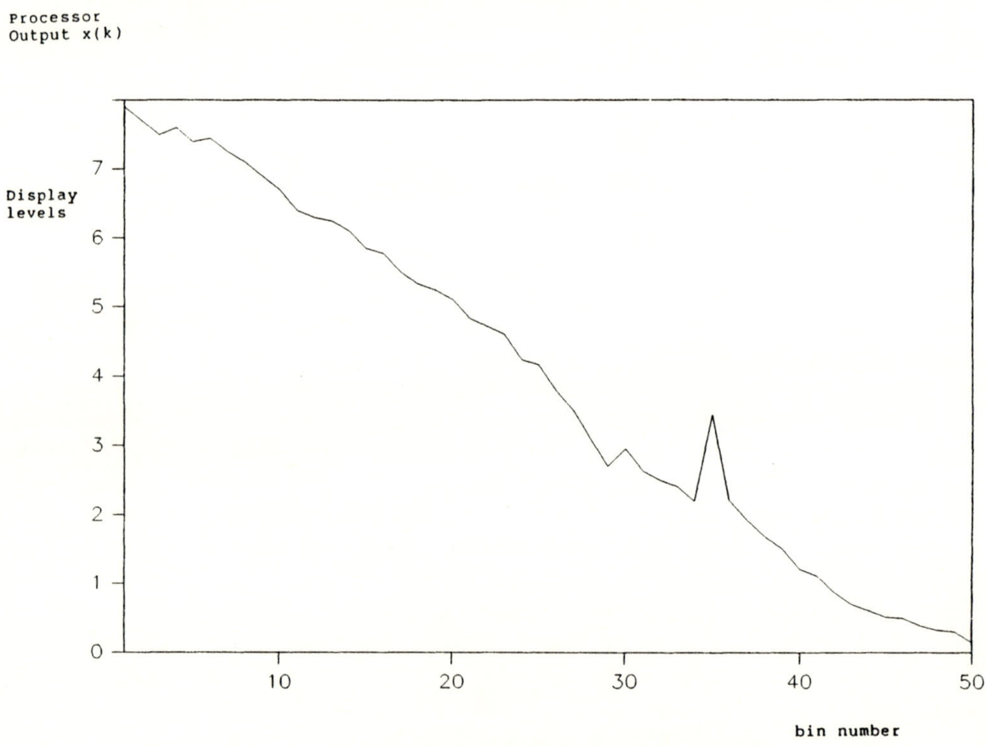

Figure 4.1 Output of processor before background normalization.

Example 4.2.1

To obtain a qualitative appreciation for the effect of a nonuniform background on the detection of signals, we assume that only three bits (eight levels) can be displayed, as shown on the vertical axis in Figure 4.1. The output of this display quantization process is

77777	77766 66665	55555 44444	33322 22223	21111 10000	00000
	↑	↑	↑	↑	↑
	bin 10	bin 20	bin 30	bin 40	bin 50

The weak signal in bin 30 is lost and the relatively strong signal in bin 35 is hardly discernible.

We would like to implement what is effectively a high-pass filter that would respond to the signal-like spikes in Figure 4.1 and attenuate the slowly varying noise background. References for this section include papers by Struzinski and Lowe [1, 2] and Wolcin [3].

4.2.1 Noise-Mean Estimation

The input to the noise-mean estimator is a set of processor output, $x(k)$, $k = 0, \dots, N - 1$. These can be the output of narrowband processing, so that $x(k)$ is the estimate of spectral level for the kth frequency bin. Alternatively, $x(k)$ can be the magnitude of the kth delay bin (correlation processing). Two techniques that can be used to estimate the noise mean are the two-pass split-window algorithm and the order-truncate-average algorithm. Both techniques use a region in the vicinity of each processor bin as the basis for calculating an estimate of the local mean.

Two-Pass Split-Window Algorithm

A two-pass split-window algorithm often is used to estimate the mean of the noise background by calculating a local mean for each sample, $x(k)$. The region, or window, used for this calculation is centered on bin k and can be specified by [1–3]

$$R_k = \{k - M, k - M + 1, \dots, k - L, k + L, \dots, k + M\} \tag{4.1}$$

where $0 \leq L < M$. The number of bins in the window is

$$K = \begin{cases} 2M + 1, & L = 0 \\ 2M + 2 - 2L, & L \neq 0 \end{cases}$$

By selecting $L \neq 0$, the window contains a gap of width $(2L - 1)$ centered on the kth bin. The expression for the window in (4.1) must be altered for bin values near either end of the data record (that is, for $0 \leq k < M$ or for $N - M - 1 < k \leq N - 1$). Assuming that k is not in either of these regions, a local mean is calculated by Pass 1:

$$\hat{x}(k) = \frac{1}{K} \sum_{i \in R_k} x(i)$$

This calculation is performed for all k. To avoid biasing the estimate of the local mean by the presence of a signal, each input, $x(k)$, is compared to $\hat{x}(k)$. If the input exceeds $\alpha\hat{x}(k)$, where α is a constant, the input is replaced by the local mean. Thus a new sequence, $y(k)$, is formed by

$$y(k) = \begin{cases} x(k), & \text{if } x(k) \leq \alpha\hat{x}(k) \\ \hat{x}(k), & \text{if } x(k) > \alpha\hat{x}(k) \end{cases}$$

The noise mean estimate, $\hat{m}(k)$, is then calculated by Pass 2 by using the clipped sequence, $y(k)$. The result is

$$\hat{m}(k) = \frac{1}{K} \sum_{i \in R_k} y(i)$$

The mean estimate is biased in the noise-only regions of the input data because large noise values have been removed. The noise-mean estimate therefore is lower than the actual value. This effect can be reduced if α is increased in value. Such an increase, however, would increase the bias of the mean estimate in regions where signals are present.

The two-pass split-window algorithm requires relatively few computations and is used widely. The algorithm does not perform well in regions containing multiple signals because a weak signal tends to be suppressed by the presence of a strong signal [1–3].

Order-Truncate-Average Algorithm

An approach for estimating the noise mean that performs better in the presence of multiple signals is the *order-truncate-average* (OTA) algorithm developed by Wolcin [3] and discussed in [1 and 2]. This algorithm uses all the bins in the window by setting L equal to zero in (4.1). For each value of k, a new sequence is obtained by ordering the K elements of $x(k)$ in the window by increasing value. This sequence is $\{y(1), y(2), \ldots, y(K)\}$, where $y(1)$ is equal to the smallest element of $x(k)$ in the window and $y(K)$ is equal to the largest. The sample median, Y_{sm}, is found, and all $y(i)$ greater than αY_{sm} are excluded from the calculation of the mean. The mean estimate is then

$$\hat{m}(k) = \frac{1}{I} \sum_{\substack{i=1 \\ i \varepsilon R_k}}^{I} y(i)$$

where $y(I) \leq \alpha Y_{sm}$ and $y(I + 1) > \alpha Y_{sm}$.

The exclusion of large elements of $x(k)$ from the calculation of the local mean in the OTA technique, rather than replacement as in the two-pass split-window algorithm, results in improved detectability of weak signals in the presence of strong signals. This is illustrated by Examples 4.2.2 and 4.2.3. The penalty for this improvement is that the OTA algorithm requires more computations.

4.2.2 Noise Removal

Noise normalization can be implemented by dividing the data by the estimate of the mean:

$$z'(k) = \frac{x(k)}{\hat{m}(k)}, \quad 0 \leqslant k \leqslant N - 1 \tag{4.2}$$

The value of α for either the two-pass split-window algorithm or the OTA algorithm is a function of the pdf assumed for the noise background and therefore depends on the type of processing (such as narrowband or correlation). References [1–3] treat the narrowband processing case in which magnitude detection is used, so the noise-only pdf is the Rayleigh distribution (Section 3.2.3). The width of the window defined by (4.1) is selected to correspond to the largest region over which the noise background is approximately uniform or has constant slope.

The noise normalization in (4.2) often is expanded to also remove the noise mean. Starting with (4.2), we subtract unity so that the expected value for the noise-only case is zero. The output is then $[x(k) - \hat{m}(k)]/\hat{m}(k)$. The mean and standard deviations for the Rayleigh distribution are related by a constant, β, so that the estimate of the standard deviation is $\hat{\sigma}(k) = \beta\hat{m}(k)$. The output of the noise-mean-removal operation therefore can be written as

$$\frac{x(k) - \hat{m}(k)}{\hat{m}(k)} = \beta\frac{x(k) - \hat{m}(k)}{\hat{\sigma}(k)} \tag{4.3}$$

It is interesting to note that (4.3) has the form of a deflection, as defined by (3.7) and repeated here:

$$\sqrt{d} = \frac{E\{\lambda_T|\mathrm{H}_1\} - E\{\lambda_T|\mathrm{H}_0\}}{\sigma\{\lambda_T|\mathrm{H}_0\}}$$

The data generated by (4.3) are scaled by a positive constant, K_A. This allows the data to be expanded or compressed to correspond to the available display levels. A second positive constant, K_B, is used to shift the scaled data so that they are centered on a positive value, rather than about zero as in (4.3). This is required because display levels in general are positive. The output of the noise-mean-estimation-and-removal algorithm then is

$$z(k) = K_A \frac{x(k) - \hat{m}(k)}{\hat{m}(k)} + K_B \tag{4.4}$$

Values of $z(k)$ that exceed the largest threshold are set equal to that threshold. Similarly, if $z(k)$ is negative, $z(k)$ is set equal to the smallest value that can be represented on the display. The implementation of (4.4) is provided in Figure 4.2.

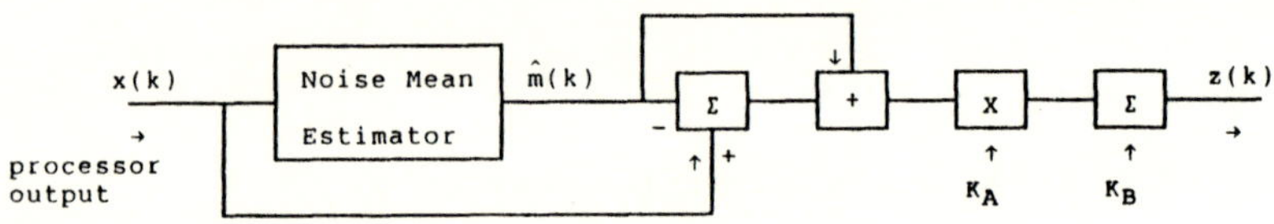

Figure 4.2 Noise-mean estimation-and-removal processing.

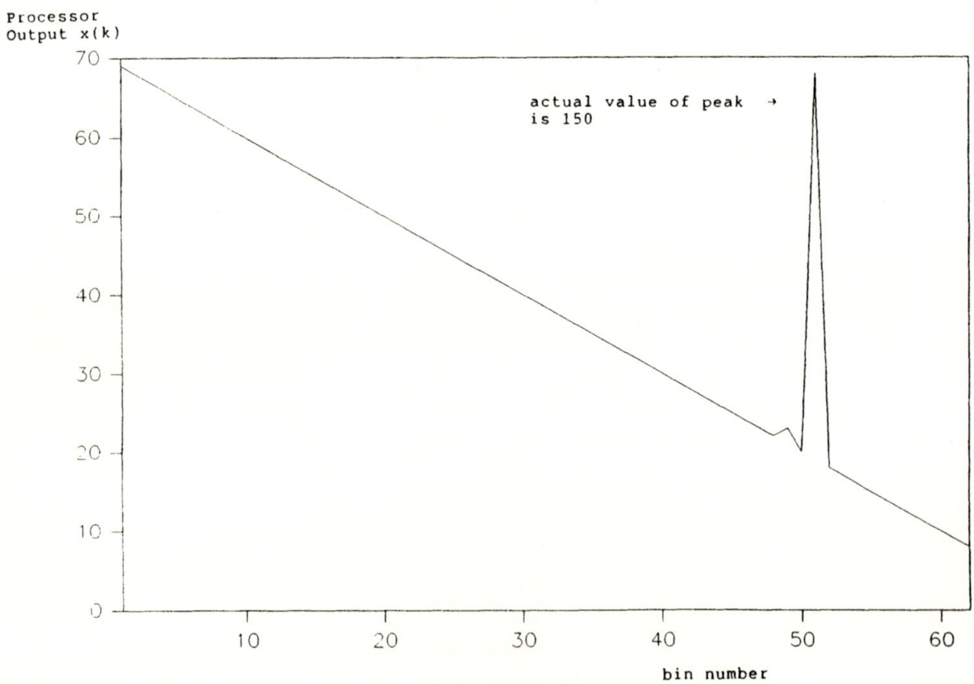

Figure 4.3 Linearly sloped background with a weak signal and a strong signal.

Example 4.2.2

We assume that the output of a processor consists of a background of constant slope, a weak signal in bin 49, and a strong signal in bin 51. This is shown in Figure 4.3. Note that there is no uncertainty in the background in this simple example. The two-pass split-window algorithm with α set equal to 2 is used to estimate the local mean, $\hat{m}(k)$. The window length is $K = 10$ bins, and because L in (4.1) is set to equal 1, the center bin is not included in the calculation. The output, $[x(k) - \hat{m}(k)]/\hat{m}(k)$, is plotted in Figure 4.4. It is apparent that the weak signal in bin 49 is not detectable.

Example 4.2.3

The OTA algorithm is now used to estimate $\hat{m}(k)$ for the combination of noise background and signals in Figure 4.3. The constant, α, is again set equal to 2 and the output, $[x(k) - \hat{m}(k)]/\hat{m}(k)$, is plotted in Figure 4.5. In this example, the weak signal is discernible.

Struzinski and Lowe [1, 2] compare other normalization techniques in terms of their ability to detect a weak sinusoidal signal in the presence of a strong sinusoidal signal.

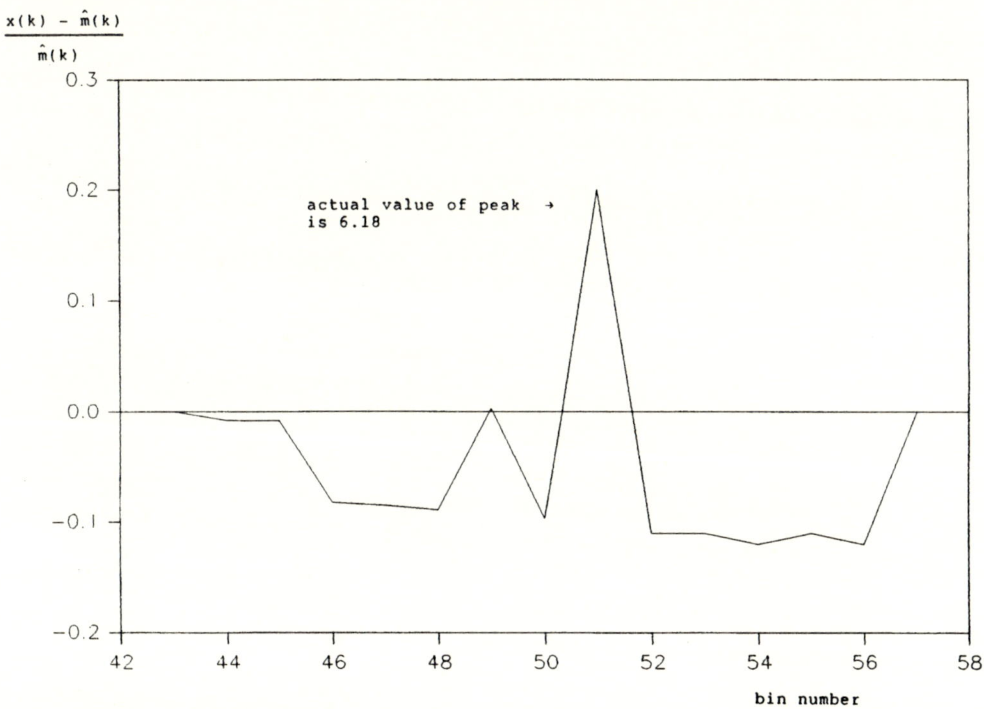

Figure 4.4 Output of the two-pass split-window noise-mean estimator (input is given in Figure 4.3).

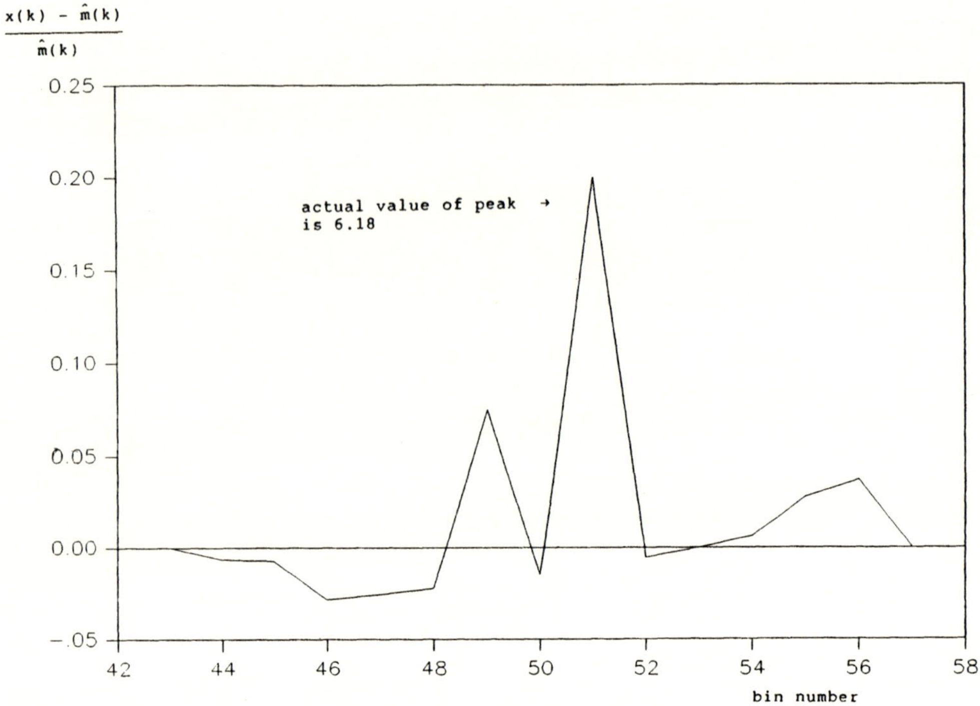

Figure 4.5 Output of the order-truncate-average noise-mean estimator (input is given in Figure 4.3).

4.3 BROADBAND PROCESSING

As shown in Figure 3.8, the NP broadband detector incorporates knowledge of the noise spatial correlation matrix, $\mathbf{Q}(k)$, in the definition of the sensor filters, $H_i(k)$, $i = 1, \ldots, M$. In addition, the signal and noise spectra are required in the expression for $F(k)$, the filter that follows the beamformer. Adaptive weights based on knowledge of the environment (such as the $H_i(k)$ filters in the NP detector) are discussed in Section 7.1, where we consider nonisotropic noise fields, including plane wave interferences.

4.3.1 Effect of $F(k)$ on Broadband Performance

In this section, we examine the effect of $F(k)$ on the performance of the broadband detector. This filter is defined by (3.16):

$$F^2(k) = \frac{S_s(k)/S_g^2(k)}{1 + S_s(k)\mathbf{p}^+(k)\mathbf{Q}^{-1}(k)\mathbf{p}(k)/S_g(k)}$$

where, from Section 3.2.2, the array gain is equal to $\mathbf{p}^{+}(k)\mathbf{Q}^{-1}(k)\mathbf{p}(k)$. Assuming a spatially uncorrelated noise field, so that $\mathbf{Q}(k) = \mathbf{I}$,

$$F^2(k) = \frac{S_s(k)/S_g^2(k)}{1 + M\,S_s(k)/S_g(k)} \tag{4.5}$$

We are interested in determining the effect of this filter on the probability of detection, P_D, where from (3.8)

$$P_D = \operatorname{erfc}\left[\frac{\operatorname{erfc}^{-1}(P_{fa}) - \sqrt{d}}{\nu}\right] \tag{4.6}$$

From Section 3.2.2, the deflection is

$$\sqrt{d} = \frac{M \sum_{k=0}^{N/2-1} F^2(k)\, S_s(k)}{\left[\sum_{k=0}^{N/2-1} F^4(k)\, S_g^2(k)\right]^{1/2}} \tag{4.7}$$

and ν^2 is given by

$$\nu^2 = \frac{\sum_{k=0}^{N/2-1} F^4(k)\,[MS_s(k) + S_g(k)]^2}{\sum_{k=0}^{N/2-1} F^4(k)\, S_g^2(k)} \tag{4.8}$$

To assess the effect of $F(k)$ on P_D, we assume some simple signal and noise spectra and evaluate (4.6) by using (1) no filter (that is, $F(k) = 1$ for all values of k); and (2) $F^2(k)$ defined by (4.5). Expressions for deflection (4.7) and ν^2 (4.8) for these two cases are given next.

No filter:

$$\sqrt{d} = \frac{M \sum_{k=0}^{N/2-1} S_s(k)}{\left[\sum_{k=0}^{N/2-1} S_g^2(k)\right]^{1/2}}$$

$$\nu^2 = \frac{\sum_{k=0}^{N/2-1} [MS_s(k) + S_g(k)]^2}{\sum_{k=0}^{N/2-1} S_g^2(k)}$$

NP filter (4.5):

$$\sqrt{d} = \frac{M \sum_{k=0}^{N/2-1} \frac{S_s^2(k)}{S_g(k)[MS_s(k) + S_g(k)]}}{\left[\sum_{k=0}^{N/2-1} \frac{S_s^2(k)}{[MS_s(k) + S_g(k)]^2} \right]^{1/2}}$$

$$v^2 = \frac{\sum_{k=0}^{N/2-1} \frac{S_s^2(k)}{S_g^2(k)}}{\sum_{k=0}^{N/2-1} \frac{S_s^2(k)}{[MS_s(k) + S_g(k)]^2}}$$

Example 4.3.1

The three sets of signal and noise spectra given in Figure 4.6 are used to calculate P_D in (4.6) for cases (a) and (b), using the appropriate expressions for $\sqrt{d}$ and v. The signal spectrum in Figure 4.6(c) is one-half cycle of a sine wave so that $S_s(k) = 0.01 \sin [2\pi(k - 1)/50]$, $k = 0, \ldots, 25$. An array of $M = 50$ sensors is assumed, and the desired value for P_{fa} is 0.01. The results are given in Table 4.1. For the spectra in Figure 4.6, it can be seen from this table that the improvement in P_D due to using the NP filter as opposed to no filter is quite pronounced. This improvement in P_D is consistent with the expectation that it should be advantageous, for example, to exclude that part of the spectrum in Figure 4.6(a) in which the signal is absent.

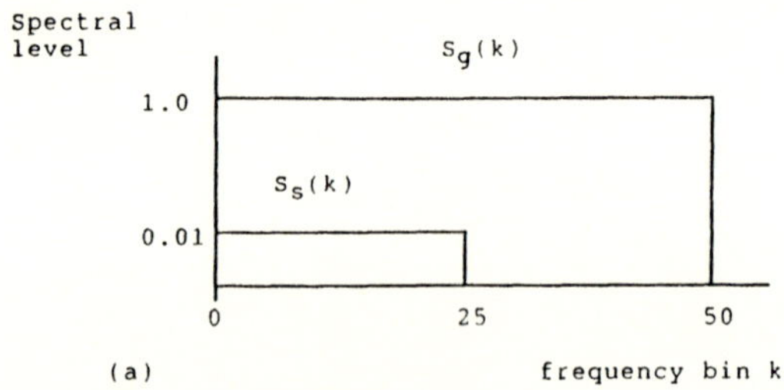

Figure 4.6 Signal spectra and noise spectra used to illustrate the effect of $F(k)$ on the performance of the broadband processor.

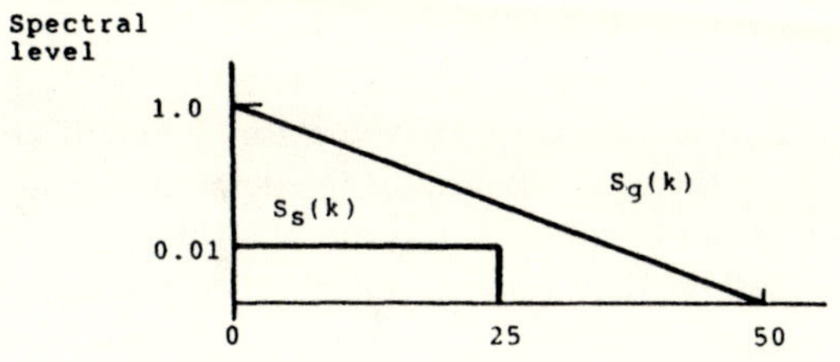

(b)

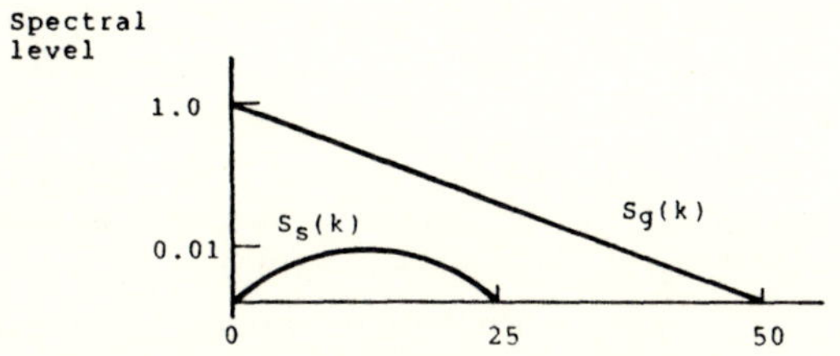

(c)

Figure 4.6 (cont'd)

Table 4.1
Broadband Detection Performance Using Different Filters

Spectra		*Deflection* ($\sqrt{d}$)	P_D
Figure 4.6 (a)			
	no filter	1.8	0.33
	NP filter	2.5	0.55
Figure 4.6 (b)			
	no filter	3.0	0.67
	NP filter	3.5	0.75
Figure 4.6 (c)			
	no filter	1.9	0.38
	NP filter	2.4	0.52

We can determine that a signal spectrum level of 0.014 is required in Figure 4.6(a) using no filter to achieve $P_D = 0.55$. (This is the value of P_D obtained using the NP filter with the original signal spectral level of 0.01.) Equivalently, assuming spherical spreading, an increase in detection range by a factor of $(0.014/0.01)^{1/2} = 1.18$ is achieved by using the NP filter.

Example 4.3.1 is intended to provide insight into the effect of filtering in the broadband processor. Actual signal and noise spectra would have to be used to assess the improvement for a specific case of interest.

In practice, $F(k)$ usually is implemented without determining the actual signal and noise spectra in real time. Instead it may be approximated with either a bandpass filter or a filter matched to expected spectra. The bandpass filter approach corresponds to assuming flat signal and noise spectra as shown in Figure 4.6(a). A set of such filters, each with a separate center frequency and passband width, can be implemented in parallel to process different spectral regions in which signals of interest are expected to occur. As shown in Figure 1.13 and repeated here in Figure 4.7, a bandpass filter can be implemented using mixing and low-pass filter operations.

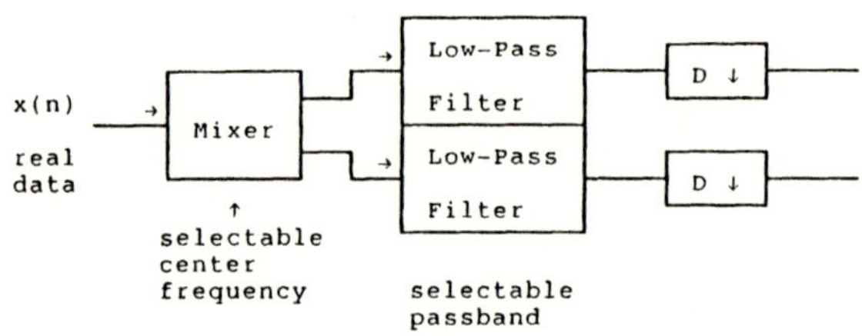

Figure 4.7 Bandpass filter implemented with mixer and low-pass filters.

The matched filter approach involves specifying a set of Eckart filters with frequency response of $F(k) = \sqrt{S_s(k)/S_g(k)}$. The expected signal spectra for signals of interest are used for $S_s(k)$. For ambient noise-limited conditions, a set of spectra, $S_g(k)$, corresponding to different ocean regions and seasons can be specified (Figure 3.3). In a flow noise-limited environment, a second set of $S_g(k)$ can be defined corresponding to different platform speeds (Figure 3.4). Assuming two signal spectra, four ambient noise spectra, and two flow noise spectra, this approach would require sixteen different filters. Detection performance would be enhanced in the presence of the hypothesized signal and noise spectra. Losses would be incurred for other signals, but because these signals are of less interest, such performance losses may be deemed acceptable.

4.3.2 Broadband Processor Implementation

A broadband processor block diagram is given in Figure 4.8. The outputs of an M-sensor array are processed using a conventional beamformer, such as one of the interpolation structures in Section 2.5.1. Following the beamformer, the data from each beam are filtered with an approximation to the NP filter. Squaring is performed using the *square-law detector* (SLD) function.

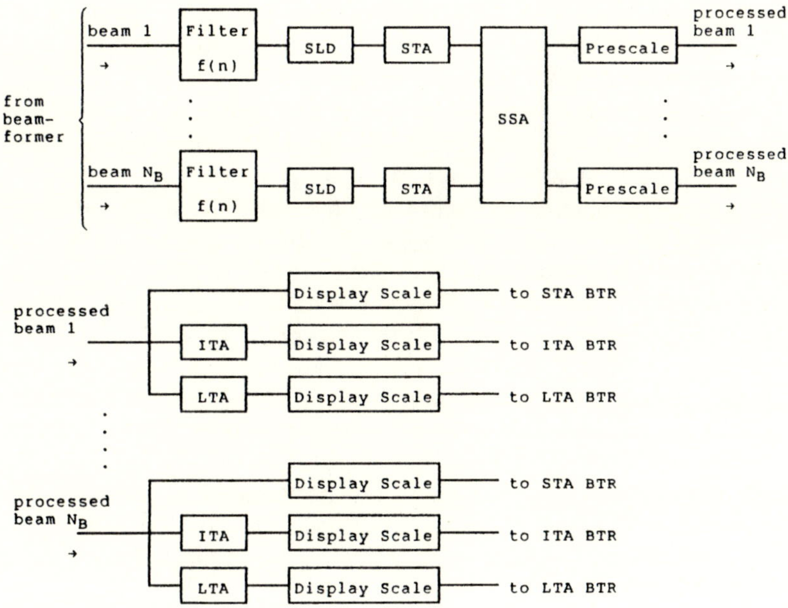

Figure 4.8 Block diagram of a broadband processor.

The observation, or integration, time should be matched to the expected duration of the signal. This duration is affected by (1) the intrinsic duration of the signal; (2) the period of time that the signal remains in a beam in a dynamic environment; and, (3) the period of time that propagation conditions support detection. To match the observation time to signal duration, a number of different averaging times are provided. These are traditionally called *short-time average* (STA), *intermediate-time average* (ITA), and *long-time average* (LTA).

The processing shown in Figure 4.8 provides a multiple alternative hypotheses test with the hypotheses:

$$\mathrm{H}_i: \quad x(n) = s(n, \psi_i) + g(n)$$

$$\mathrm{H}_0: \quad x(n) = g(n)$$

where $s(n, \psi_i)$ and $g(n)$ are sample sequences of zero-mean uncorrelated broadband Gaussian random processes. The signal is assumed to be propagating as a plane wave with direction ψ_i, where i is the index of available beam directions. (For a linear array of M equally spaced sensors, i would be in the range of values, $i = 1, \ldots, M + 1$.) The MLE for arrival angle (Figure 3.15) involves a search over the continuous angle variable. This is approximated in Figure 4.8 by providing a set of beam directions that can be equally spaced in $\sin\psi$ space. A detection is declared if the output of one of these beams exceeds a threshold. An estimate of arrival angle is obtained by observing which beam contains the signal.

After the STA function in Figure 4.8, the noise mean can be removed using a function called *sector-space averaging* (SSA). In Figure 3.17, an estimate of noise power was used to estimate the power of the signal. Due to the fact that multiple beams are formed, another approach uses a beam-to-beam comparison to detect the presence of a signal as a departure from the background mean. Because of the small number of beams typically available, a background noise estimation algorithm often is used that is simpler than the techniques discussed in Section 4.2. This can have the form of a length-K moving window, where K is odd. For example, with $K = 5$, the window has the coefficients $\{-1, -1, 4, -1, -1\}$. It is easy to verify that the output of convolving this window with a background of constant slope is zero. To do so, we assume a spatial background function across beam space defined by $x(i) = \alpha i + \beta$, where α and β are constants and i again is the beam index. The output of the SSA function then is

$$\begin{aligned} y(i) &= -\alpha(i-2) - \beta - \alpha(i-1) - \beta + 4\alpha i + 4\beta \\ &\quad - \alpha(i+1) - \beta - \alpha(i+2) - \beta \\ &= 0 \end{aligned}$$

We determined in Section 2.3 that the response of an array of sensors to isotropic noise is a function of beam direction. This response can be found by integrating the noise output, $\mathbf{w}^+(k)\mathbf{Q}(k)\mathbf{w}(k)$, in (2.10) with respect to frequency for each beam direction. The output of the SSA function for each beam in Figure 4.8 is scaled to compensate for this variation in noise gain. This is referred to as prescaling.

Table 4.2 presents the noise output as a function of beam direction for a linear array of 50 equally spaced sensors. The noise spectrum is assumed to be flat from

Table 4.2
Noise Power and Spatial Prescalars as a Function of Beam Direction (relative to the broadside beam)

Beam Angle Ψ *(degrees)*	*Relative Noise Power*	*Spatial Prescalars*
0	1.000	1.000
±10	0.9966	1.003
±20	0.9861	1.014
±30	0.9680	1.033
±40	0.9408	1.063
±50	0.9021	1.109
±60	0.8470	1.181
±70	0.7659	1.306
±80	0.6372	1.569
±90	0.5575	1.794

zero to $c/(2d)$Hz, the upper design frequency of the array. The results in this table are normalized by the noise power of the broadside beam ($\psi = 0°$). The reciprocals of these noise powers can be used as the prescalars.

The data at this point are scaled for display (STA) or integrated further and then scaled (ITA and LTA). The display format is termed the *bearing-time record* (BTR). A horizontal line of data consists of one output from each beam, and it is available for display each time an averaged output has been generated. On a CRT, the most recent line of data is placed at the top of the display and, to maintain a constant display size, the oldest line is removed from the bottom. This "waterfall" effect is commonly used to display the output of other types of processing as well. An example of a BTR is shown in Figure 4.9. The appearance of a high display

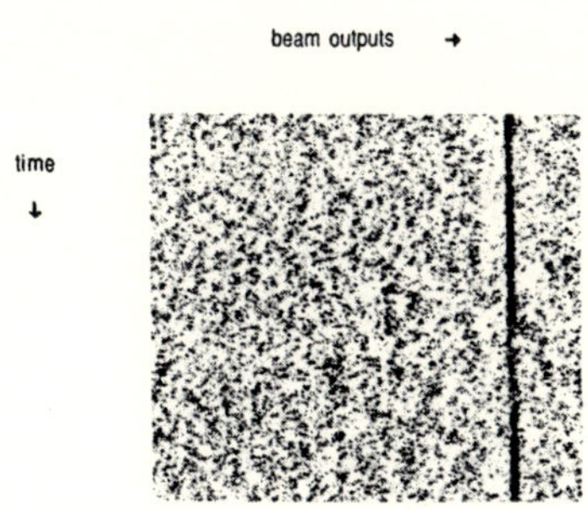

Figure 4.9 Bearing-time record generated with simulated data.

level in a beam over several averaged outputs is an indication that a signal in fact is present in the direction corresponding to that beam.

4.4 NARROWBAND PROCESSING

We showed in Section 3.2.3 that the probability of detection of the narrowband detector in Figure 3.11 increases with integration time (that is, with the number of samples, N, that are summed). It therefore is of interest to coherently integrate the signal for as long as possible. Performance losses are encountered if only *one* integration period is used. These losses occur if the signal is of shorter duration than the integration period or if the bandwidth of the signal exceeds that of the integrator. In addition, the source and platform dynamics may be such that the signal remains in a given beam for a time interval shorter than the integration period. Also, doppler shifts may cause the signal to change frequency during the integration period.

4.4.1 Narrowband Processor Implementation

A practical implementation involves providing several coherent and incoherent integration times to accommodate different signal bandwidths and durations. Coherent integration is provided by a DFT. Here, an attempt is made to match the frequency width of the DFT bin to the expected bandwidth of the signal. Incoherent integration is provided by the postdetection averaging of consecutive DFT outputs. The overall integration time of the processor is selected to match the expected duration of the signal.

There are two basic approaches for providing several coherent integration times. The first uses DFTs of different lengths. The resolution of a DFT with a rectangular window is f_s/N, where f_s is the sampling frequency of the data at the input to a length-N DFT. Thus, by providing various values of N, several resolutions can be provided. The second, and more common, approach uses a single-length DFT and obtains different resolutions by changing the input sampling frequency. Table 4.3 is a representative list of typical resolutions. These are obtained by using a length-2048 DFT.

Figure 4.10 illustrates the result of using the bandpass filter as implemented in Figure 4.7. A center frequency, f_m, is specified for the region of the spectrum to be analyzed as shown in Figure 4.10(a). The complex demodulation, or mixing, operation, shifts the spectrum by this frequency (Figure 4.10(b)). The low-pass filter in Figure 4.10(c) is used to specify the width of the spectral region to be analyzed, which is equal to $2f_p$. After the filter, the sampling frequency can be reduced. A convenient value is $f_{so} = 2.56f_p$. Using a length-N DFT, the number of frequency bins that correspond to the region of interest, $2f_p$, is $N/1.28$. Assuming that N can

Table 4.3
Typical Resolutions in Narrowband Processing (length-2048 DFT)

Sampling Frequency at DFT, f_{so} (Hz)	*Resolution (Hz)*	*Decimation (D)*
2500	1.22	1
1250	0.610	2
625	0.305	4
312.5	0.153	8
156.3	0.0763	16
78.13	0.0381	32

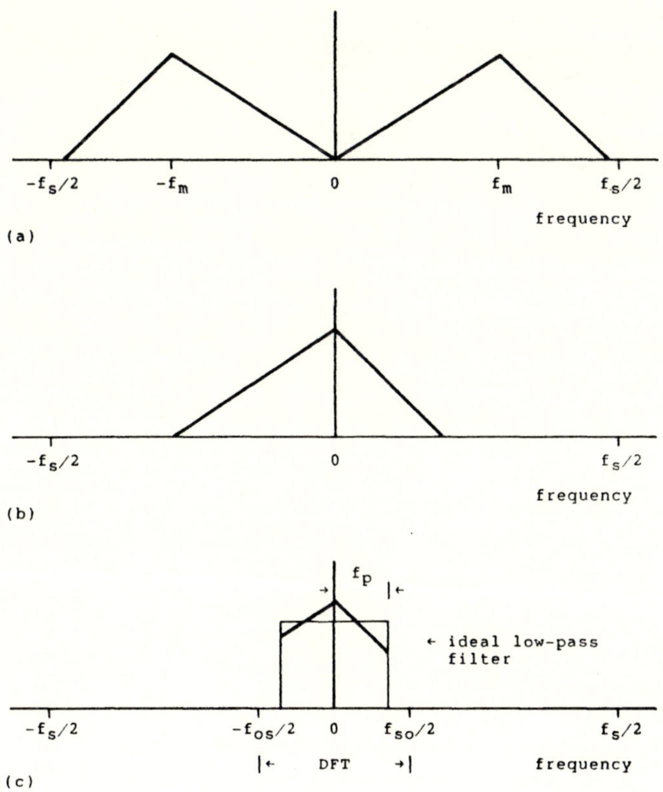

Figure 4.10 Selection of a segment of the spectrum by using a bandpass filter.

be expressed as 2^n, where n is an integer greater than 4, the number of frequency bins to be displayed, $N/1.28$, is equal to an integer, hence the factor 2.56.

Compromise between the Computational Requirements of the Low-Pass Filter and the FFT

A compromise can be made between the amount of calculations performed by the low-pass filter function and by the FFT to achieve a given resolution and processing bandwidth. Such a compromise may be of value if we are dealing with hardware that is more efficient at executing one algorithm relative to the other. The desired resolution is taken to be $2.56f_p/N$ and the processing bandwidth is $2f_p$. The two approaches are as follows:

1. The input sampling frequency to the FFT is $2.56f_p$. This requires a filter transition region of $f_a - f_p = (2.56f_p - f_p) - f_p = 0.56f_p$. A length-$N$ FFT is required to achieve the desired resolution.
2. The input sampling frequency to the FFT is $5.12f_p$. The filter transition region has increased to $(5.12f_p - f_p) - f_p = 3.12f_p$. However, a length-$2N$ FFT is now required.

Assuming that a single FIR filter is used, approach 1 requires about $3.12f_p/0.56f_p$, or 5.5 times as many stages as approach 2. This follows because the length (that is, the number of stages) of the FIR filter is inversely proportional to the transition region (Section 1.2.1). Because the filter in approach 2 is computed at twice the rate of that in approach 1, we would expect a $5.5/2 = 2.75$ times higher computational load for the filter in approach 1. In general, we would not implement a single filter but rather a cascade structure as described in Section 1.2.3. Using such a cascade structure, we still expect that the number of operations per second required to implement the filter function will be higher for approach 1.

The FFT rate is the same for each approach, but the FFT in approach 2 is twice as long, thus requiring $5[2N \log_2(2N)]/[5N \log_2(N)] = 2[1 + 1/(\log_2 N)]$ more computations. Again the choice of approach may depend on the nature of the signal processing architecture in question. The following example illustrates the idea for a particular set of parameters.

Example 4.4.1

We wish to analyze a frequency region of width 200 Hz (f_p = 100 Hz) with a resolution of 0.125 Hz. The input sampling frequency to the mixer is 2048 Hz. A redundancy of 2 is assumed, and the filter specifications are a 0.5 dB passband ripple ($\delta_1 = 2.88 \cdot 10^{-2}$) and a 70 dB stopband attenuation ($\delta_2 = 3.16 \cdot 10^{-4}$). These filter parameters are defined in Section 1.1.1.

Approach 1: Based on the parameters, the sampling frequency at the input to the FFT is $2.56f_p = 256$ Hz and the length of the FFT is $2.56f_p/0.125 = 2048$. The required decimation factor is 8. This can be achieved efficiently by implementing a 20-stage FIR filter with a decimation factor of 4, followed by a 27-stage FIR filter with a decimation factor of 2. Each of these filters provides a passband ripple of 0.25 dB, instead of 0.5 dB, to ensure that the composite response satisfies the 0.5 dB requirement. The computational rate for the filter function is

$$\begin{aligned}\text{Filter computational rate} &= 2\ rs/cs \\ &\quad \times 2\ \text{real ops/stage}/rs \\ &\quad \times (20\ \text{stages} \times 512\ cs/\text{s} \\ &\quad + 27\ \text{stages} \times 256\ cs/\text{s}) \\ &= 68{,}608\ \text{real ops/s}\end{aligned}$$

where *rs* denotes real samples, *cs* denotes complex samples, and a real op (operation) denotes a real addition or a real multiplication. Based on a radix-2 FFT algorithm,

$$\begin{aligned}\text{FFT computational rate} &= 5N \log_2(N)\ \text{real ops/FFT} \\ &\quad \times 1\ \text{FFT}/1024\ cs \times 256\ cs/\text{s} \\ &= 28{,}160\ \text{real ops/s}, \quad \text{for } N = 2048\end{aligned}$$

Approach 2: The FFT length is 4096. A decimation factor of 4 is now required to generate a 512 Hz sampling frequency at the input to the FFT. A single FIR filter consisting of 18 stages is adequate.

$$\begin{aligned}\text{Filter computational rate} &= 2\ rs/cs \times 2\ \text{real ops/stage}/rs \\ &\quad \times 18\ \text{stages} \times 512\ cs/\text{s} \\ &= 36{,}864\ \text{real ops/s}\end{aligned}$$

$$\begin{aligned}\text{FFT computational rate} &= 5 \times 4096 \times \log_2(4096)\ \text{real ops/FFT} \\ &\quad \times 1\ \text{FFT}/2048\ cs \times 512\ cs/\text{s} \\ &= 61{,}440\ \text{real ops/s}\end{aligned}$$

We see from this example that, of the combined computational requirements for filter and FFT algorithms, the filter requires 71 percent in approach 1 and 38 percent in approach 2.

Narrowband Processor Block Diagram

A narrowband processor block diagram is provided in Figure 4.11. The input can be data from a sensor or a beam. A segment of the spectrum is selected by specifying a center frequency in the digital mixer block and a low-pass filter passband. After decimation and windowing, the quadrature receiver in Section 3.2.3 is approximated with a DFT bin and SLD. Here the MLE for frequency due to Rife and Boorstyn (Section 3.3.3), which requires a search over the continuous frequency variable, is approximated with the set of discrete frequencies provided by the DFT. This is a multiple alternative hypotheses test with the hypotheses:

$$H_j: \quad x(n) = A\sin(2\pi f_j n/f_s) + g(n), \quad j = 1, \ldots, N$$

$$H_0: \quad x(n) = g(n)$$

A detection is declared if the value of a DFT filter, after SLD, integration, and noise mean estimation and removal, exceeds a threshold. An estimate of frequency is obtained by observing which bin contains the signal.

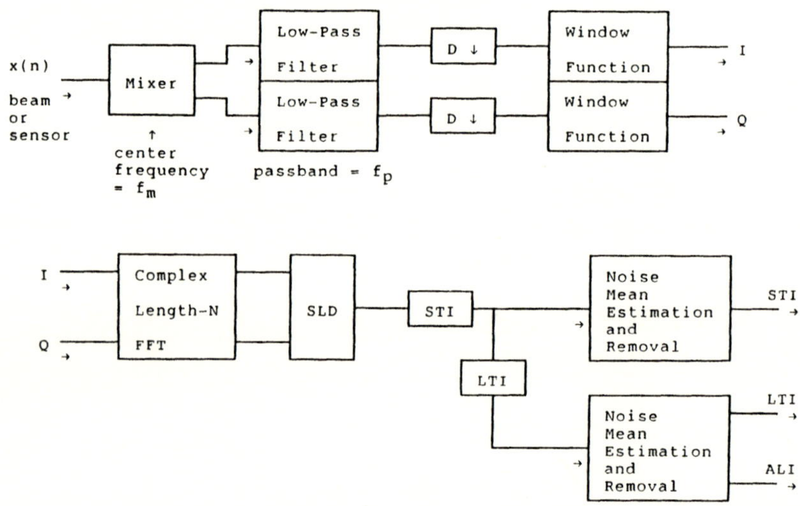

Figure 4.11 Block diagram of a narrowband processor.

Different postdetection integration times are provided in an attempt to match the characteristics of the signal. These are termed *short-time integration* (STI) and *long-time integration* (LTI). The display format of STI and LTI data is termed the *lofargram*, where the horizontal axis is frequency and the vertical axis is time. An example is given in Figure 4.12. This display uses the same waterfall approach as the BTR display in Section 4.3. By integrating the lofargram display over time, an *automatic line integration* (ALI) display is obtained. The block marked *noise mean estimation and removal* in Figure 4.11 implements the algorithm in (4.4).

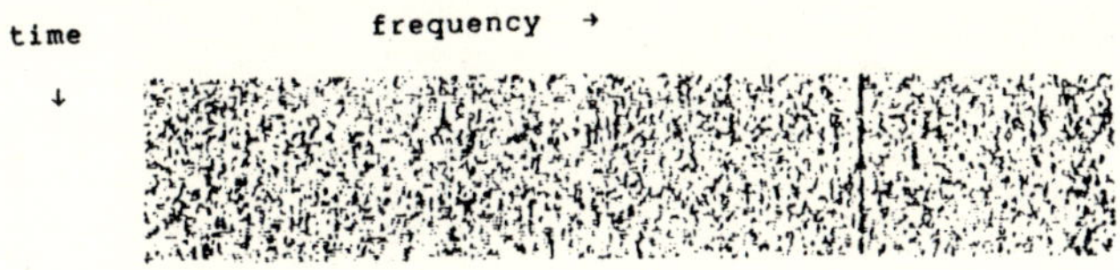

Figure 4.12 Lofargram generated with simulated data.

To achieve different DFT resolutions or, equivalently, different coherent integration times such as those in Table 4.3, we must provide a number of the processing structures, of the type shown in Figure 4.11, in parallel. Each such structure often is called a *vernier band*. Thus, different segments of the spectrum of the input data are processed in parallel with different resolutions.

Another approach to selecting the spectral regions for narrowband analysis is shown in Figure 4.13. The frequency bands to be processed are fixed, in contrast to the vernier processing in Figure 4.11. The output of a sensor or beam is analyzed with a set of frequency bands of different resolution. Because of the manner in which the frequency limits of these bands are defined, this is called *octave processing*. Shown in Figure 4.13 are five such octave bands providing the resolutions given in Table 4.4. A high degree of parallelism is possible with such a structure. Each of the five octave bands utilizes precisely the same processing. Successively finer resolutions are obtained because the input sampling frequency to each octave band is half of that to the preceding band. The input to the octave-band processor in Figure 4.13 consists of a set of N_B beams. Note that fewer beams are processed for the lower octaves because these beams are broader and thus provide adequate spatial coverage. Of course, the input to the octave-band processor also could be from a single sensor or beam.

4.4.2 Narrowband Processor Performance

In Section 3.2.3, we discussed the performance of a narrowband detector. We now include in the analysis the effect of postdetection integration, shown in Figure 4.11 as the STI and LTI functions. The goal is to determine the S/N in a 1 Hz band at the input to the DFT required to achieve the desired values of P_D and P_{fa}. Sets of ROC curves, relating d, P_D, and P_{fa}, are available that include the effect of integration. We start by obtaining an expression for d. This analysis will display the

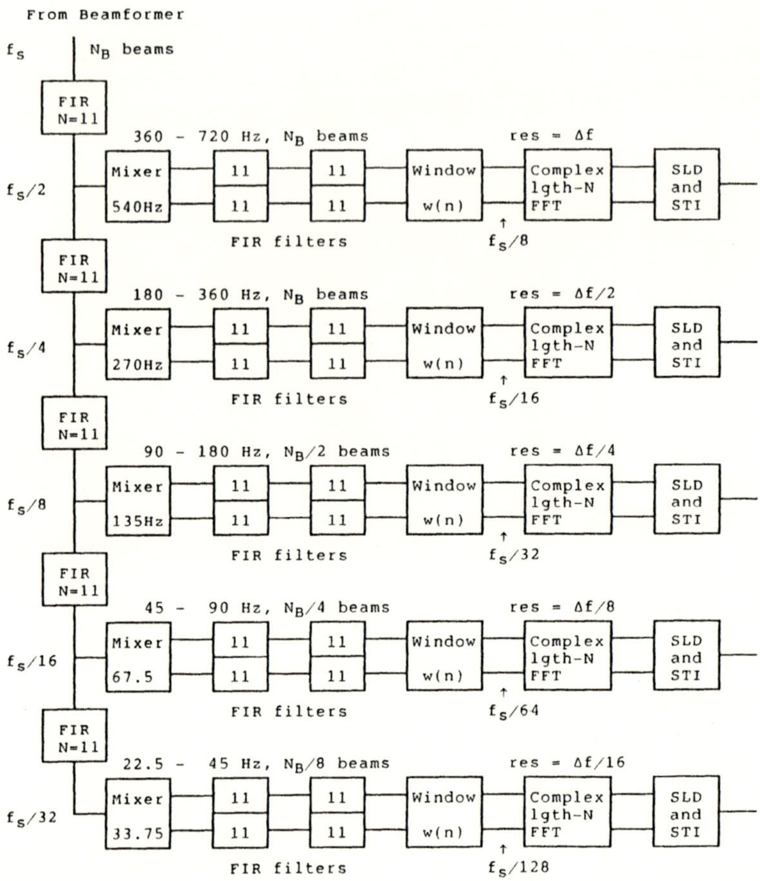

Figure 4.13 Octave-band implementation of a narrowband processor.

effects of the window function, $w(n)$, FFT redundancy, R, and scalloping loss. The processing to be analyzed is shown in Figure 4.14. The deflection, $\sqrt{d}$, is

$$\sqrt{d} = \frac{E\{I(k)|\mathrm{H}_1\} - E\{I(k)|\mathrm{H}_0\}}{\sigma\{I(k)|\mathrm{H}_0\}} \tag{4.9}$$

where $I(k)$ is the output for the kth frequency bin of the integrator. This integrator is implemented as an averaging function where an output is provided after K spectral samples have been summed. Thus $I(k)$ is defined by

$$I(k) = \frac{1}{K} \sum_{i=1}^{K} |Y_i(k)|^2$$

Table 4.4
Octave Processing Parameters for Figure 4.13

Octave Band	*Frequency Range (Hz)*	*Mixing Frequency (Hz)*	*Resolution*
1	360–720	540	Δf
2	180–360	270	$\Delta f/2$
3	90–180	135	$\Delta f/4$
4	45–90	67.5	$\Delta f/8$
5	22.5–45	33.75	$\Delta f/16$

Note: $\Delta f = f_s/(8N)$

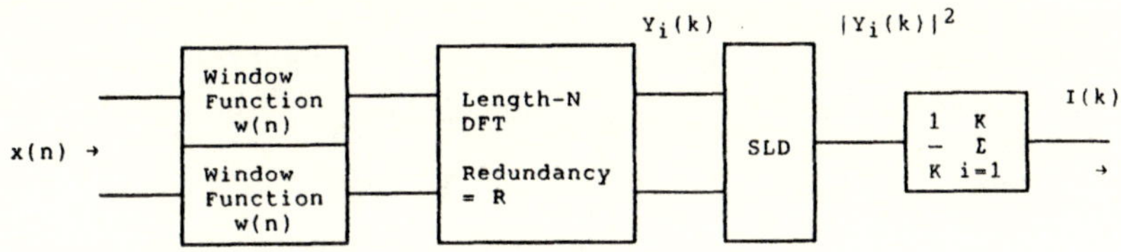

Figure 4.14 The narrowband processing function analyzed in Section 4.7 to calculate deflection.

The ith spectral estimate, $|Y_i(k)|^2$, is calculated by using the input sequence

$$w(n)\, x[n + (i - 1)N/R], \quad n = 0, \ldots, N - 1$$

where $w(n)$ is the window function and R is the DFT redundancy. Because $R = N/N_{\text{new}}$, each input sequence is offset by N_{new} samples. For the two hypotheses, the sequence, $x(n)$, is defined by

$$\text{H}_1\colon x(n) = A\, e^{j2\pi f n/f_s} + g(n)$$

$$\text{H}_0\colon x(n) = g(n)$$

We assume that $g(n)$ is a sample sequence of a zero-mean white Gaussian random process. The output of the kth frequency bin of the ith DFT is

$$Y_i(k) = \sum_{n=0}^{N-1} w(n)\, x[n + (i - 1)N/R]\, e^{-j2\pi kn/N}$$

The quantities $E\{I(k)|H_1\}$, $E\{I(k)|H_0\}$, and $\sigma\{I(k)|H_0\}$ are calculated in Section 4.7 (below), expressions (4.29), (4.22), and (4.28), respectively. The deflection is found by substituting these expressions into (4.9). This yields

$$\sqrt{d} = \frac{\left| A \sum_{n=0}^{N-1} w(n) \exp[j2\pi n(f/f_s - k/N)] \right|^2}{\dfrac{\sigma_g^2 \sqrt{\alpha}}{\sqrt{K}} \sum_{n=0}^{N-1} w^2(n)} \tag{4.10}$$

It is instructive to rewrite (4.10) in the following form:

$$\sqrt{d} = \frac{A^2}{\sigma_g^2/f_s f_s/N} \frac{1}{\text{ENBW}} \frac{\sqrt{K_I}}{1} \frac{1}{10^{SL/10}} \tag{4.11}$$

where $K_I = K/\alpha$, the effective number of independent samples available for integration. The *equivalent noise bandwidth* (ENBW) of the window function, in terms of bins (Table 1.3), is

$$\text{ENBW} = \frac{N \sum_{n=0}^{N-1} w^2(n)}{\left[\sum_{n=0}^{N-1} w(n) \right]^2}$$

The *scalloping loss* (SL) is defined by

$$\text{SL} = 10 \log_{10} \left\{ \frac{\left[\sum_{n=0}^{N-1} w(n) \right]^2}{\left| \sum_{n=0}^{N-1} w(n) \exp[j2\pi n(f/f_s - k/N)] \right|^2} \right\} \tag{4.12}$$

The frequency of the sinusoidal signal is in the range $f_s(k - 1/2)/N \leqslant f \leqslant f_s(k + 1/2)/N$. Expression (4.12) gives the scalloping loss at a particular frequency as opposed to the average scalloping loss, SL_{AV}, defined in Section 1.3.1. The worst-case scalloping loss given in Table 1.3 occurs if the signal frequency is equal to $f_s(k \pm 1/2)/N$ in (4.12).

The *minimum detectable signal* (MDS) is defined as that input S/N in a 1-Hz band, in dB, required to achieve the desired values of P_D and P_{fa}. The MDS therefore is equal to

$$\text{MDS} = 10 \log_{10} \left(\frac{A^2}{\sigma_g^2 / f_s} \right)$$

because the power spectral density of white noise is σ_g^2/f_s, and A^2 is the power of the complex sinusoidal signal. Solving (4.11) for MDS yields

$$\begin{aligned} \text{MDS} = {} & 10 \log_{10}(f_s/N) + 10 \log_{10}\text{ENBW} \\ & + \text{SL} + PL - 5 \log_{10}K_I + 5 \log_{10}d \end{aligned} \tag{4.13}$$

where PL represents an implementation loss consisting of A/D converter noise, improper noise mean estimation and removal, filter passband ripple, quantization errors, and other effects. The ROC curves for the narrowband processor are used to determine d based on P_D and P_{fa}.

In our performance analysis leading to (4.13), we assumed that spectral estimates were formed by calculating the squared magnitude of the DFT outputs in Figure 4.14 (as well as Figures 4.11 and 4.13). The analysis of this type of detector, provided in Section 4.7, is relatively straightforward. In practice, however, a magnitude or linear detector often is implemented. This detector requires a smaller dynamic range. A second advantage is that an approximation exists for the linear detector function, $|Y(k)| = |Y_R(k) + jY_I(k)|$, that is simple to implement. This approximation, by Robertson [4], is given by

$$|Y(k)| \approx \max\{|Y_R(k)|, |Y_I(k)|\} + \beta \min\{|Y_R(k)|, |Y_I(k)|\}$$

where β is a constant and $Y_R(k)$ and $Y_I(k)$ denote the real and imaginary parts of $Y(k)$, respectively.

Robertson [5] has analyzed the linear detector and obtained a set of ROC curves relating d, P_D, P_{fa}, and the number of independent samples, K_I, that are integrated. The ROC curves for an integration of one sample is given in Figure 4.15. (This set of curves is identical to the curves for the incoherent detector in Figure 3.13, in which d is plotted *versus* P_D for various values of P_{fa}.) A set of Robertson's curves for values of K_I equal to powers of 2 can be found in [5] as well as in Whalen [6].

The difference in the value of d, required for a specific set of P_D, P_{fa}, and K_I, between the linear detector and the square-law detector is quite small (less than 0.2 dB) as shown in Figure 4.16. As a result, Robertson's curves often are used to determine the performance for a square-law detector even though they were derived for the linear detector. Albersheim [8] has developed the following closed-form

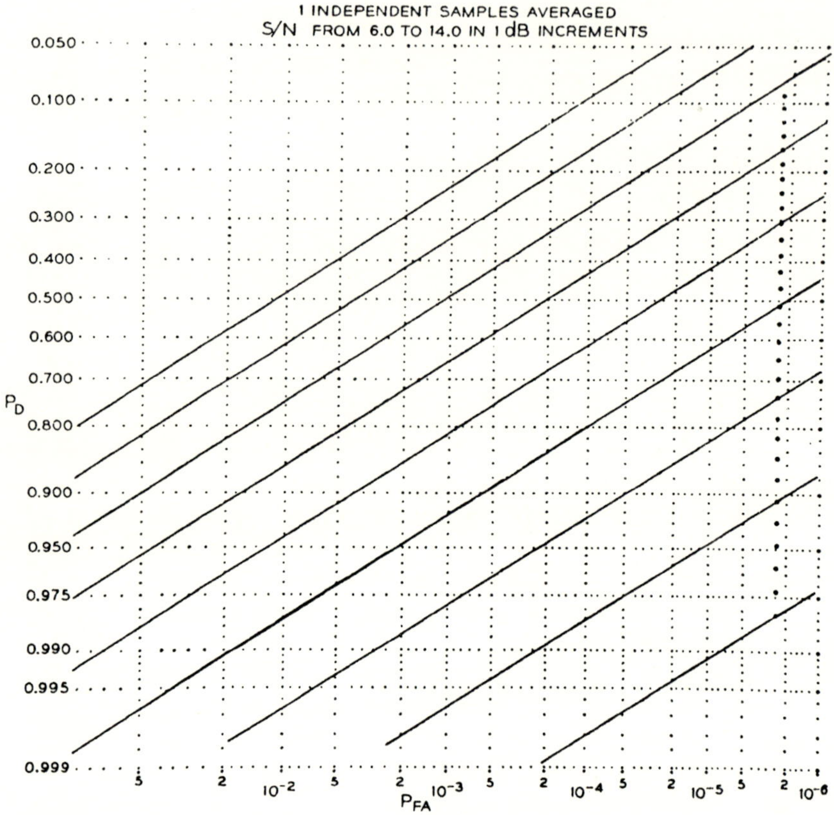

Figure 4.15 Robertson's curve for one sample (top diagonal line corresponds to $S/N = 6.0$ dB; bottom line corresponds to 14.0 dB).
Source: [5] G.H. Robertson, "Operating Characteristics for a Linear Detector of CW Signals in Narrow-Band Gaussian Noise," *Bell Syst. Tech. J.*, Vol. 46, No. 4, April 1967, pp. 755–774. (© AT&T 1967. Reprinted with special permission).

expression for Robertson's curves, which is accurate to within 0.2 dB over a wide range of parameters:

$$5 \log_{10} d = \left[6.2 + \frac{4.54}{(K_I + 0.44)^{1/2}} \right] \log_{10}(A + 0.12AB + 1.7B) \qquad (4.14)$$

where $A = \ln(0.62/P_{fa})$ and $B = \ln[P_D/(1 - P_D)]$. Expression (4.14) is plotted in Figure 4.17 for $P_{fa} = 10^{-4}$ and for values of 0.5 and 0.9 for P_D. We can see from (4.14) that the output S/N, or d, has a weak dependence on the number of independent samples integrated, and for large values of K_I, d is approximately indepen-

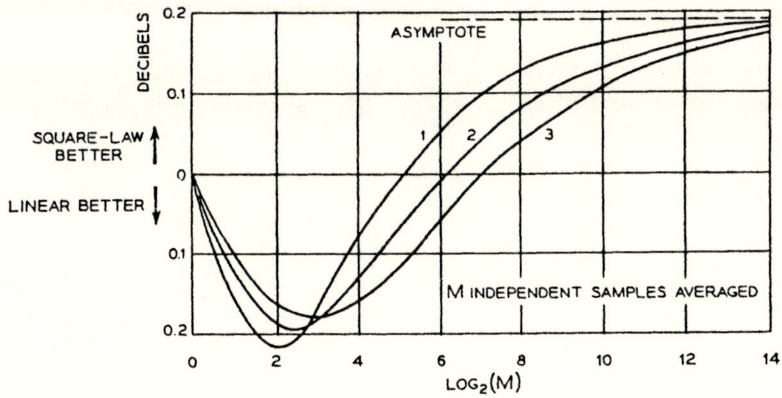

Figure 4.16 Comparison of linear and square-law detectors (P_{fa} = 10^{-6}; curve 1, P_D = 0.01; curve 2, P_D = 0.5; curve 3, P_D = 0.995).
Source: [7] G.H. Robertson, "Performance Degradation by Postdetector Nonlinearities," *Bell Syst. Tech. J.*, Vol. 47, No. 3, March 1968, pp. 407–414. (© AT&T 1968. Reprinted with special permission).

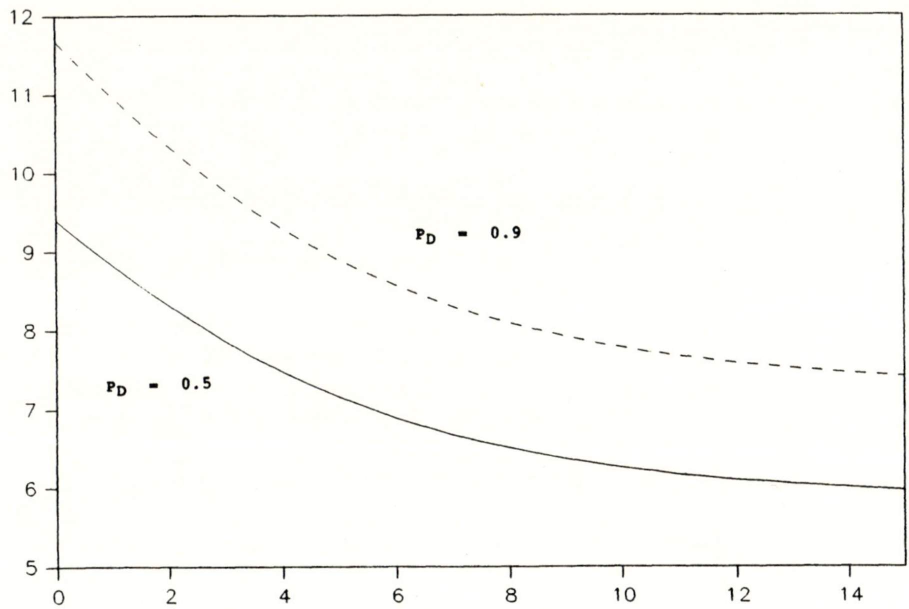

Figure 4.17 Plot of Albersheim's equation (4.14) for P_{fa} = 10^{-4} and P_D = 0.5 (solid line) and P_D = 0.9 (dashed line) (K_I is the number of independent samples for integration).

dent of this parameter. In this case, doubling the number of samples reduces the required input S/N, or MDS, in (4.13) by the familiar value of 5 $\log_{10}2$ or 1.5 dB.

We again examine expression (4.28) from Section 4.7 (below) for the variance of the integrated spectral estimate, $\sigma^2\{I(k)|H_0\}$, and rewrite it in the form:

$$\sigma^2\{I(k)|H_0\} = \frac{\sigma^2\{|Y_i(k)|^2\}}{K}\left[1 + \frac{2}{K}\sum_{i=1}^{K-1}(K-i)\,c^2(i/R)\right] \tag{4.15}$$

where

$$c(i/R) = \frac{\left[\sum_{n=0}^{N-1} w(n)\,w(n+iN/R)\right]}{\sum_{n=0}^{N-1} w^2(n)}$$

The correlation coefficient, $c(i/R)$, is a measure of redundancy, or overlap. For the rectangular window, it is easy to show that $c(0.25) = 0.75$, $c(0.5) = 0.5$, and $c(0.75) = 0.25$. For the Hann window, we can show that the correlation coefficient is approximately equal to

$$c(i/R) = \frac{[1 - i/R][1 + 0.5\cos(2\pi i/R)] + (0.75/\pi)\sin(2\pi i/R)}{1.5} \tag{4.16}$$

where $i/R < 1$. (Because any length-N window function is defined only over the interval $0 \le n \le N-1$, $c(i/R)$ is zero for values of $i/R \ge 1$.) Using (4.16), we obtain $c(0.25) = 0.65915$, $c(0.5) = 0.16667$, and $c(0.75) = 0.0075117$ for the Hann window. Values for $c(i/R)$ for other window functions are found in Harris [9, Table 1] and DeFatta, Lucas, and Hodgkiss [10, Table 8.1].

For example, with the rectangular window and $R = 1$, the variance of the output of the integrator (4.15) is reduced by a factor of K so that the number of independent samples is $K_I = K$. If we now use $R = 2$, we have $2K$ samples to integrate. The only nonzero correlation coefficient in (4.15) is $c(0.5) = 0.5$. The variance is found from (4.15) by setting the number of samples integrated equal to $2K$:

$$\sigma^2\{I(k)|H_0\} = \frac{\sigma^2\{|Y_i(k)|^2\}}{(2K)}\left[1 + \frac{2}{2K}(2K-1)\,(0.5)^2\right]$$

For large values of K, this is approximately

$$\sigma^2\{I(k)|\mathrm{H}_0\} = \frac{\sigma^2\{|Y_i(k)|^2\}}{4K/3}$$

Therefore, the variance for $R = 2$ is three-quarters of that for $R = 1$, or equivalently, the number of independent samples is $K_I = 4K/3$. Table 4.5 provides the number of independent samples available by using the rectangular and Hann windows. These are normalized by the number of independent samples available with no overlap (that is, with $R = 1$).

Table 4.5
Comparison of the Number of Independent Samples Available (K_I)

	Redundancy		
Window Function	1	2	4
Rectangular	1.0	1.33	1.45
Hann	1.0	1.89	2.08

Increasing the redundancy from 2 to 4 results in only a modest increase in K_I. The penalty associated with using $R = 4$ is that the window, FFT, and subsequent processing functions must be executed at twice the rate. Typically, a value of $R = 2$ is specified as a compromise between detection performance and computational load.

Average Scalloping Loss

The maximum scalloping loss for a given window function can be used to calculate MDS in (4.13). As discussed in Section 1.3.1, this corresponds to a signal frequency located halfway between a pair of DFT frequencies so that $f = f_s(k \pm 1/2)/N$. A more reasonable value to use may be the average scalloping loss, also defined in Section 1.3.1. This corresponds to the fact that we generally would expect the frequency of a signal to be uniformly distributed over the interval from $(k - 1/2)f_s/N$ to $(k + 1/2)f_s/N$. The average scalloping loss is about 1.25 dB for the rectangular window and 0.5 dB for the Hann window [11].

Example 4.4.2

We wish to evaluate (4.13) for the case of a 300 s observation time, a DFT filter bandwidth of 0.1 Hz, $P_D = 0.5$, and $P_{fa} = 10^{-4}$. A DFT bandwidth of $f_s/N = 0.1$ Hz corresponds to a coherent integration of 10 s. Therefore, 300/10 = 30 samples are available for incoherent integration with no overlap ($R = 1$). We assume that the implementation loss is equal to 1 dB. The following four cases are considered:

1. Rectangular window and $R = 1$: The number of independent samples for integration is $K_I = 30$ ($5 \log_{10} K_I = 7.39$ dB). Based on the desired values of P_D and P_{fa} and with $K_I = 30$, $5 \log_{10} d = 6.61$ dB from (4.14), and ENBW from Table 1.3 is equal to 1 ($10 \log_{10}$ENBW = 0 dB). The scalloping loss is 1.25 dB [11], and, from (4.13), the MDS is

$$\begin{aligned} \text{MDS} &= 10 \log_{10}(0.1) + 0 + 1.25 + 1.0 - 7.39 + 6.61 \\ &= -8.53 \text{ dB} \end{aligned}$$

   Therefore, an S/N of $-$ 8.53 dB in a 1-Hz band is required at the input to the DFT to achieve the required P_D and P_{fa} at the output of the integrator.

2. Rectangular window and $R = 2$: From Table 4.5, the number of independent samples is $K_I = 1.33 \times 30 = 40$ (8.01 dB). From (4.14), $5 \log_{10} d = 6.51$ dB, and the MDS for this case is

$$\begin{aligned} \text{MDS} &= 10 \log_{10}(0.1) + 0 + 1.25 + 1.0 - 8.01 + 6.51 \\ &= -9.25 \text{ dB} \end{aligned}$$

   We thus have improved the detection performance by about 0.7 dB by increasing the redundancy to 2, at the cost of twice the computational load for the FFT.

3. Hann window and $R = 1$: In this case, the ENBW is 1.5 (1.76 dB) from Table 1.3, and the average scalloping loss is 0.5 dB [11]. The MDS then is

$$\begin{aligned} \text{MDS} &= 10 \log_{10}(0.1) + 1.76 + 0.5 + 1.0 - 7.39 + 6.61 \\ &= -7.52 \text{ dB} \end{aligned}$$

   Thus the use of the Hann window results in a loss in performance of about 1 dB relative to case (1).

4. Hann window and $R = 2$: From Table 4.5, the number of independent samples is $K_I = 1.89 \times 30 = 57$ (8.77 dB). From (4.14), $5 \log_{10} d = 6.40$ dB, and the

MDS is equal to

$$\text{MDS} = 10 \log_{10}(0.1) + 1.76 + 0.5 + 1.0 - 8.77 + 6.40$$
$$= -9.11 \text{ dB}$$

We conclude that by using redundancy of 2 with the Hann window, detection performance is decreased by only 0.14 dB relative to that provided by the rectangular window and $R = 2$. The benefit is that we have substantially reduced the sidelobes and, therefore, the effect of spectral leakage. We would expect that the ability of the narrowband processor to detect a weak sinusoidal signal in the presence of a strong signal has been enhanced, provided that the separation of these signals is greater than the width of a DFT bin.

The technique described in this section for performing narrowband analysis was popularized by Welch [12]. In summary, it consists of the following steps:

1. Partition the data record into a number of possibly overlapped segments ($R \geq 1$).
2. Multiply each segment by a window function in the time domain. (Equivalently, the output of the FFT can be convolved with the transform of the window function.)
3. Calculate an FFT for each segment.
4. Calculate the autospectrum estimate by using a linear or square-law detector.
5. Average over the available segments.

This technique, referred to as the *Welch periodogram*, is used extensively in sonar signal processing.

4.5 MODULATION PROCESSING

A block diagram of a typical modulation processor is shown in Figure 4.18. This is a practical implementation of the NP detector in Figure 3.14. The output of a sensor or beam is filtered by using an approximation to the filter derived in Section 3.2.4 (3.36). As was discussed in Section 4.3, dealing with the broadband processor, this filter can be approximated with a bandpass filter or a filter based on assumed signal and noise spectra, $S_s(k)$ and $S_g(k)$, respectively. Returning to Figure 4.18, the filter output is then squared. At this point, the optimum receiver requires that the signal be applied to a quadrature receiver, matched to modulation frequency, f_m. We again implement a multiple alternative hypotheses test to detect the presence of the modulating component and simultaneously estimate its frequency. The hypotheses are

$$\text{H}_j\text{:} \quad x(n) = [1 + m \cos(2\pi f_j n/f_s)]\, s(n) + g(n)$$
$$\text{H}_0\text{:} \quad x(n) = s(n) + g(n)$$

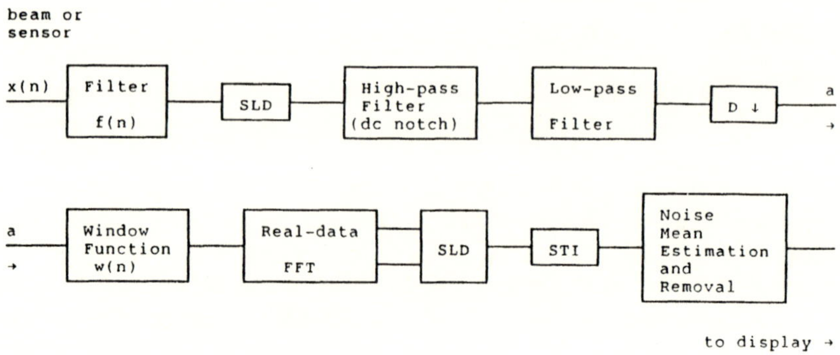

Figure 4.18 Block diagram of a modulation processor.

where the index, j, is in the range $j = 1, \ldots, j_{\max}$, and $j_{\max}$ corresponds to the highest modulation frequency expected in the data. The DFT provides a set of narrowband filters. A detection is declared if the output of one of these filters exceeds a threshold, and an estimate of frequency is obtained by observing which DFT bin contains the signal.

The spectrum of the output of the first SLD in Figure 4.18 contains a dc component in addition to narrowband components at frequencies f_m and $2f_m$. This dc component can be removed with a filter that has zero response to dc. A typical transfer function for such a high-pass, or dc notch, filter is

$$H(z) = (z - 1)/(z - \alpha)$$

where α is real and less than unity. The spectrum at the output of this filter may be hundreds or several thousands of Hertz in width. The range of modulation frequencies usually is considerably smaller. To provide high-resolution frequency bins using a DFT of reasonable length, it therefore is desirable to process the data with a low-pass filter followed by a reduction in sampling frequency. The passband of this filter is selected to be larger than the maximum expected value of modulating frequency.

The processing at the output of the decimator is identical to that described in Section 4.4 for the narrowband processor. The output is presented in the same waterfall format as the lofargram in Figure 4.12. Note that the data at the input to the FFT in Figure 4.18 are real. This presents an opportunity to apply the real-data FFT algorithm in Figure 1.33 to reduce the computational requirements.

As was the case with both the broadband and narrowband processors, the modulation processor in Figure 4.18 may be repeated a number of times so that the same sensor or beam data are processed in parallel. For the case where $F(k)$ is approximated with a bandpass filter, each implementation would examine a different portion of the broadband spectrum for modulation. If the design of the filter, $F(k)$, is based on expected signal and noise spectra, then each implementation would correspond to a different set of assumed spectra.

4.6 CORRELATION PROCESSING

Correlation processing may be used to form an estimate of an autocorrelation function for a single sensor or beam. Correlation peaks correspond to differences in times of arrival of a signal due to multiple propagation paths, or multipath. For simple propagation geometries, a display of the autocorrelation function over time can be used to determine the *closest point of approach* (CPA) of the source.

It also is of interest to calculate an estimate of a cross-correlation function for a pair of sensors or perhaps for a pair of beams from different arrays (coherent interarray processing). Correlation peaks in this case would correspond to different times of arrival from a source due to the physical separation of the pair of sensors. The range then can be estimated by triangulation.

As was discussed in Section 3.3.4, the calculation of the correlation function can be performed in the frequency domain (Figure 3.19) or the time domain (Figure 3.20). We consider the computational requirements for each. An unnormalized estimate of the cross-correlation function of two length-N complex sequences, $x(n)$ and $y(n)$, is

$$\hat{R}_{xy}(m) = \sum_{n=0}^{N-m-1} x^*(n)\, y(n+m), \quad \text{for} \quad m = 0, \ldots, L-1 \qquad (4.17)$$

We are interested in evaluating (4.17) for L values for m. These results often are called *lag products*. That is, $y(n)$ is shifted by L different values of m relative to $x(n)$ and the corresponding values of $\hat{R}_{xy}(m)$ are obtained. We assume that $L \ll N$.

The cross-correlation function also can be computed by using the DFT by implementing IDFT{DFT*[$x(n)$] DFT[$y(n)$]}, where for the moment we ignore the issue of circular correlation. The time-domain approach requires approximately NL complex multiplications and NL complex additions. A complex addition requires two real additions, and a complex multiplication can be computed with four real multiplications and two real additions. The time-domain approach then is implemented with $8NL$ real operations. We use a standard radix-2 FFT algorithm requiring $5N \log_2 N$ real operations to calculate the DFT in the frequency-domain approach. Therefore, $15N \log_2 N$ real operations are required for the 3 FFTs and $6N$ real operations are used to multiply the outputs of the two FFTs. We thus determine for this simple example that the frequency-domain approach requires fewer computations if the inequality, $L > (15/8) \log_2 N + 0.75$ is satisfied. For $N = 2048$, the transition occurs at $L = 22$ lag products. We often must calculate a considerably larger number of lag products to accommodate an anticipated range of delays. We expect, therefore, that the frequency-domain approach would require fewer computations in cases of practical interest.

We return to the issue of circular correlation. The cross-correlation function obtained via the frequency domain is denoted by $R'_{xy}(m) = \text{IDFT}\{X^*(k)Y(k)\}$, where $X(k)$ and $Y(k)$ are the DFTs of $x(n)$ and $y(n)$, respectively. Thus,

$$R'_{xy}(m) = \frac{1}{N}\sum_{k=0}^{N-1} X^*(k)\, Y(k)\, e^{j2\pi km/N}$$

or, after substituting for $X^*(k)$ and $Y(k)$,

$$R'_{xy}(m) = \frac{1}{N}\sum_{k=0}^{N-1}\left[\sum_{n=0}^{N-1} x^*(n)\, e^{j2\pi kn/N}\right]\left[\sum_{p=0}^{N-1} y(p)\, e^{-j2\pi kp/N}\right] e^{j2\pi km/N}$$

Rearranging the orders of summation yields

$$R'_{xy}(m) = \sum_{n=0}^{N-1} x^*(n) \sum_{p=0}^{N-1} y(p) \left\{\frac{1}{N}\sum_{k=0}^{N-1} \exp[-j2\pi k(p-n-m)/N]\right\} \qquad (4.18)$$

The factor in braces is equal to $\delta(p + rN - n - m)$, where $\delta(n)$ is the discrete delta function and r is an integer. Because p, n, and m are nonnegative integers, we observe that $\delta(p + rN - n - m)$ equals unity for

$$p = \begin{cases} n + m, & \text{if } n + m < N \quad (r = 0) \\ n + m - N, & \text{if } n + m \geq N \quad (r = 1) \end{cases}$$

With this result, (4.18) is equal to

$$R'_{xy}(m) = \sum_{n=0}^{N-m-1} x^*(n)\, y(n + m) + \sum_{n=N-m}^{N-1} x^*(n)\, y(n + m - N) \qquad (4.19)$$

The first sum in (4.19) represents a linear correlation of $x^*(n)$ and $y(n)$ and is equivalent to $\hat{R}_{xy}(m)$ in (4.17). The second sum is incorrect and represents the effect of circular correlation as illustrated in Figure 4.19. That is, the first m elements of $y(n)$ are multiplied by the last m elements of $x(n)$, which is an erroneous contribution to the estimate of the correlation function. Therefore, $R'_{xy}(m)$, obtained by the DFT approach, does not equal the desired $\hat{R}_{xy}(m)$ in (4.17). For example, with $N = 8$ and

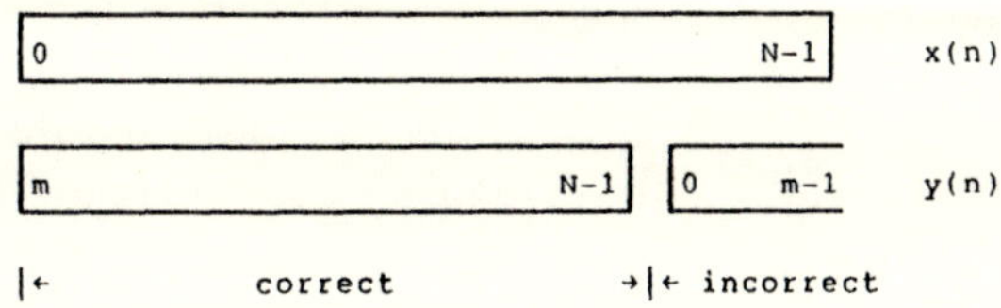

Figure 4.19 Circular correlation resulting from the DFT approach.

$m = 3$, (4.19) gives

$$R'_{xy}(3) = [x^*(0)y(3) + x^*(1)y(4) + x^*(2)y(5) + x^*(3)y(6) + x^*(4)y(7)] \\ + [x^*(5)y(0) + x^*(6)y(1) + x^*(7)y(2)]$$

The first group of terms is the desired estimate of the correlation function, whereas the second group consists of incorrect circular correlation terms. Apparently we can use the DFT approach and convert the circular correlation in (4.19) to a linear correlation (4.17) by setting the last $(L - 1)$ samples of $x(n)$ equal to zero. Again, $L - 1$ is the maximum value attained by the lag product index, m. In our example, we would set $x(5)$, $x(6)$, and $x(7)$ equal to zero if the desired number of lag products is $L = 4$.

To this point, we have used a DFT of length equal to that of the two sequences, $x(n)$ and $y(n)$. In the event that N is too large to be implemented with a single DFT, DFTs of shorter length can be used by partitioning these sequences. Oppenheim and Schafer [13, pp. 556–562] describe an approach that uses length-$2L$ FFTs, where again L satisfies the inequality, $L \ll N$. Subsequences are formed by partitioning $x(n)$:

$$x_i(n) = \begin{cases} x[n + (i - 1)L], & n = 0, \ldots, L - 1 \\ 0, & n = L, \ldots, 2L - 1 \end{cases}$$

That is, half of the elements of each subsequence of $x(n)$ equal zero. The number of such subsequences is $K = N/L$, so that $i = 1, \ldots, K$. Subsequences of $y(n)$ are formed by

$$y_i(n) = y[n + (i - 1)L], \quad n = 0, \ldots, 2L - 1$$

A cross-spectral estimate,

$$S_{xy}(k) = \sum_{i=1}^{K} X_i^*(k)\, Y_i(k), \quad k = 0, \ldots, 2L - 1$$

is computed by summing the K products, $X_i^*(k)Y(k)$. The desired cross-correlation function is found by calculating $R'_{xy}(m) = \text{IDFT}[S_{xy}(k)]$, for $m = 0, \ldots, L - 1$. We can show that, for large N/L, this approach is more efficient than the time-domain implementation if the inequality $2L - 5 \log_2 L > 9$ is satisfied. This occurs for $L \geq 16$.

Equation (4.17) provides cross-correlation estimates for $m \geq 0$. It might be of interest to evaluate $\hat{R}_{xy}(m)$ for negative values of m. This can be achieved by calculating

$$\hat{R}_{xy}(-m) = \sum_{n=0}^{N-m-1} x^*(n+m)\, y(n), \quad m = 0, \ldots, L-1$$

Using the DFT, this is accomplished by calculating

$$R'_{xy}(-m) = \Big(\text{IDFT}\,\{\text{DFT}[x(n)]\ \text{DFT}^*[y(n)]\}\Big)^*, \quad m = 0, \ldots, L-1$$

Here the last $(L - 1)$ samples of $y(n)$ are set equal to zero to obtain a linear correlation.

In addition to offering a reduction in the required computations, the frequency-domain implementation allows the removal of discrete frequency components. Such components result in sinusoidal interference in the correlation domain that may mask the desired correlation peaks.

A block diagram of a typical cross-correlation processor is shown in Figure 4.20. This represents a practical implementation of the frequency-domain MLE of

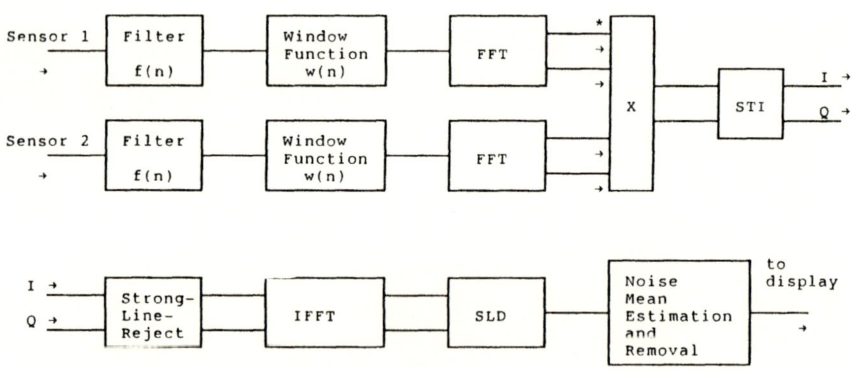

Figure 4.20 Block diagram of a cross-correlation processor.

delay given in Figure 3.19. Each channel is processed by a filter that approximates the desired response

$$F(k) = \left\{ \frac{\alpha S_s(k)}{[S_s(k) + S_{g1}(k)][S_s(k) + S_{g2}(k)] - \alpha^2 S_s^2(k)} \right\}^{1/2}$$

derived in Section 3.3.4. Here $S_{g1}(k)$ and $S_{g2}(k)$ are the noise spectra at the two sensors. It is interesting to observe that for low S/N, $\alpha = 1$, and equal noise spectra at the two sensors (that is, $S_{g1}(k) = S_{g2}(k)$), this filter is approximately equal to the Eckart or matched filter, $\sqrt{S_s(k)}/S_g(k)$, encountered in the broadband and modulation processors.

As shown in Figure 4.20, the cross-spectral estimate at the output of the averaging, or STI, block is applied to the "strong-line-reject" function. This function detects the presence of narrowband components by estimating the local mean, $\hat{m}(k)$, for each frequency bin. An approach such as the two-pass split-window algorithm or the OTA algorithm, described in Section 4.2, can be used to obtain $\hat{m}(k)$. If the magnitude of a spectral bin exceeds a constant times $\hat{m}(k)$, that value is replaced by $\hat{m}(k)$, thus decreasing the effect of the narrowband component. The output of the SLD provides an estimate of the squared magnitude of the cross-correlation function, $|R'_{xy}(m)|^2$, for a set of delay indices, $m = 0, \ldots, L - 1$. The background-noise-estimation-and-removal algorithm, described in Section 4.2, is used to remove slowly varying trends in the correlation function.

The IFFT in Figure 4.20 provides a set of delay bins. A detection of correlated components at the two sensors is declared if the squared magnitude of a delay bin exceeds a threshold. An estimate of delay is obtained by observing which IFFT bin contains the peak. The data are displayed in the standard waterfall format as shown on Figure 4.21.

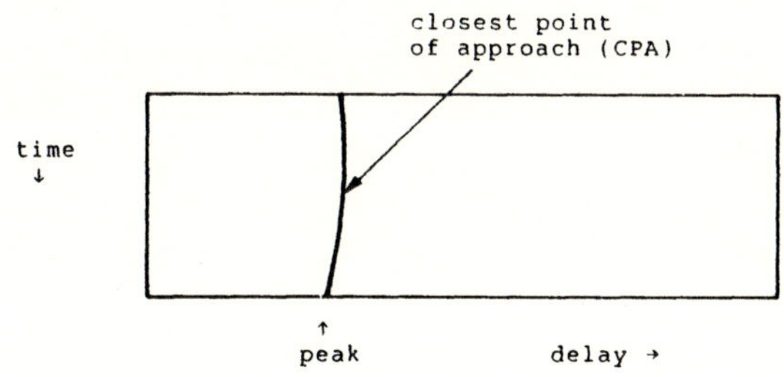

Figure 4.21 Correlation display format.

Autocorrelation Processing

A cross-correlation processor is shown in Figure 4.20. For a single sensor, autocorrelation processing can be implemented by calculating the autospectrum at the output of the single FFT. A real-data IFFT then can be employed (Sec. 1.3.3).

Simple two-path geometry, taken from Burdic [14], is shown in Figure 4.22. Burdic shows that if the receiver depth, h, and the water depth, d, are small compared to the horizontal range between source and receiver, r, the difference between the two paths is equal to

$$\delta r = \frac{3h\,(4d - h)}{2r} \tag{4.20}$$

It follows from this expression that δr attains a maximum value when the horizontal range is minimal. Therefore, the largest value of δr occurs at the CPA in a dynamic environment. Because the delay, D, between the two paths is equal to $\delta r f_s/c$ samples, D also attains maximum value at the CPA. This is illustrated in Figure 4.21. An estimate of horizontal range, r, can be obtained by using (4.20) for the simple geometry in Figure 4.22, assuming that water and receiver depths are known.

Note that the sign associated with a correlation peak can be used to infer the types of propagation paths present. Such types include the direct path, the bottom

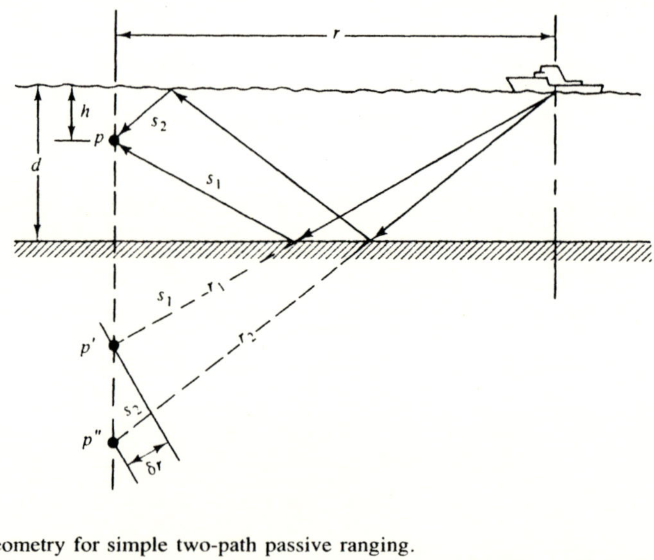

Figure 4.22 Geometry for simple two-path passive ranging.
Source: [14] W.S. Burdic, *Underwater Acoustic System Analysis*, Prentice-Hall, Englewood Cliffs, NJ; p. 426. (© 1984. Reprinted with permission.)

reflection path, and the bottom-surface reflection path. A reflection from a perfectly rigid bottom has no phase reversal, whereas a reflection from a perfectly free surface imparts a 180° phase reversal. Therefore, the multipath environment shown in Figure 4.22 should generate a negative correlation peak. The input to the SLD block in Figure 4.20 can be examined to determine the signs of the correlation peaks in the data.

Interference Pattern

Although the presence of multiple paths is of interest in autocorrelation processing, these paths can create an interference pattern on the narrowband, or lofargram, display that may hinder the detection of a narrowband line. This can be seen readily by examining the spectrum of a broadband signal, $s(n)$, in a two-path environment. Neglecting noise, the received sequence is

$$x(n) = s(n) + \alpha\, s(n - D)$$

The output of a length-N DFT is approximately equal to

$$X(k) = S(k) + \alpha\, S(k)\, e^{-j2\pi kD/N}$$

and the spectrum is

$$|X(k)|^2 = |S(k)|^2[1 + \alpha^2 + 2\,\alpha \cos(2\pi kD/N)]$$

where $S(k)$ is the DFT of $s(n)$. Therefore, a sinusoidal interference pattern is present on the lofargram. This is shown in Figure 4.23 by using simulated data. The narrowband line clearly discernible in Figure 4.12 now is obscured by this interference pattern.

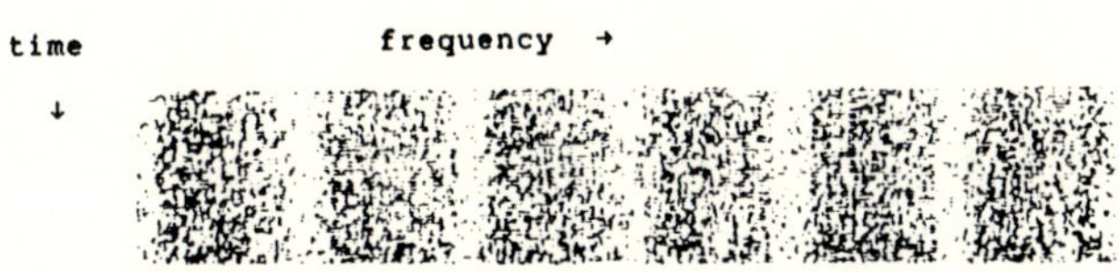

Figure 4.23 Multipath interference pattern on lofargram (simulated data).

4.7 APPENDIX

This appendix provides derivations of the quantities required for the deflection (4.9) at the output of the narrowband processor in Section 4.4.2. These quantities are

$E\{I(k)|H_0\}$, $\sigma\{I(k)|H_0\}$, and $E\{I(k)|H_1\}$, where $I(k)$ is the integrator output as shown in Figure 4.14. Note that $I(k)$ is based on noise-only data in the derivations of $E\{I(k)|H_0\}$ and $\sigma\{I(k)|H_0\}$, whereas $I(k)$ is based on signal plus noise in the calculation of $E\{I(k)|H_1\}$.

Find $E\{I(k)|H_0\}$

From the definition of the integrator,

$$E\{I(k)|H_0\} = \frac{1}{K}\sum_{i=1}^{K} E\{|Y_i(k)|^2\} \tag{4.21}$$

By substituting for $Y_i(k)$, we obtain

$$E\{|Y_i(k)|^2\} = \sum_{n=0}^{N-1}\sum_{m=0}^{N-1} w(n)w(m)$$
$$E\{x[n + (i - 1)N/R]x^*[m + (i - 1)N/R]\}\, e^{-j2\pi k(n-m)/N}$$

For the H_0 hypothesis, $x(n) = g(n)$. Because $g(n)$ is a sample sequence of a zero-mean white Gaussian process, $E\{x(n)x^*(m)\} = \sigma_g^2\, \delta(n - m)$ and

$$E\{|Y_i(k)|^2\} = \sigma_g^2 \sum_{n=0}^{N-1} w^2(n)$$

Finally, we have that

$$E\{I(k)|H_0\} = \sigma_g^2 \sum_{n=0}^{N-1} w^2(n) \tag{4.22}$$

because of the manner in which $I(k)$ is normalized.

Find $\sigma^2\{I(k)|H_0\}$

From the definition of $I(k)$, it follows that

$$E\{I^2(k)|H_0\} = \frac{1}{K^2}\sum_{i=1}^{K}\sum_{p=1}^{K} E\{Y_i(k)Y_i^*(k)Y_p(k)Y_p^*(k)\}$$

Using a result for complex Gaussian signals found by Reed [15],

$$E\{Y_i(k)Y_i^*(k)Y_p(k)Y_p^*(k)\} = E\{Y_i(k)Y_i^*(k)\}\, E\{Y_p(k)Y_p^*(k)\} + E\{Y_i(k)Y_p^*(k)\}\, E\{Y_i^*(k)Y_p(k)\}$$

the expression for $E\{I^2(k)|H_0\}$ becomes

$$E\{I^2(k)|H_0\} = \frac{1}{K^2}\sum_{i=1}^{K}\sum_{p=1}^{K} E\{Y_i(k)Y_i^*(k)\}\, E\{Y_p(k)Y_p^*(k)\}$$
$$+ \frac{1}{K^2}\sum_{i=1}^{K}\sum_{p=1}^{K} E\{Y_i(k)Y_p^*(k)\}\, E\{Y_i^*(k)Y_p(k)\}$$

From (4.21), we see that the first term in this expression is equal to $E^2\{I(k)|H_0\}$. Thus, the variance of $I(k)$ is equal to the second term,

$$\sigma^2\{I(k)|H_0\} = \frac{1}{K^2}\sum_{i=1}^{K}\sum_{p=1}^{K} E\{Y_i(k)Y_p^*(k)\}\, E\{Y_i^*(k)Y_p(k)\}$$

which can be rewritten as

$$\sigma^2\{I(k)|H_0\} = \frac{1}{K^2}\sum_{i=1}^{K} E^2\{|Y_i(k)|^2\} + \frac{1}{K^2}\sum_{\substack{i=1\\ i\neq p}}^{K}\sum_{p=1}^{K} E\{Y_i(k)Y_p^*(k)\}\, E\{Y_i^*(k)Y_p(k)\} \quad (4.23)$$

We obtain the following expression for $E\{Y_i(k)Y_p^*(k)\}$ by substituting for $Y_i(k)$ and rearranging the summations:

$$E\{Y_i(k)Y_p^*(k)\} = \sum_{n=0}^{N-1}\sum_{m=0}^{N-1} w(n)w(m)$$
$$\cdot E\{x[n + (i-1)N/R]x^*[m + (p-1)N/R]\}\, e^{-j2\pi k(n-m)/N}$$

Because we are dealing with white noise, $E\{Y_i(k)Y_p^*(k)\}$ can be written as

$$E\{Y_i(k)Y_p^*(k)\} = \sum_{n=0}^{N-1}\sum_{m=0}^{N-1} w(n)w(m)\, \sigma_g^2$$
$$\cdot \delta\{[n + (i-1)N/R] - [m + (p-1)N/R]\}\, e^{-j2\pi k(n-m)/N} \quad (4.24)$$

The discrete delta function equals unity when $m = n + N(i-p)/R$. Using this result in (4.24) yields

$$E\{Y_i(k)Y_p^*(k)\} = \sigma_g^2 \sum_{n=0}^{N-1} w(n)w[n + (i-p)N/R]\, \exp[-j2\pi k(p-i)/R] \quad (4.25)$$

By substituting (4.25) into the expression for the variance (4.23), we obtain

$$\sigma^2\{I(k)|\mathrm{H}_0\} = \frac{1}{K}\left[\sigma_g^2 \sum_{n=0}^{N-1} w^2(n)\right]^2 + \frac{1}{K^2} \sum_{\substack{i=1 \\ i\neq p}}^{K} \sum_{p=1}^{K} \left\{\sigma_g^2 \sum_{n=0}^{N-1} w(n)w[n + (i-p)N/R]\right\}^2 \tag{4.26}$$

It is straightforward to show that for $i > p$, the second summation in (4.26) is equal to

$$\frac{1}{K^2} \sum_{i=1}^{K} (K - i) \left[\sigma_g^2 \sum_{n=0}^{N-1} w(n)w(n + iN/R)\right]^2 \tag{4.27}$$

Furthermore, the second summation in (4.26) also equals (4.27) for $i < p$. Then (4.26) can be written as

$$\sigma^2\{I(k)|\mathrm{H}_0\} = \frac{1}{K}\left[\sigma_g^2 \sum_{n=0}^{N-1} w^2(n)\right]^2 + \frac{2}{K^2} \sum_{i=1}^{K} (K - i) \left[\sigma_g^2 \sum_{n=0}^{N-1} w(n)w(n + iN/R)\right]^2$$

After simplifying, we obtain the desired result:

$$\sigma^2\{I(k)|\mathrm{H}_0\} = \frac{\sigma^2\{|Y_i(k)|^2\}}{K}\,\alpha \tag{4.28}$$

where

$$\alpha = 1 + \frac{\dfrac{2}{K} \displaystyle\sum_{i=1}^{K-1} (K - i) \left[\sum_{n=0}^{N-1} w(n)w(n + iN/R)\right]^2}{\left[\displaystyle\sum_{n=0}^{N-1} w^2(n)\right]^2}$$

and

$$\sigma^2\{|Y_i(k)|^2\} = \left[\sigma_g^2 \sum_{n=0}^{N-1} w^2(n)\right]^2$$

Expression (4.28) is equivalent to Welch's result [12] and relates the window function and the redundancy to the reduction in variance of the spectral estimate. With $R = 1$ in (4.28), so that there is no overlap, the reduction in variance due to integrating K outputs of the SLD in Figure 4.14 is equal to K. The increase in variance due to $R > 1$ is given by α.

Find $E\{I(k)|H_1\}$

It remains now to find $E\{I(k)|H_1\}$ to be able to evaluate the expression for deflection in (4.9). From the definition of $I(k)$, this is equivalent to finding $E\{|Y_i(k)|^2\}$. For this hypothesis, $x(n) = A\ e^{j2\pi fn/f_s} + g(n)$. It is straightforward to show that

$$E\{I(k)|H_1\} = E\{|Y_i(k)|^2\}$$

$$= \left| A \sum_{n=0}^{N-1} w(n) \exp[j2\pi n(f/f_s - k/N)] \right|^2 + \sigma_g^2 \sum_{n=0}^{N-1} w^2(n) \qquad (4.29)$$

REFERENCES

1. Struzinski, W.A., and E.D. Lowe, "A Performance Comparison of Four Noise Background Normalization Schemes Proposed for Signal Detection Systems," *J. Acoustical Society of America,* Vol. 76, No. 6, December 1984, pp. 1738–1742.
2. Struzinski, W.A., and E.D. Lowe, "The Effect of Improper Normalization on the Performance of an Automated Energy Detector," *J. Acoustics Society of America,* Vol. 78, No. 3, September 1985, pp. 936–941.
3. Wolcin, J.J., "On the Statistical Properties of Noise Background Equalization Schemes," *NUSC Technical Memo.* No. 781159, Naval Underwater Systems Center, New London, CT, July 31, 1978.
4. Robertson, G.H., "A Fast Amplitude Approximation for Quadrature Pairs," *Bell System Technical J.,* Vol. 50, No. 8, October 1971, pp. 2849–2852.
5. Robertson, G.H., "Operating Characteristics for a Linear Detector of CW Signals in Narrow-Band Gaussian Noise," *Bell System Technical J.,* Vol. 46, No. 4, April 1967, pp. 755–774.
6. Whalen, A.D., *Detection of Signals in Noise,* Academic Press, New York, 1971.
7. Robertson, G.H., "Performance Degradation by Postdetector Nonlinearities," *Bell Syst. Technical J.,* Vol. 47, No. 3, March 1968, pp. 407–414.
8. Albersheim, W.J., "A Closed-Form Approximation to Robertson's Detection Characteristics," *Proc. IEEE,* Vol. 69, No. 7, July 1981, p. 839.
9. Harris, F.J., "On the Use of Windows for Harmonic Analysis with the Discrete Fourier Transform," *Proc. IEEE,* Vol. 66, No. 1, January 1978, pp. 51–83.
10. DeFatta, D.J., J.G. Lucas, and W.S. Hodgkiss, *Digital Signal Processing: A System Design Approach,* John Wiley & Sons, New York, 1988.
11. Pryor, C.N., "Calculation of the Minimum Detectable Signal for Practical Spectrum Analyzers," *NOLTR* 71–92, Naval Ordnance Laboratory, Silver Spring, MD, August 2, 1971.
12. Welch, P.D. "The Use of the Fast Fourier Transform for the Estimation of Power Spectra: A Method Based on Time Averaging over Short, Modified Periodograms," *IEEE Trans. on Audio Electroacoustics,* Vol. AU-15, No. 2, June 1967, pp. 70–73.
13. Oppenheim, A.V., and R.W. Schafer, *Digital Signal Processing,* Prentice-Hall, Englewood Cliffs, NJ, 1975.
14. Burdic, W.S., *Underwater Acoustic System Analysis,* Prentice-Hall, Englewood Cliffs, NJ, 1984.
15. Reed, I.S., "On a Moment Theorem for Complex Gaussian Processes," *IRE Trans. on Information Theory,* Vol. IT-8, No. 3, April 1962, pp. 194–195.

Chapter 5
Active Sonar

A look at the characteristics of the sonar channel is followed by an investigation of types of transmitted pulses that reduce the impact of the channel. The optimum receiver is presented for range-doppler-shift processing in a background-noise–limited environment. The effect of reverberation is treated, followed by implementations based on the fast Fourier transform.

5.1 CHANNEL MODELING

We assume that the channel, or medium, consists of surface, bottom, and volume scatterers. The velocities of the surface and volume scatterers are assumed to be random variables. In addition, the strengths of the returns from these scatterers are random variables. The distribution of volume scatterers in space is assumed to be inhomogeneous. Volume scattering arises from thermal layers, biological sources, and reflections from particles suspended in the channel.

As a result of these scatterers, the channel gives rise to dispersion or spreading in time, frequency, and angle. The characteristics of the received signal depend on the transmitted waveform, so in active sonar the effects of the channel can be reduced by the appropriate selection of transmitted waveforms.

The effects of the channel are discussed in Urick [1], Burdic [2], Van Trees [3], and numerous other references such as papers appearing in the *Journal of the Acoustical Society of America*. Figure 5.1(a) illustrates these effects for passive sonar. The signal received from a source suffers time, frequency, and angle spreading. In addition, the receiver must operate in the presence of background noise consisting of ambient noise and flow noise. For active sonar, Figure 5.1(b) shows a bistatic system; that is, the source and the receiver are located at separate points. In a monostatic system, the functions of transmission and reception share the same array of sensors. As illustrated in Figure 5.1(b), the transmitted pulse is spread by the channel and undergoes further change as a result of the interaction with the contact. The

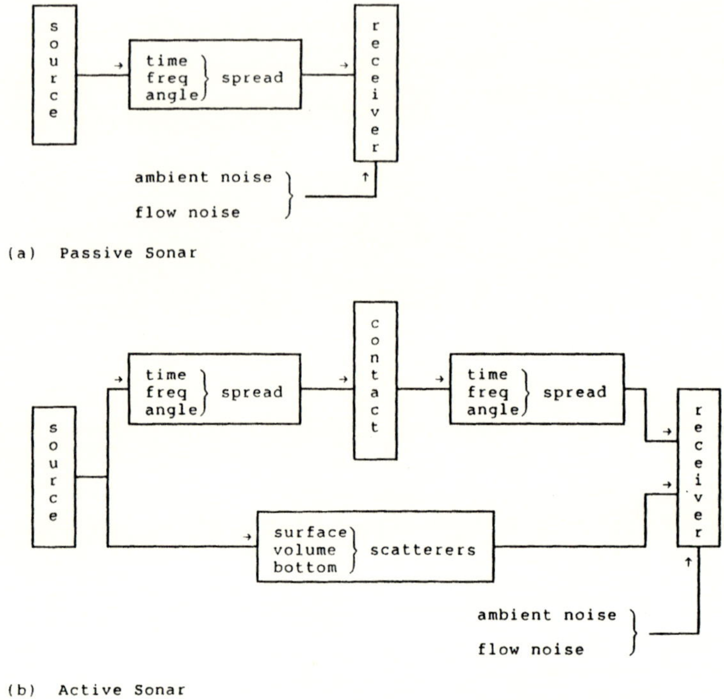

Figure 5.1 Effects of the channel: (a) passive sonar; (b) active sonar.

return also is spread by the channel. Here, the receiver again must operate in the presence of background noise. An additional component of the noise field is due to returns from surface, volume, and bottom scatterers. This component is called *reverberation noise*. The sonar system is said to be background noise-limited if the sum of ambient noise and flow noise dominates the reverberation component, and reverberation noise limited if the converse is true.

Time Dispersion

One effect of the channel on a transmitted pulse is to disperse or spread the pulse in time due to the presence of a number of different propagation paths. Because different arrival times are associated with these paths, the pulse arrives at the receiver spread in the time domain. The extent of this time spread often is denoted by L seconds. Such time spreading gives rise to frequency-selective fading, as indicated in Example 5.1.1.

Example 5.1.1

A sine wave of frequency f_0 is transmitted. In a two-path environment, the received signal has the form:

$$s_r(n) = \alpha_1 \sin[2\pi f_0(n + n_1)/f_s] + \alpha_2 \sin[2\pi f_0(n + n_2)/f_s] \qquad (5.1)$$

where α_i and n_i represent the attenuation and delay, respectively, associated with the ith path. Using the identity

$$\sin A + \sin B = 2 \sin[(A + B)/2] \cos[(A - B)/2]$$

it follows that $s_r(n)$ in (5.1) has a component of the form

$$\cos[\pi f_0(n_1 - n_2)/f_s] \sin[\pi f_0(2n + n_1 + n_2)/f_s]$$

The modulating function, $\cos[\pi f_0(n_1 - n_2)/f_s]$, equals zero for values of transmission frequency given by

$$f_0 = \frac{(2k - 1)\, f_s}{2\,|n_1 - n_2|}$$

where k is an integer.

We can extend Example 5.1.1 to the case of multiple paths where the difference in propagation times between the longest and the shortest paths is equal to L s.

Frequency Dispersion

A transmitted sine wave of frequency f_0 is shifted after being reflected from a scatterer, and the frequency at the receiver is f. Because each scatterer moves with respect to the transmitter, the received signal consists of a sum of sine waves with a distribution of frequencies related to the distribution of the velocities of the scatterers. The width of this distribution is denoted by B Hz.

Measurements of time and frequency dispersion have been made in the Florida Straits by DeFerrari and Nghiem-Phu [4] and by Veenkant [5]. In the work cited in [4], time spreads of up to 600 ms were observed for a 19 ms pulse with a center frequency of 420 Hz. Frequency spreads of several tenths of Hz were also observed, due largely to interactions with the surface. The shape of the spectral sidebands of the received signal closely resembled the shape of the spectrum of the height of the ocean waves.

Angle Dispersion

At long ranges, transmitted signals are received from a number of different directions due to the existence of multiple refracted and reflected paths. For an array that is steered to one of these paths, the signals arriving from the other paths appear as noise. An important contributor to the refractive effects is the variation of the speed of sound as a function of depth [1, Chapter 6; 2, Chapter 5].

5.2 SIGNAL CONSIDERATIONS

We transmit a waveform, $s(n)$, with a length equal to N samples and receive a sequence

$$x(n) = A\, s[n - \tau(n)] + g(n) \tag{5.2}$$

where A is a range-dependent attenuation, $\tau(n)$ is the two-way delay in samples experienced by the portion of the sequence received at time nT, and $g(n)$ is a noise sequence. For simplicity, we assume that the contact is moving with a constant radial velocity of v relative to the source, so that the range as a function of time is $R(nT) = R_0 + vnT$. R_0 is the range for $n = 0$, the time at which the waveform is transmitted, and $T = 1/f_s$ is the interval of time between samples.

A sketch of the geometry associated with the constant-velocity assumption is provided by Figure 5.2. As shown in this figure, a positive value of v indicates that the distance between source and contact is increasing with time. We assume that $|v| \ll c$, where c is the speed of sound in the medium. For a part of the waveform

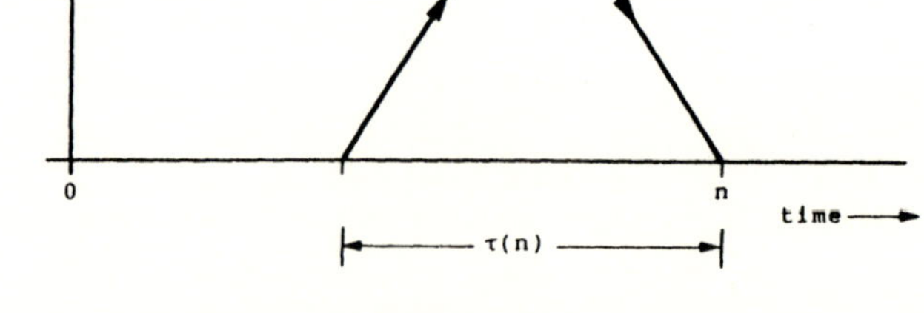

Figure 5.2 Geometry for the assumption of constant relative velocity.

received at time nT, the two-way delay is then approximately equal to

$$\tau(n) = 2\left(\frac{vn}{c} + \frac{R_0}{cT}\right) \tag{5.3}$$

An expression for the received waveform is found by substituting (5.3) into (5.2):

$$x(n) = A\, s\left(n - \frac{2vn}{c} - \frac{2R_0}{cT}\right) + g(n)$$

which can be written in the form

$$x(n) = A\, s[n(1 - \delta) - n_0] + g(n) \tag{5.4}$$

where $\delta = 2v/c$, the time compression parameter, provides information concerning doppler shift, or velocity. The quantity $n_0 = 2R_0/(cT)$ in (5.4) is the time delay parameter that corresponds to range. The assumption that $v \ll c$ is typically valid because c is about 3000 knots and $|v|$ may be on the order of 30 knots.

We can see from (5.4) that the waveform in fact is compressed or expanded as a result of the constant radial velocity between source and contact. The transmitted pulse, $s(n)$, is nonzero only for values of the argument in the range $0 \leqslant n \leqslant N - 1$. We apply this constraint to (5.4) and find that n must satisfy $0 \leqslant n(1 - \delta) - n_0 \leqslant N - 1$. Rewriting this inequality yields $n_0/(1 - \delta) \leqslant n \leqslant (N - 1)/(1 - \delta) + n_0/(1 - \delta)$. We therefore see that the original length-N waveform is received as a length-$[N/(1 - \delta)]$ waveform. For positive values of v, the waveform therefore is expanded, whereas for negative values of v, $s(n)$ is compressed.

5.2.1 Signal Ambiguity Function

In Section 3.2.1, we saw that the optimum detector for a known signal in white Gaussian noise is the correlation receiver, described by $\lambda = \mathbf{s}^+\mathbf{R}_g^{-1}\mathbf{x}$. The vectors $\mathbf{s}$ and $\mathbf{x}$ are the replica and the received waveforms, respectively, and $\mathbf{R}_g$ is the noise correlation matrix. We assume for the moment that $g(n)$ in (5.4) is a sample sequence of a zero-mean white Gaussian noise process so that $\mathbf{R}_g = \sigma_g^2\mathbf{I}$. The output of the correlation receiver at time $\hat{n}T$ is

$$\lambda(\hat{n}) = \sum_{n=n_1}^{n_2} x(n)\, s^*(n - \hat{n})$$

or, after substituting for $x(n)$ from (5.4),

$$\lambda(\hat{n}) = \sum_{n=n_1}^{n_2} \{A\, s[n(1 - \delta) - n_0] + g(n)\}\, s^*(n - \hat{n}) \tag{5.5}$$

For convenience, we have omitted the factor $1/\sigma_g^2$ in (5.5), and the limits of summation will be defined later. The output of the correlation receiver is calculated using N samples so that this detector in fact is matched to the nominal length of the waveform.

It is of interest to determine the response of the correlation receiver to mismatches in both the compression (or doppler shift) parameter, δ, and the delay (or range) parameter, $\hat{n}$. This can be done by considering the receiver output (5.5) in the absence of noise. The resulting function is

$$\Psi(\hat{n}, \delta) = \sum_{n=n_1}^{n_2} s[n(1 - \delta) - n_0]\, s^*(n - \hat{n}) \tag{5.6}$$

where A in (5.5) has been set equal to unity. The function given in (5.6) sometimes is referred to as the *uncertainty function* for $s(n)$. We assume for convenience that $s(n)$ is nonzero only for $-N/2 \leqslant n \leqslant N/2 - 1$. The summation limits in (5.6) are then defined by

$$\begin{aligned} n_1 &= \max\,[-N/2 + \hat{n}, \quad (-N/2 + n_0)/(1 - \delta)] \\ n_2 &= \min\,[N/2 - 1 + \hat{n}, \quad (N/2 - 1 + n_0)/(1 - \delta)] \end{aligned} \tag{5.7}$$

where $-N/2 + \hat{n}$ and $N/2 - 1 + \hat{n}$ are associated with the replica, and the other pair of limits in (5.7) correspond to the return. The relationship of the limits in (5.7) is sketched in Figure 5.3 for the case where δ is positive, so that the return pulse is longer than the replica. A similar figure can be drawn for $\delta < 0$.

The squared magnitude of the uncertainty function often is referred to as the *ambiguity function*, $|\Psi(\hat{n}, \delta)|^2$. A sketch of an ambiguity function, centered at $\hat{n} = n_0$ and $\delta = 0$ is shown in Figure 5.4. As in Section 3.2.1, we are interested in setting a threshold and comparing the receiver output to that threshold. If the threshold is exceeded, a detection is declared. Such a detection threshold, as illustrated in Figure 5.4, consists of a plane parallel to the $(\hat{n}, \delta)$ plane and located at a distance of B above this plane. The intersection of this threshold plane and the ambiguity function defines an ambiguity diagram, as shown in Figure 5.5. The vertical axis in this figure represents the doppler shift $\beta = \delta f_0$ in Hz rather than the compression parameter as in Figure 5.4.

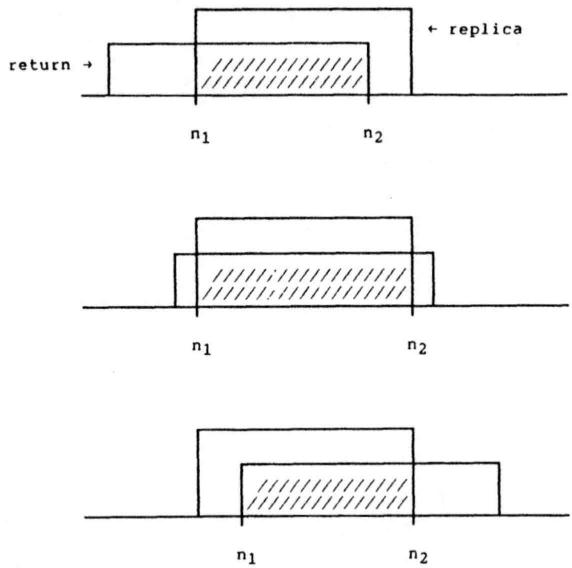

Figure 5.3 Regions of overlap between the return pulse and the replica ($\delta > 0$).

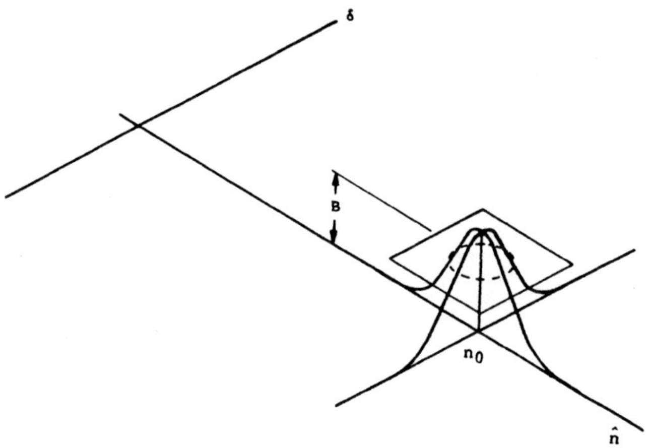

Figure 5.4 Typical ambiguity function, centered at $\hat{n} = n_0$ and $\delta = 0$.

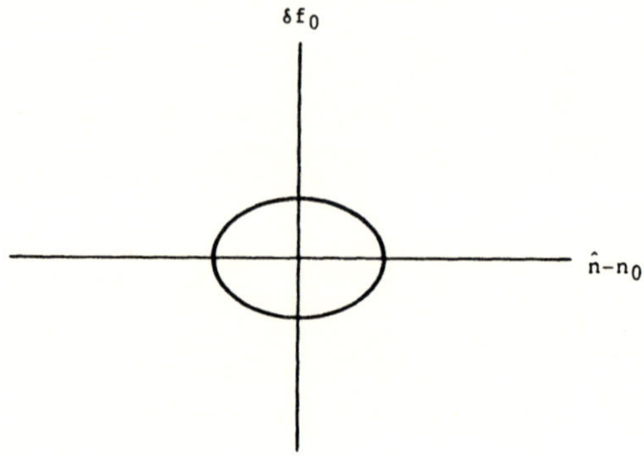

Figure 5.5 Hypothetical ambiguity diagram.

The ambiguity diagram provides an indication of the accuracy with which range and doppler shift can be determined, using a given waveform, $s(n)$. Because the only information provided by the correlation receiver is simply that the threshold has been exceeded, any pair of doppler shift and delay parameters $(\beta, \hat{n})$ inside the ambiguity diagram constitutes a possible estimate of the true velocity and range of the contact.

5.2.2 Continuous Wave Pulse

We illustrate the ideas of the ambiguity function and ambiguity diagram by considering the simple *continuous wave* (CW) pulse, where $s(n)$ is defined by

$$s(n) = \begin{cases} \exp(j2\pi f_0 n/f_s), & -N/2 \leq n \leq N/2 - 1 \\ 0, & \text{otherwise} \end{cases}$$

From (5.6), the uncertainty function for the CW pulse is

$$\Psi(\hat{n}, \delta) = \sum_{n=n_1}^{n_2} \exp\{j2\pi f_0[n(1-\delta) - n_0]/f_s\} \cdot \exp[-j2\pi f_0(n - \hat{n})/f_s]$$

or, equivalently,

$$\Psi(\hat{n}, \delta) = \sum_{n=n_1}^{n_2} \exp[j2\pi f_0(\hat{n} - n_0)/f_s] \exp(-j2\pi \delta f_0 n/f_s) \tag{5.8}$$

We ignore $1/(1 - \delta)$ factors in (5.7) and obtain the summation limits:

$$\begin{aligned} n_1 &= \max(-N/2 + \hat{n}, \quad -N/2 + n_0) \\ n_2 &= \min(N/2 - 1 + \hat{n}, \quad N/2 - 1 + n_0) \end{aligned} \tag{5.9}$$

The ambiguity function is found from (5.8):

$$|\Psi(\hat{n}, \delta)|^2 = \left| \sum_{n=n_1}^{n_2} \exp(-j2\pi\delta f_0 n/f_s) \right|^2 \tag{5.10}$$

We assume first that $\hat{n} \leq n_0$. Then from (5.9) the limits of the summation are $n_1 = -N/2 + n_0$ and $n_2 = N/2 - 1 + \hat{n}$. After simplifying (5.10), we obtain

$$|\Psi(\hat{n}, \delta)|^2 = \left| \frac{\sin[\pi\delta f_0(N + \hat{n} - n_0)/f_s]}{\sin(\pi\delta f_0/f_s)} \right|^2 \tag{5.11}$$

Next we assume that $\hat{n} > n_0$, so that the limits of summation are $n_1 = -N/2 + \hat{n}$ and $n_2 = N/2 - 1 + n_0$. With these limits, the ambiguity function in (5.10) can be written in the form

$$|\Psi(\hat{n}, \delta)|^2 = \left| \frac{\sin[\pi\delta f_0(N + n_0 - \hat{n})/f_s]}{\sin(\pi\delta f_0/f_s)} \right|^2 \tag{5.12}$$

Both (5.11) and (5.12) are equivalent to

$$|\Psi(\hat{n}, \delta)|^2 = \left| \frac{\sin[\pi\delta f_0(N - |\hat{n} - n_0|)/f_s]}{\sin(\pi\delta f_0/f_s)} \right|^2 \tag{5.13}$$

The ambiguity diagram next is obtained for the CW pulse to assess the expected accuracy of the estimates of doppler shift and range. We select a detection-threshold plane in Figure 5.4 that is located a distance above the $(\hat{n}, \delta)$ plane equal to one-half the height of the ambiguity function. (The height of the ambiguity function is $|\Psi(n_0, 0)|^2 = N^2$.) To calculate the δf_0-axis intercepts, we set $\hat{n} = n_0$ in (5.13), solve the equation:

$$|\Psi(n_0, \delta)|^2 = \left| \frac{\sin(\pi\delta f_0 N/f_s)}{\sin(\pi\delta f_0/f_s)} \right|^2 = N^2/2$$

and obtain $\delta f_0 \approx \pm 0.44 f_s/N$. Similarly, the $\hat{n}$-axis intercepts are found by solving

$$|\Psi(\hat{n}, 0)|^2 = (N - |\hat{n} - n_0|)^2 = N^2/2$$

These intercepts occur for $\hat{n} - n_0 = \pm 0.29\ N$. The continuous-time equivalents of the δf_0 and $\hat{n}$ intercepts are $\pm 0.44/T$ and $\pm 0.29\ T$, respectively, where T is the duration of the pulse in seconds.

Plotted in Figure 5.6 is the ambiguity diagram for the CW pulse. As indicated in Figure 5.6(a), a long pulse (that is, a pulse with a large value of N) should provide good resolution of doppler shift at the expense of range resolution. Conversely, the use of a short pulse should give a good estimate of range but a poor estimate of doppler shift, as shown in Figure 5.6(b). Ideally, it would be desirable to transmit a waveform with an ambiguity function equal to N^2 at the origin of the $(\hat{n}, \delta f_0)$ plane and equal to zero everywhere else. Such a waveform would provide estimates of

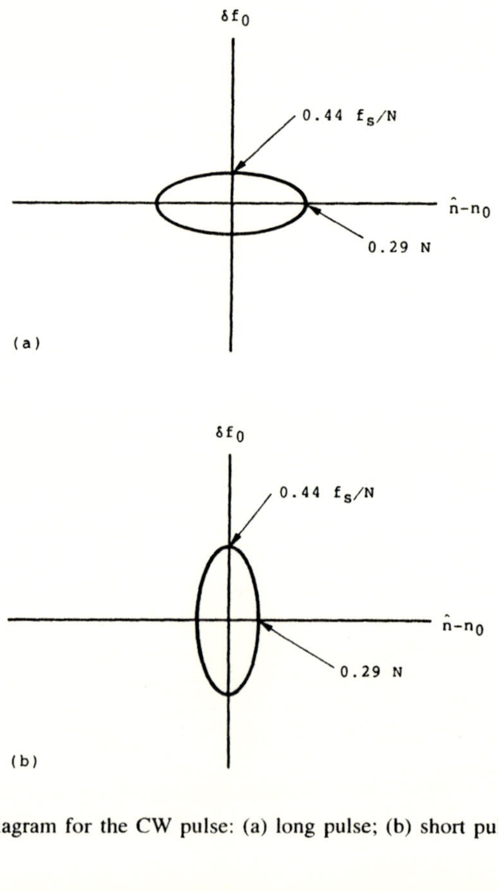

Figure 5.6 Ambiguity diagram for the CW pulse: (a) long pulse; (b) short pulse.

range and doppler shift with zero uncertainty. As shown next, the volume under the ambiguity function is equal to the square of the signal energy for the CW pulse so this ideal waveform cannot be achieved.

Volume under the Ambiguity Function

The volume, V, under the ambiguity function is found by summing over $\hat{n}$ and δ. We again choose to express the doppler shift in Hz and therefore use $\beta = \delta f_0$, where β can assume values in the range $-f_s/2 \leq \beta \leq f_s/2$. Using M frequency points for the summation, β assumes the discrete values of kf_s/M, where k is in the range $-M/2 \leq k \leq M/2 - 1$. The volume then is equal to

$$V = \sum_{\hat{n}=-\infty}^{\infty} \sum_{k=-M/2}^{M/2-1} |\Psi(\hat{n}, \delta)|^2$$

Using the expression for the ambiguity function of the CW pulse (5.10) with $\delta f_0/f_s$ replaced by k/M, we obtain

$$V = \sum_{\hat{n}=-\infty}^{\infty} \sum_{k=-M/2}^{M/2-1} \sum_{n=n_1}^{n_2} \exp(-j2\pi kn/M) \sum_{i=n_1}^{n_2} \exp(j2\pi ki/M)$$

Reordering the summations yields

$$V = \sum_{\hat{n}=-\infty}^{\infty} \sum_{n=n_1}^{n_2} \sum_{i=n_1}^{n_2} \sum_{k=-M/2}^{M/2-1} \exp[-j2\pi k(n-i)/M] \tag{5.14}$$

The last summation in (5.14) is simply equal to M for $n = i$ and equal to zero otherwise. Thus (5.14) is equivalent to

$$V = M \sum_{\hat{n}=-\infty}^{\infty} \sum_{n=n_1}^{n_2} 1 \tag{5.15}$$

We rewrite (5.15) in the form

$$V = M \sum_{\hat{n}=-\infty}^{n_0} \sum_{n=n_1}^{n_2} 1 + M \sum_{\hat{n}=n_0+1}^{\infty} \sum_{n=n_1}^{n_2} 1 \tag{5.16}$$

Using limits from (5.9), the first term in (5.16) is

$$V_- = M \sum_{\hat{n}=-\infty}^{n_0} \sum_{n=-N/2+n_0}^{N/2-1+\hat{n}} 1 \tag{5.17}$$

We can see from (5.17) that $\hat{n}$ must be in the range $-N + 1 + n_0 \leqslant \hat{n} \leqslant n_0$ to obtain nonzero contributions to V_-. The second sum in (5.17) equals 1 for $\hat{n} = -N + 1 + n_0$, 2 for $\hat{n} = -N + 2 + n_0$, and so forth. For $\hat{n} = n_0$, the second sum is equal to N. Therefore (5.17) is equal to

$$V_- = M \sum_{i=1}^{N} i = MN(N + 1)/2 \tag{5.18}$$

Similarly, the second term in (5.16) is equal to

$$V_+ = M \sum_{\hat{n}=n_0+1}^{\infty} \sum_{n=-N/2+\hat{n}}^{N/2-1+n_0} 1$$

which is equivalent to

$$V_+ = M \sum_{i=1}^{N-1} i = MN(N - 1)/2 \tag{5.19}$$

The total volume from (5.18) and (5.19) is $V = V_- + V_+ = MN^2$. The energy in the CW pulse is simply

$$\sum_{n=0}^{N-1} |s(n)|^2 = N$$

To obtain an approximation to the continuous volume, we scale V by the time interval, $1/f_s$, and the frequency interval, f_s/M. Thus the volume under the ambiguity function is $V(1/f_s)(f_s/M) = MN^2/M = N^2$, the square of the signal energy. It follows therefore that the volume under the ambiguity function is equal for CW waveforms of equal energy.

Because the volume is constant, an attempt to simultaneously limit the extent of the main lobe in both time and frequency therefore will cause sidelobes to appear. These are undesirable because, in the presence of noise, they may be mistaken for the main lobe. Erroneous estimates of range and doppler shift then would result. In addition, as the main lobe is narrowed in one dimension, it expands in the other. It therefore is not possible to simultaneously obtain highly accurate estimates of both range and doppler shift with a single type of pulse. This well-known uncertainty principle is mentioned in Section 5.3.1.

5.2.3 Linear FM Pulse

Another common waveform is the *linear frequency-modulated* (LFM) pulse, defined by

$$s(n) = \begin{cases} \exp\{j2\pi[f_0 n/f_s + mn^2/(2f_s^2)]\}, & -N/2 \leq n \leq N/2 - 1 \\ 0, & \text{otherwise} \end{cases} \tag{5.20}$$

where m, the sweep or chirp rate, is equal to Wf_s/N, and W is the sweep extent in Hz. The frequency behavior of this pulse is plotted in Figure 5.7 for $m > 0$.

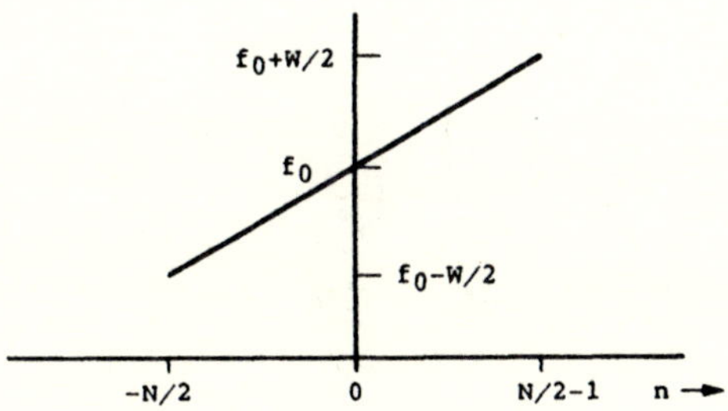

Figure 5.7 Frequency-time characteristics for the LFM pulse.

We develop the ambiguity function for the LFM pulse by substituting (5.20) into the definition of the uncertainty function (5.6). This yields

$$\begin{aligned} \Psi(\hat{n}, \delta) = \sum_{n=n_1}^{n_2} & \exp\{j2\pi f_0[n(1-\delta) - n_0]/f_s\} \\ & \cdot \exp\{j2\pi m[n(1-\delta) - n_0]^2/(2f_s^2)\} \\ & \cdot \exp[-j2\pi f_0(n-\hat{n})/f_s] \cdot \exp[-j2\pi m(n-\hat{n})^2/(2f_s^2)] \end{aligned} \tag{5.21}$$

It is convenient to multiply (5.21) by

$$1 = \exp(j2\pi mnn_0\delta/f_s^2)\exp(-j2\pi mnn_0\delta/f_s^2)\cdot\exp(j2\pi mn_0^2\delta/f_s^2)\exp(-j2\pi mn_0^2\delta/f_s^2)$$

After simplification, we obtain

$$|\Psi(\hat{n}, \delta)| = \left| \begin{aligned} & \sum_{n=n_1}^{n_2} \exp(-j2\pi\delta f_0 n/f_s) \\ & \quad \cdot \exp\{j2\pi mn[\hat{n} - n_0(1+\delta)]/f_s^2\} \\ & \quad \cdot \exp[-j2\pi m\delta(n-n_0)^2/f_s^2] \end{aligned} \right| \tag{5.22}$$

where quantities in δ^2 have been neglected. The difference in slope between the transmitted LFM waveform and the time-compressed return is $\Delta m = 2m\delta(1 - \delta/2)$, which is approximately equal to $2m\delta$ for $\delta \ll 1$. This difference appears in the last exponent of (5.22).

Mismatches between the replica and the return in frequency (δf_0), in slope, and in time contribute to the total correlation loss so that $|\Psi(\hat{n}, \delta)|$ in (5.22) attains a value less than the maximum value of N. If Δm is small, we ignore the last exponential factor in (5.22) and obtain

$$|\Psi(\hat{n}, \delta)|^2 = \left| \sum_{n=n_1}^{n_2} \exp(-j2\pi n\{\delta f_0 - m[\hat{n} - n_0(1 + \delta)]/f_s\}/f_s) \right|^2$$

This expression is the same as the ambiguity function for the CW pulse (5.10) if δf_0 is replaced by $\delta f_0 - m[\hat{n} - n_0(1 + \delta)]/f_s$. We therefore obtain

$$|\Psi(\hat{n}, \delta)|^2 = \left| \frac{\sin\{\pi[\delta f_0 - m(\hat{n} - n_0)/f_s](N - |\hat{n} - n_0|)/f_s\}}{\sin\{\pi[\delta f_0 - m(\hat{n} - n_0)/f_s]/f_s\}} \right|^2 \tag{5.23}$$

for $\delta \ll 1$. We can show that along the line defined by $\delta f_0 = m(\hat{n} - n_0)/f_s$, (5.23) is equal to one-half the maximum value of $|\Psi(n_0, 0)|^2 = N^2$ if $[(N - |\hat{n} - n_0|)/N]^2 = 0.5$. This occurs for $\hat{n} - n_0 = \pm(1 - 1/\sqrt{2})\, N = \pm 0.29\, N$. The corresponding δf_0 values are $m(\hat{n} - n_0)/f_s = \pm 0.29\, mN/f_s = \pm 0.29\, W$. Next, the 3-dB intercepts for the δf_0 axis in the ambiguity diagram are found by solving

$$|\Psi(n_0, \delta)|^2 = \left| \frac{\sin(\pi N \delta f_0/f_s)}{\sin(\pi \delta f_0/f_s)} \right|^2 = N^2/2$$

which yields $\delta f_0 = \pm 0.44\, f_s/N$. Finally, the 3-dB intercepts for the $\hat{n}$ axis are found by solving

$$|\Psi(\hat{n}, 0)|^2 = \left| \frac{\sin[\pi m(\hat{n} - n_0)(N - |\hat{n} - n_0|)/(f_s^2)]}{\sin[\pi m(\hat{n} - n_0)/(f_s^2)]} \right|^2 = N^2/2$$

If we assume that the $\hat{n}$-axis intercepts are small compared to N, which typically is valid, these intercepts are found by solving $m|\hat{n} - n_0| = 0.44 f_s^2/N$. The result is $\hat{n} - n_0 = \pm 0.44\, f_s/W$. The ambiguity diagram for the LFM pulse is shown in Figure 5.8. Again, this ambiguity function was derived with the assumption that the difference in slope between return and stored replica is small.

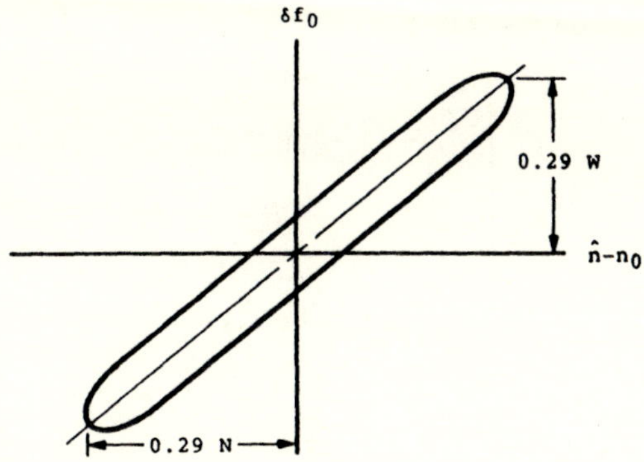

Figure 5.8 Ambiguity diagram for the narrowband LFM pulse.

Narrowband Approximation

A pulse often is said to satisfy the narrowband approximation if a compression in time can be accurately modeled as a translation in frequency [6, p. 39]. That is,

$$s[n(1-\delta)] \approx s(n)\exp(-j2\pi\delta f_0 n/f_s)$$

This relationship holds exactly for the CW pulse in Section 5.2.2. We tacitly invoked the narrowband approximation in the development of the ambiguity function for the LFM pulse (5.23) by assuming that Δm could be neglected. Therefore, the ambiguity diagram in Figure 5.8 applies for the narrowband LFM pulse.

Velocity Tolerance

We are interested here in determining the correlation loss that results when the change in slope between the replica and the return cannot be neglected. We start with (5.6) but set the frequency of the replica equal to $\hat{f}_0$. Then the uncertainty function is

$$\Psi(\hat{n},\delta) = \begin{array}{l} \displaystyle\sum_{n=n_1}^{n_2} \exp\{j2\pi f_0[n(1-\delta)-n_0]/f_s\} \\ \quad\cdot \exp\{j2\pi m[n(1-\delta)-n_0]^2/(2f_s^2)\} \\ \quad\cdot \exp[-j2\pi\hat{f}_0(n-\hat{n})/f_s] \\ \quad\cdot \exp[-j2\pi m(n-\hat{n})^2/(2f_s^2)] \end{array}$$

which, after simplification and taking the magnitude, is equivalent to

$$|\Psi(\hat{n}, \delta)| = \left| \begin{array}{l} \sum_{n=n_1}^{n_2} \exp\{j2\pi n[f_0(1 - \delta) - \hat{f}_0]/f_s\} \\ \quad \cdot \exp\{j2\pi m[n(1 - \delta) - n_0]^2/(2f_s^2)\} \\ \quad \cdot \exp[-j2\pi m(n - \hat{n})^2/(2f_s^2)] \end{array} \right| \tag{5.24}$$

By selecting $\hat{f}_0 = f_0(1 - \delta)$ and $\hat{n} = n_0/(1 - \delta)$, (5.24) becomes

$$|\Psi(\hat{n}, \delta)| = \left| \sum_{n=n_1}^{n_2} \exp[j2\pi m\delta(\delta - 2)(n - \hat{n})^2/(2f_s^2)] \right| \tag{5.25}$$

From (5.9) the limits of summation are $\hat{n} - N/2$ and $\hat{n} + N/2 - 1$ for $\delta \ll 1$, so we can rewrite (5.25) in the following form:

$$|\Psi(\hat{n}, \delta)| = \left| \sum_{n=-N/2}^{N/2-1} \exp[j2\pi m\delta(\delta - 2)n^2/(2f_s^2)] \right| \tag{5.26}$$

By selecting the parameters of the replica so that $\hat{f}_0 = f_0(1 - \delta)$ and $n_0 = \hat{n}(1 - \delta)$, we have obtained an expression (5.26) for the square root of the ambiguity function that contains only the effect of a mismatch in slope, $\Delta m = 2m\delta(1 - \delta/2)$. Expression (5.26) is equivalent to

$$|\Psi(\hat{n}, \delta)|^2 = \left\{ \sum_{n=-N/2}^{N/2-1} \cos[2\pi m\delta(\delta - 2)n^2/(2f_s^2)] \right\}^2 + \left\{ \sum_{n=-N/2}^{N/2-1} \sin[2\pi m\delta(\delta - 2)n^2/(2f_s^2)] \right\}^2$$

or, if quantities in δ^2 are ignored,

$$|\Psi(\hat{n}, \delta)|^2 = \left\{ \sum_{n=-N/2}^{N/2-1} \cos[2\pi m\delta n^2/(f_s^2)] \right\}^2 + \left\{ \sum_{n=-N/2}^{N/2-1} \sin[2\pi m\delta n^2/(f_s^2)] \right\}^2 \tag{5.27}$$

We can approximate the sums in (5.27) with integrals by noting that

$$\frac{1}{f_s} \sum_{n=-N/2}^{N/2-1} \cos[2\pi m\delta n^2/(f_s^2)] \approx \int_{-T/2}^{T/2} \cos(2\pi m\delta t^2)\, dt \tag{5.28}$$

where n/f_s is replaced by t, and $T = N/f_s$. Using the approximation in (5.28), we can rewrite (5.27) in integral form:

$$|\Psi(\hat{n}, \delta)|^2 \approx f_s^2 \left\{ \int_{-T/2}^{T/2} \cos(2\pi m \delta t^2)\, dt \right\}^2 + f_s^2 \left\{ \int_{-T/2}^{T/2} \sin(2\pi m \delta t^2)\, dt \right\}^2 \quad (5.29)$$

We let $u = 2t\sqrt{\delta m}$ in (5.29) and obtain

$$\begin{aligned} |\Psi(\hat{n}, \delta)|^2 \approx \frac{f_s^2}{\delta m} & \left\{ \int_0^{T\sqrt{\delta m}} \cos(\pi u^2/2)\, du \right\}^2 \\ & + \frac{f_s^2}{\delta m} \left\{ \int_0^{T\sqrt{\delta m}} \sin(\pi u^2/2)\, du \right\}^2 \end{aligned} \quad (5.30)$$

which is equivalent to the result obtained in [6, eq. 41].

We wish to find that value of δ such that $|\Psi(\hat{n}, \delta)|^2$ in (5.30) is equal to $N^2/2$, or one-half its peak value. That is, we solve

$$N^2/2 = f_s^2[C^2(N\sqrt{\delta m}/f_s) + S^2(N\sqrt{\delta m}/f_s)]/(\delta m) \quad (5.31)$$

where $C(x)$ and $S(x)$ are the Fresnel integrals:

$$C(x) = \int_0^x \cos(\pi y^2/2)\, dy$$

$$S(x) = \int_0^x \sin(\pi y^2/2)\, dy$$

From [7, Table 7.7], (5.31) is satisfied if the argument of the Fresnel integrals is equal to approximately 1.32. Thus,

$$\frac{N^2\, \delta m}{f_s^2} = \frac{N^2}{f_s^2} \frac{2v}{c} \frac{W f_s}{N} \approx (1.32)^2 \quad (5.32)$$

We solve (5.32) for v to find the value of velocity that corresponds to a 3-dB correlation loss due to slope mismatch. This value is

$$v = \frac{2614}{W(N/f_s)} \text{ knots} \quad (5.33)$$

where c is assumed to equal 3000 knots. For a frequency sweep of $W = 200$ Hz and a pulse length of $N/f_s = 1$ s, the velocity tolerance is about 13.1 knots. That is, the correlation loss due to slope mismatch is no greater than 3 dB for any value of velocity in the interval $-13.1 \leq v \leq 13.1$ knots.

We can see from the ambiguity diagram in Figure 5.8 that the doppler-shift tolerance of the narrowband LFM pulse is equal to 0.29 W. The corresponding velocity tolerance is found by solving $\delta f_0 = (2v/c)f_0 = 0.29\,W$ for v, which yields

$$v = \frac{435\,W}{f_0}\ \text{knots} \tag{5.34}$$

where again c is assumed to equal 3000 knots.

The wideband velocity tolerance in (5.33) was obtained on the assumption that the replica and the return pulse were aligned in time and frequency and differed only in slope. Although correlation loss results from mismatches in slope and in temporal overlap, the mismatch in slope is dominant for the wideband pulse [6, p. 45], [8, p. 631]. Thus (5.33) still provides a reasonably accurate estimate of the velocity tolerance in the presence of temporal mismatch. On the other hand, the decorrelation between the replica and the return for the narrowband LFM pulse is due primarily to temporal overlap loss. This is illustrated in Figure 5.9.

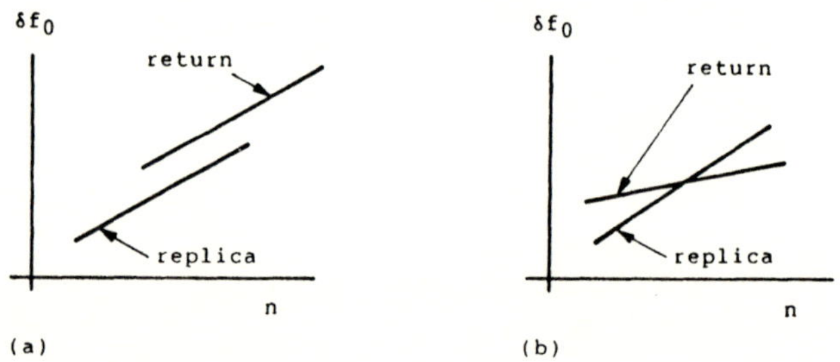

Figure 5.9 Decorrelation between the return pulse and the replica: (a) narrowband LFM pulse; (b) wideband LFM pulse.

We note that a correlation loss also arises from a mismatch in frequency for both narrowband and wideband LFM signals. This problem is solved by providing a set of correlation receivers closely spaced in frequency (Section 5.3.3).

In summary, the velocity tolerance of the narrowband pulse is equal to 435 W/f_0 knots whereas that of the wideband LFM signal is 2614 $f_s/(WN) = 2614/(WT)$ knots, where T is the duration of the pulse in seconds. If the wideband velocity tolerance (5.33) is smaller than the narrowband tolerance (5.34) for a particular set of pulse parameters, the actual velocity tolerance must be predicted using the wideband expression.

It follows from the ambiguity diagram of the LFM pulse (Figure 5.8) that, if the doppler shift is known beforehand, the uncertainty in the range estimate can be reduced. This occurs if the contact is known to be stationary so that the relative velocity, v, is just equal to the velocity of the platform. Furthermore, if the range is known, then the uncertainty in the estimate of doppler shift can be reduced.

5.2.4 Echo Excess Equation

In active sonar, as in passive sonar, we stipulate desired values for probability of detection and probability of false alarm. As in Chapters 3 and 4, these values of P_D and P_{fa} then are used to establish the value of S/N (or d) required at the output of the processor by using a set of ROC curves.

For this section, the processor shall consist of an array of sensors, a beamformer, and a correlation receiver. The output of the processor is compared to a threshold to determine if a signal is present.

The S/N at the output of the beamformer required to achieve an S/N in dB of $10 \log_{10} d$ at the processor output, is called the *detection threshold* (DT). It is of interest to determine the *echo excess* (EE), which is the difference between the actual beamformer output S/N and DT. For the background noise-limited case, EE in dB in a 1-Hz band is given by

$$\mathrm{EE} = \mathrm{SL} - 2\mathrm{TL} + \mathrm{TS} - (\mathrm{NL} - \mathrm{DI}) - \mathrm{DT} \tag{5.35}$$

where

EE = echo excess,
SL = source level (dB//μPa at 1 yd),
TL = transmission loss,
TS = target strength,
NL = background noise level,
DI = directivity index of the receiver array,
DT = detection threshold (required S/N at the beamformer output).

The source level is defined by $\mathrm{SL} = 171.5 + 10 \log_{10} P + \mathrm{DI}_t$, where P is the radiated power in watts and DI_t is the directivity index of the transmitter array [1, p. 69]. The DI of the receiver array is defined in Section 2.3.2 as the ratio of S/N at the output of the beamformer to the S/N at a single sensor.

For the reverberation noise-limited environment, (5.35) becomes

$$\mathrm{EE} = \mathrm{SL} - 2\mathrm{TL} + \mathrm{TS} - \mathrm{RL} - \mathrm{DT} \tag{5.36}$$

where RL is the equivalent plane-wave reverberation level observed at the receiver array. Also useful is an expression for passive sonar relating the difference between the S/N at the output of the beamformer and DT. This difference, called the *signal excess* (SE), is given by

$$\text{SE} = \text{SL} - \text{TL} - (\text{NL} - \text{DI}) - \text{DT}$$

We can rewrite (5.35) without logarithms and obtain

$$\text{ee} = \frac{\text{sl ts di}}{\text{tl}^2 \text{ nl dt}} \tag{5.37}$$

where, for instance, $\text{sl} = 10^{\text{SL}/10}$.

5.2.5 Center Frequency Selection

The choice of a center frequency for the transmitted pulse is influenced by opposing concerns. For instance, ambient noise decreases with frequency (in the frequency range of interest); and, for a fixed-size transmitter array with constant radiated power, the source level increases with frequency. These considerations would suggest the use of a high center frequency. On the other hand, the absorption of the medium increases with frequency. It therefore is possible to find a frequency that maximizes the echo excess in (5.37) for a given range. This type of analysis is discussed in [9].

Optimum Frequency as a Function of Range

We can determine an optimum center frequency for any range of interest by maximizing the echo excess in (5.37) with respect to frequency. We write the frequency dependencies for the factors in (5.37) as follows:

$$\begin{aligned} \text{Source level, sl} &= c_1 f^i, \\ \text{Transmission loss, tl} &= c_2(r)\, 10^{[\alpha(f) r/10]} \ (r \text{ in kyds; } f \text{ in kHz}), \\ \text{Target strength, ts} &= c_3 f^j, \\ \text{Ambient noise, nl} &= c_4 f^k, \\ \text{Directivity index, di} &= c_5 f^m, \\ \text{Detection threshold, dt} &= c_6 f^p, \end{aligned} \tag{5.38}$$

where $c_2(r)$ is a frequency-independent spreading loss. The absorption loss is estimated using (3.1), which is repeated here:

$$\alpha(f) = \frac{0.1f^2}{1+f^2} + \frac{40f^2}{4100+f^2} + 2.75 \cdot 10^{-4}f^2$$

where α is in dB per kyd and f is in kHz. Substituting (5.38) into (5.37) yields

$$\text{ee} = Af^{i+j+m-k-p}\,10^{-[2\alpha(f)r/10]} \tag{5.39}$$

where

$$A = \frac{c_1\,c_3\,c_5}{c_2^2\,c_4\,c_6}$$

We make the following assumptions to obtain values for the exponents of frequency in (5.39):

$$\begin{aligned}
i &= 2, && \text{fixed-size transmitter array with constant radiated power,}\\
j &= 0, && \text{constant target strength,}\\
k &= -2, && \text{ambient noise spectrum,}\\
m &= 0, && \text{DI of receiver array independent of } f,\\
p &= 0, && \text{receiver gain independent of } f.
\end{aligned} \tag{5.40}$$

The choice of values in (5.40) is discussed next. Because we are assuming a fixed-size transmitter array with constant radiated power, the frequency dependence of the source level is found in DI_t. We further assume that a two-dimensional transmitter array is used. For such an array, the DI_t increases as the square of the frequency. We are operating in that portion of the ambient noise curve in Figure 3.3 where the noise level varies approximately as $1/f^2$. The size of the receiver array is allowed to change to maintain constant angular resolution with frequency. Therefore, the receiver DI is not a function of frequency.

We set the composite exponent of frequency in (5.39) equal to n, so that $n = i + j + m - k - p$, and calculate the partial derivative of echo excess with respect to frequency. Equation (5.39) becomes

$$\text{ee} = Af^n\,10^{-[2\alpha(f)r/10]}$$

and the partial derivative is

$$\frac{\partial\,\mathrm{ee}}{\partial f} = A\,nf^{n-1}\,10^{-2[\alpha(f)r/10]} + Af^{n}\,10^{-[2\alpha(f)r/10]}\,\ln(10)\,(-2r/10)\frac{\partial\alpha(f)}{\partial f} \tag{5.41}$$

The partial derivative of absorption loss with respect to frequency is

$$\frac{\partial\alpha(f)}{\partial f} = \frac{0.2f}{(1+f^2)^2} + \frac{3.28\cdot 10^5 f}{(4100+f^2)^2} + 5.5\cdot 10^{-4}f \tag{5.42}$$

We set (5.41) equal to zero and solve for r:

$$r = \frac{5\,n}{f\,\partial\alpha(f)/\partial f\,\ln(10)} \tag{5.43}$$

where $\partial\alpha(f)/\partial f$ is given in (5.42). We thus have in (5.43) an equation relating range

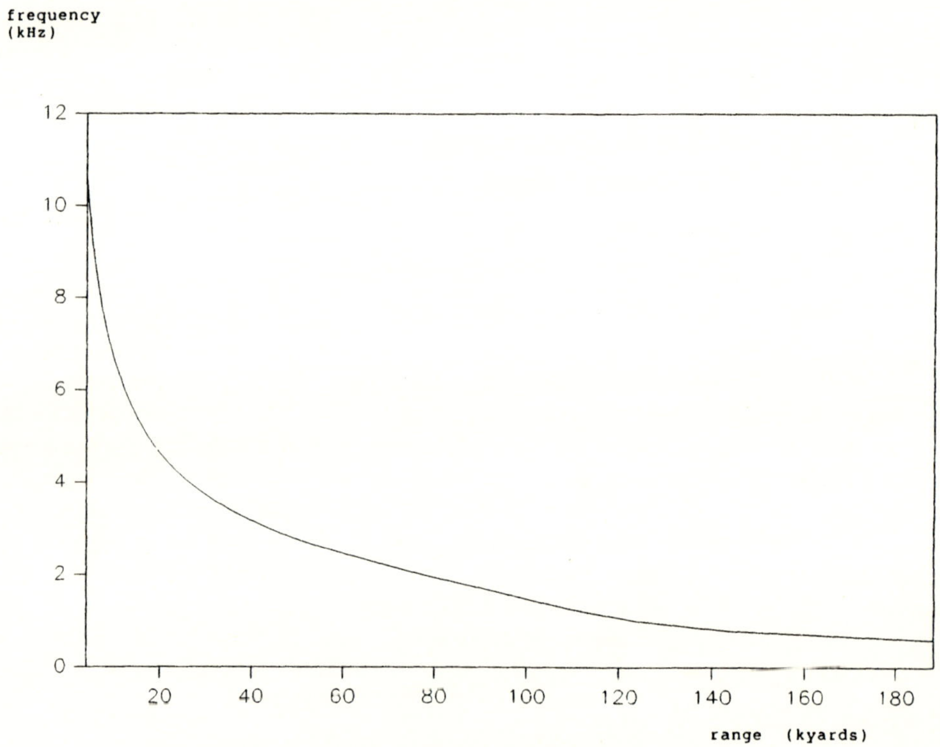

Figure 5.10 Optimum center frequency as a function of range (for the assumptions listed in (5.40)).

in kyds and optimum frequency in kHz, as a function of the system design parameters contained in n. Expression (5.43) is evaluated in Figure 5.10, where, based on the assumptions listed in (5.40), $n = 4$. It is apparent from this figure that low values of center frequency are required for long-range active sonar operation.

Performance Loss Due to Nonoptimum Center Frequency

For a desired range, we wish to calculate the loss in performance (that is, the reduction in echo excess) associated with a center frequency that differs from the optimum value as determined by (5.43). Substituting the values for the exponents of frequency (5.40) into (5.39) gives

$$\text{ee} = Af^4\, 10^{-[2\alpha(f)r/10]} \tag{5.44}$$

Figure 5.11 was generated by assuming a desired range of 100 kyds in (5.44) and illustrates the reduction in ee as a function of frequency. The optimum value of center

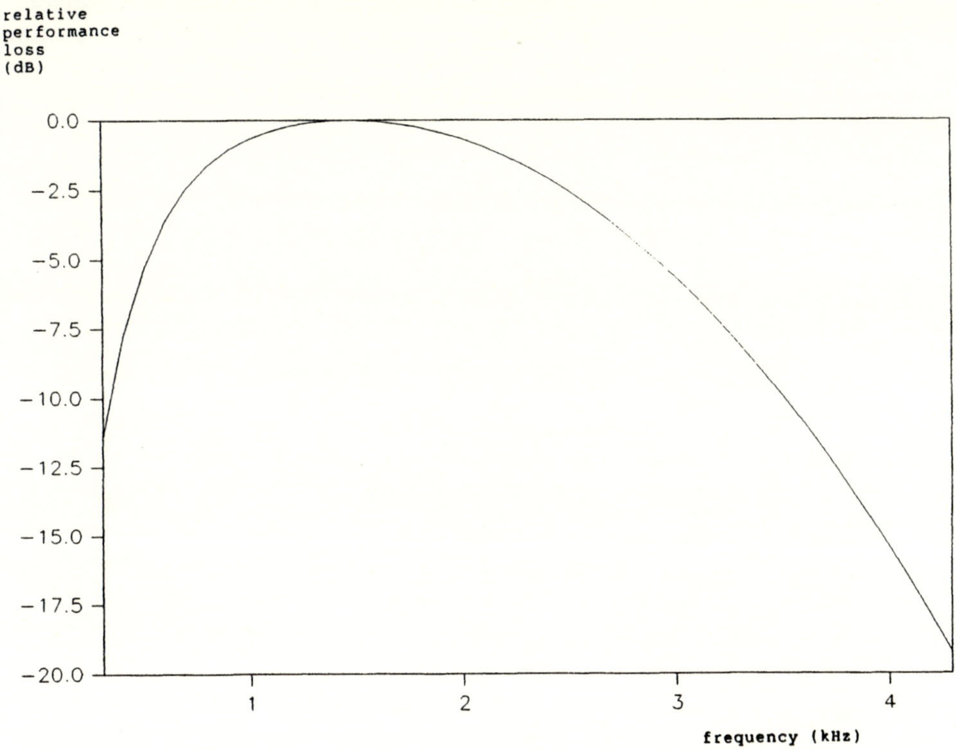

Figure 5.11 Reduction in echo excess as a function of center frequency (range = 100 kyds).

frequency for 100 kyds can be obtained from Figure 5.10 and is approximately equal to 1500 Hz.

Effect of Reverberation Noise on Optimum Center Frequency

Up to this point, the effects of reverberation noise in the calculation of the optimum center frequency have not been considered. In a reverberation noise-limited environment, we would use the appropriate expression for ee (5.36) and include the frequency dependence of $10^{RL/10}$. This dependence is a function of the type of reverberation noise (surface, volume, or bottom) encountered. Thus, for a desired range, a frequency different from the value provided by Figure 5.10 now would be optimum.

Ownship Doppler Nullification

The velocities of scatterers are assumed to be samples of a zero-mean random process so that the returns from these scatterers are centered at zero frequency in the absence of platform motion. Sometimes we want to arrange for these returns to be centered at a fixed frequency in the presence of platform motion, regardless of the direction of the transmitted beam. The function that removes the component of doppler shift due to platform (or ownship) motion is called *ownship doppler nullification* (ODN). Assuming that we send a signal with a center frequency equal to f_0, the received component due to reverberation noise from surface or volume scatterers is centered at

$$f = f_0[1 + (2v/c)\cos\theta] \tag{5.45}$$

where v is the velocity of the platform and θ is the angle between the platform heading and the beam direction, as shown in Figure 5.12. To center the reverberation

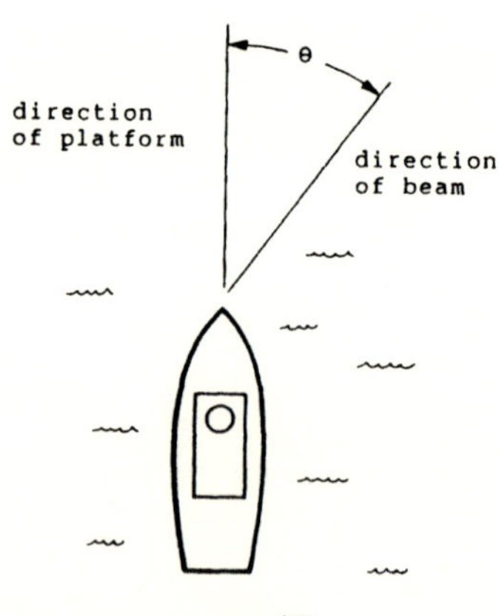

Figure 5.12 Direction of transmitted beam.

noise at f_1 for all beam directions, we arrange for the center frequency to be a function of the beam direction. It follows from (5.45) that the center frequency for a transmitted beam at angle θ should be

$$f_0 = \frac{f_1}{[1 + (2v/c)\cos\theta]}$$

5.3 RANGE-DOPPLER-SHIFT PROCESSING

In this section, we consider the optimum detectors for the CW pulse and the LFM pulse in a background noise-limited environment. The Cramer-Rao lower bounds (Section 3.3.1) for the variances of the range and doppler-shift estimates also are discussed.

A statistical characterization of reverberation noise is required to derive the optimum detector in a reverberation noise-limited environment. Because of the difficulties in obtaining such a characterization in practice, we often use the detector optimized for the background noise-limited case and attempt to minimize the effect of reverberation noise on performance. Parameters under the control of the system designer include pulse type, pulse length, sweep rate (for the LFM pulse), and center frequency (Section 5.2.5). Also discussed in Section 5.3 are FFT-based approximations to optimum detectors.

5.3.1 Background Noise-Limited Environment

We first consider the CW pulse. Assuming that we are operating in zero-mean white Gaussian noise (such as flow noise or isotropic ambient noise, where the sensor spacing is equal to $\lambda/2$), the optimum detector is just the quadrature receiver that we derived in Section 3.2.3. This is shown again in Figure 5.13, where the number

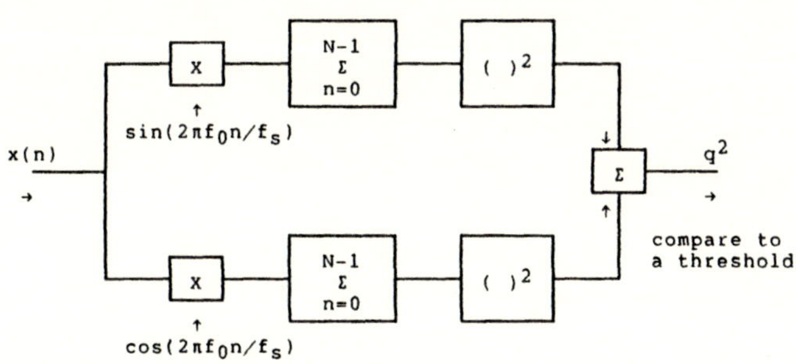

Figure 5.13 Quadrature receiver (number of samples summed equals the length of the CW pulse).

of samples that is summed is set equal to the length of the pulse. The fact that the quadrature receiver is the optimum detector is reasonable because the problem of detecting a CW pulse is equivalent to that of detecting a sinusoidal signal of finite duration in noise.

For the LFM pulse, the optimum detector has the same form as the quadrature receiver with the exception that

$$\sin(2\pi f_0 n/f_s) \text{ is replaced by } \sin\{2\pi[f_0 n/f_s + mn^2/(2f_s^2)]\}$$

$$\cos(2\pi f_0 n/f_s) \text{ is replaced by } \cos\{2\pi[f_0 n/f_s + mn^2/(2f_s^2)]\}$$

in Figure 5.13. That is, we are correlating the received signal with a stored replica of the LFM pulse.

Estimation

We are interested in the CRLBs for the variances of the joint estimates of range and doppler shift. A general waveform

$$s(n) = b(n)\exp\{j[2\pi f_0 n/f_s + \phi(n) + \theta]\} \tag{5.46}$$

is considered where $b(n)$ and $\phi(n)$ are real functions. The CW pulse treated in Section 5.2.2 has the form shown in (5.46) if we set $b(n)$ equal to unity for $-N/2 \leq n \leq N/2 - 1$ and zero otherwise; and if the angle modulation function, $\phi(n)$, is zero. For the LFM pulse, $\phi(n)$ in (5.46) is equal to $\pi mn^2/f_s^2$.

The complex envelope of (5.46) has the form

$$\tilde{s}(n) = b(n)\exp[j\phi(n)]\exp(j\theta) \tag{5.47}$$

We neglect the factor $\exp(j\theta)$ and obtain the continuous-time domain equivalent of (5.47):

$$\tilde{s}(t) = b(t)\exp[j\phi(t)] \tag{5.48}$$

Whalen [10, Section 10.7] derives the following *Fisher information matrix* (FIM) for the joint estimates of delay (τ) and angular frequency $\omega = 2\pi f$ for the signal in (5.48):

$$\mathbf{J} = \begin{bmatrix} -E\left\{\dfrac{\partial^2 \ln p(\mathbf{x}|\omega, \tau)}{\partial \tau^2}\right\} & -E\left\{\dfrac{\partial^2 \ln p(\mathbf{x}|\omega, \tau)}{\partial \tau\, \partial \omega}\right\} \\ -E\left\{\dfrac{\partial^2 \ln p(\mathbf{x}|\omega, \tau)}{\partial \omega\, \partial \tau}\right\} & -E\left\{\dfrac{\partial^2 \ln p(\mathbf{x}|\omega, \tau)}{\partial \omega^2}\right\} \end{bmatrix}$$

$$= \begin{bmatrix} \dfrac{2\,Ef_d^2}{N_0} & -\dfrac{2\,E(\omega t)}{N_0} \\ -\dfrac{2\,E(\omega t)}{N_0} & \dfrac{2\,E\,t_d^2}{N_0} \end{bmatrix} \tag{5.49}$$

where

E = signal energy,
$N_0/2$ = noise power spectral density of $g(n)$,
f_d^2 = a measure of the bandwidth of $\tilde{s}(t)$ and
= $\int(2\pi f)^2\, |S(f)|^2\, \mathrm{d}f$,
$S(f)$ = Fourier transform of $\tilde{s}(t)$,
t_d^2 = a measure of the time duration of $\tilde{s}(t)$ and
= $\int t^2 |\tilde{s}(t)|^2\, \mathrm{d}t$,
(ωt) = a measure of the average frequency-time product:

$$(\omega t) = \int t\, \mathrm{d}\phi(t)/\mathrm{d}t\; b^2(t)\, \mathrm{d}t \tag{5.50}$$

We assume that $\tilde{s}(t)$ in (5.48) is normalized so that

$$\int |\tilde{s}(t)|^2\, \mathrm{d}t = \int |S(f)|^2\, \mathrm{d}f = 1$$

and the time and frequency coordinates are selected so that

$$\int t |\tilde{s}(t)|^2\, \mathrm{d}t = \int f |S(f)|^2\, \mathrm{d}f = 0$$

Inverting the FIM in (5.49) yields the following CRLBs for the variances of the estimates of delay and doppler shift:

$$\hat{\tau}: \quad \frac{t_d^2}{2(E/N_0)\,[f_d^2 t_d^2 - (\omega t)^2]} \tag{5.51}$$

$$\hat{\omega}: \quad \frac{f_d^2}{2(E/N_0)\,[f_d^2 t_d^2 - (\omega t)^2]} \tag{5.52}$$

The CRLBs given in (5.51) and (5.52) are determined for the case of high S/N to produce an unambiguous peak at the output of the optimum detector in the $(\tau,\ \omega)$ domain that is close to the true values of τ and ω. As we can see from the definition of (ωt) in (5.50), the estimates of range and doppler shift are uncoupled if $\tilde{s}(t)$ contains no angle modulation factor (that is, $d\phi(t)/dt = 0$). The off-diagonal elements in (5.49) therefore are equal to zero. The CW pulse is an example of a signal where $d\phi(t)/dt = 0$. With zero angle modulation, (5.51) and (5.52) reduce to

$$\hat{\tau}: \quad \frac{1}{2(E/N_0)f_d^2}$$

$$\hat{\omega}: \quad \frac{1}{2(E/N_0)\,t_d^2}$$

As we saw in Figure 5.6, a pulse with a long time duration (large t_d^2) provides an estimate of frequency with low variance. Alternatively, a pulse with a large frequency bandwidth (large f_d^2) provides a good estimate of range. (A pulse with a large frequency bandwidth equivalently has a short duration in the time domain.)

Uncertainty Relation

Whalen [10, p. 348] shows that the time-bandwidth product of a signal is bounded by

$$t_d^2 f_d^2 \geq (\omega t)^2 + 1/4 \tag{5.53}$$

Therefore, we cannot design a single pulse that has small durations in both the time and the frequency domains. We showed in Section 5.3.5 that equality is obtained in (5.53) using the Gaussian pulse: (1) with no angle modulation and (2) with angle modulation.

5.3.2 Reverberation Noise-Limited Environment

As we mentioned in Section 5.3.1, the optimum detector in the presence of zero mean wide-sense stationary Gaussian noise is the correlation receiver. From Example 3.2.2, the test statistic is

$$\lambda_T = \mathbf{s}^+\, \mathbf{R}_g^{-1}\, \mathbf{x} \tag{5.54}$$

where $\mathbf{s}$ is the known signal vector, $\mathbf{x}$ is the vector of received samples, and $\mathbf{R}_g$ is the noise correlation matrix. For white noise, $\mathbf{R}_g = \sigma_g^2\, \mathbf{I}$ and

$$\sigma_g^2 = \int_{-f_s/2}^{f_s/2} S_g(f)\,df = N_0 f_s/2$$

The noise spectrum, $S_g(f)$, is shown in Figure 5.14, where f_s is the sampling frequency. Substituting $\mathbf{R}_g = \sigma_g^2\,\mathbf{I}$ into (5.54) yields

$$\lambda_T = \frac{\mathbf{s}^+\,\mathbf{x}}{\sigma_g^2} = \frac{1}{\sigma_g^2}\sum_{n=1}^{N} s^*(n)\,x(n) \tag{5.55}$$

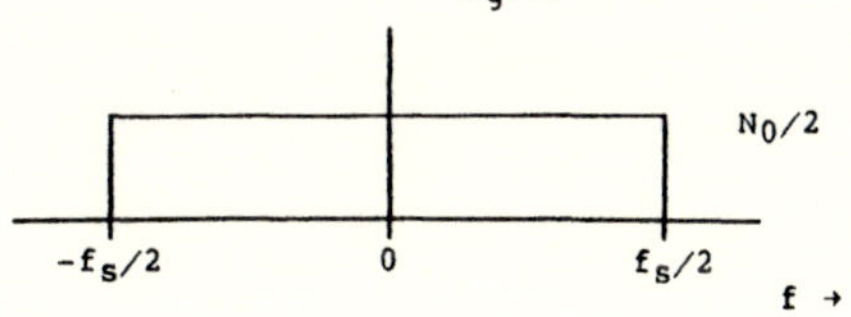

Figure 5.14 Spectrum of white noise.

We use the correlation receiver (5.55) in the presence of reverberation noise and obtain an expression for the S/N, or d, at the output of the receiver. Thus $\mathbf{R}_g$ will consist of a component from background noise and a second component from reverberation noise. The output S/N is equal to the square of the deflection (3.7):

$$d = \frac{(E\{\lambda_T|\mathrm{H}_1\} - E\{\lambda_T|\mathrm{H}_0\})^2}{\sigma^2\{\lambda_T|\mathrm{H}_0\}} \tag{5.56}$$

The hypotheses are

$$\mathrm{H}_1 : \mathbf{x} = \sqrt{E_r}\,\mathbf{s} + \mathbf{g}$$
$$\mathrm{H}_0 : \mathbf{x} = \mathbf{g}$$

where E_r is the received energy and $\mathbf{g}$ contains contributions from flow noise, ambient noise, and reverberation noise. Expanding (5.56), we find that d is equal to

$$d = \frac{E_r \left(\sum_{n=1}^{N} |s(n)|^2 \right)^2}{\sum_{n=1}^{N} \sum_{m=1}^{N} s^*(n)\, s(m)\, R_g(n, m)} \tag{5.57}$$

where $R_g(n, m)$ denotes the (n, m)th element of $\mathbf{R}_g$. We note that the reverberation noise in general is nonstationary because the scatterers are assumed to be nonuniformly distributed in range. This is reflected in (5.57), where the noise correlation is a function of the particular time indices in question. In the case of a wide-sense stationary noise field, the correlation can be expressed in terms of the difference in indices.

We now choose to specify the noise correlation matrix as consisting of a white noise component plus a reverberation noise component. The (n, m)th element of $\mathbf{R}_g$ is then equal to

$$\mathbf{R}_g(n, m) = \sigma_g^2\, \delta(n - m) + \mathbf{R}_r(n, m) \tag{5.58}$$

where $\mathbf{R}_r$ denotes the correlation matrix of the reverberation noise. Substituting (5.58) into (5.57) yields

$$d = \frac{E_r \left(\sum_{n=1}^{N} |s(n)|^2 \right)^2}{\sigma_g^2 \sum_{n=1}^{N} |s(n)|^2 + \sum_{n=1}^{N} \sum_{m=1}^{N} s^*(n)\, s(m)\, R_r(n, m)} \tag{5.59}$$

We assume that the reverberation noise, $g_r(n)$, is contributed by a large number of scatterers. For a transmitted waveform equal to $s(n)$, the received reverberation noise is then equal to

$$g_r(n) = \sum_k Z_k \exp(-j2\pi\beta_k n/f_s)\, s(n - n_k) \tag{5.60}$$

where the summation is computed over all scatterers. For the kth scatterer, Z_k is the complex strength of the return, β_k is the doppler shift due to the radial velocity, and n_k is the time delay. This time delay is the sum of the propagation times from the source to the kth scatterer and from the kth scatterer to the receiver. Secondary scattering is ignored in (5.60).

Van Trees [11] developed an expression for the correlation matrix for $g_r(n)$ in (5.60). By using this expression for $\mathbf{R}_r$ and replacing the summations in (5.59) with integrals, we can obtain the expression for d given in [11, eq. (78)]:

$$d = \frac{2E_r/N_0}{1 + \dfrac{E\{|Z|^2\}\,E_t}{2N_0} \displaystyle\int\!\!\int V_r(\tau, \beta)\,|\Psi(\tau - \tau_D, \beta - \beta_D)|^2\, d\tau\, d\beta} \tag{5.61}$$

where

E_r	=	received energy,
E_t	=	transmitted energy,
$N_0/2$	=	power spectral density of the ambient noise,
$E\{\|Z\|^2\}$	=	variance of the strength of the scatterers,
$V_r(\tau, \beta)$	=	reverberation scattering function of the channel,
$\|\Psi(\tau, \beta)\|^2$	=	signal ambiguity function (normalized so that the peak equals unity),
(τ_D, β_D)	=	location of the contact in the (τ, β) plane.

The double integral in (5.61) is the two-dimensional convolution of the reverberation scattering function and the signal ambiguity function. The scattering function can be viewed as the product of $p_\tau(x)$ and $p(\beta|x)$ [11, eq. (75)], where $p_\tau(x)$ is the pdf governing the distribution of scatterers in range. The conditional pdf, $p(\beta|x)$, describes the probability that a scatterer has a velocity corresponding to a doppler shift in the frequency interval $[\beta, \beta + d\beta]$, given that a scatterer is present in the range interval $[x, x + dx]$.

We observe from (5.61) that the effect of reverberation noise only serves to reduce the output S/N because the contribution of the double integral is nonnegative. The numerator in (5.61) is just the output S/N for the background noise-limited case. We further observe that increasing the transmitted energy (E_t) to increase the received energy (E_r) does not improve the performance in the reverberation noise-limited case because the denominator of (5.61) also is increased. Again, the detector in (5.54) that is optimized for the background noise-limited case often is used in the reverberation noise-limited environment because of the difficulty in estimating $\mathbf{R}_r$.

Pulse Selection

The overlap between the signal ambiguity function and the scattering function is sketched in Figure 5.15, where the one-sigma extent in frequency of the scattering function is shown. As can be seen from Figure 5.15(a), the overlap is minimized by selecting a short CW pulse or a LFM pulse for a low-velocity contact. For a

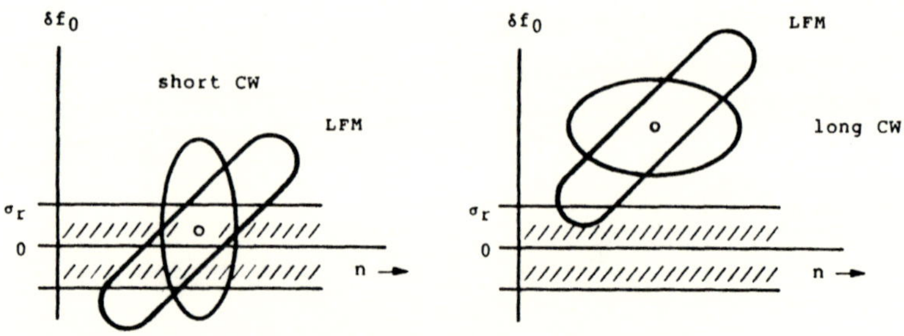

Figure 5.15 Sketch of the overlap of the scattering function and the signal ambiguity function in (5.61): (a) low-velocity contact; (b) high-velocity contact.
After: [6] T.H. Glisson, C.I. Black, and A.P. Sage, "On Sonar Signal Analysis," *IEEE Trans. on Aerospace and Electronic Systems*, Vol. AES-6, No. 1, January 1970, pp. 37–49.

contact with a large relative velocity (Figure 5.15(b)), the use of a long CW pulse minimizes the value of the double integral in (5.56). Typically a combination of such pulses is employed.

Other pulse types that can be used include the hyperbolic FM pulse and the *pseudorandom noise* (PRN) pulse. Harris and Kramer [12] discuss the doppler-invariant properties of the hyperbolic FM pulse and study its behavior in a constant-acceleration environment. The PRN pulse offers good range and doppler-shift resolution, at the cost of having a large number of range-doppler cells to search. Brief discussions of the PRN pulse can be found in [6, pp. 46–49; 13, pp. 355 and 362; and 14, pp. 293–294].

Different Transmitted and Replica Pulses

We can improve the performance of the correlation receiver by using a replica that differs from the transmitted pulse. Kincaid addressed this question for the background noise-limited case in [15]. Different waveforms for the transmitted pulse and the replica were chosen to maximize detection. These waveforms were selected to deal with the effect of random multipath on the transmitted pulse. In [16], Kincaid discussed the same problem for the reverberation noise-limited environment. Of course, to be successful, the approach of using different transmitted and replica waveforms requires a knowledge of the effects of the channel.

In place of the signal ambiguity function encountered in Section 5.2.1, we use a cross-ambiguity function. Letting the transmitted and replica sequences be denoted by $u(n)$ and $v(n)$, respectively, this cross-ambiguity function has the form

$$|\Psi_{uv}(\hat{n}, \delta)|^2 = \left| \sum_{n=n_1}^{n_2} u[n(1 - \delta) - n_0]\, v^*(n - \hat{n}) \right|^2$$

We conclude this section by observing that, for low S/N conditions, the sidelobes of the signal ambiguity function are important because spurious peaks may be generated in the range-doppler-shift domain that may exceed the true peak. We would use any *a priori* information about the contact to reject detections that occur in unreasonable range-doppler-shift cells.

5.3.3 FFT-Based Implementations

As done in Chapter 4 for the broadband, narrowband, and modulation processors, we establish a set of multiple alternative hypotheses, which, unlike those in Chapter 4, now involve two variables. These hypotheses are

$$\begin{aligned} \mathrm{H}_{ij}&: \quad x(n) = s(n - n_i, \beta_j) + g(n) \\ \mathrm{H}_0&: \quad x(n) = g(n) \end{aligned} \tag{5.62}$$

The composite receiver consists of a set of optimum detectors, adjusted to the discrete frequencies, β_j, $j = 1, \ldots, J$ and calculated at a set of discrete times, n_i, $i = 1, \ldots, I$. At each value of n_i, the maximum output of the J detectors is compared to a threshold. If this maximum output exceeds the threshold, the frequency value to which the detector is matched serves as the estimate of β, or doppler shift. The time at which the detection occurs is the estimate of range.

Processing for the CW Pulse

We consider first the CW pulse. As noted in the discussion of the narrowband approximation (Section 5.2.3), time compression and frequency translation are equivalent for this simple pulse. That is,

$$\begin{aligned} s[n(1 - \delta)] &= \exp[\mathrm{j}2\pi f_0 n(1 - \delta)/f_s] \\ &= \exp(\mathrm{j}2\pi f_0 n/f_s) \exp(-\mathrm{j}2\pi\delta f_0 n/f_s) \\ &= s(n) \exp(-\mathrm{j}2\pi\delta f_0 n/f_s) \end{aligned}$$

The optimum detector is the quadrature receiver in Figure 3.11. A bank of such detectors is implemented in parallel, where each detector is adjusted to a separate frequency, $\beta_j = \delta_j f_0$. The output of the detectors is calculated at time intervals n_1, $n_2, \ldots, n_I$. At the ith time, the output of the jth optimum detector is given by

$$q_j^2(n_i + N) = \left[\sum_{n=n_i}^{N-1+n_i} x(n) \sin(2\pi\beta_j n/f_s) \right]^2 + \left[\sum_{n=n_i}^{N-1+n_i} x(n) \cos(2\pi\beta_j n/f_s) \right]^2$$

The implementation of the multiple alternative hypotheses test (5.62) is shown in Figure 5.16. The global maximum over the indices i and j in (5.62) provides the estimates of range and doppler shift. A DFT, followed by an SLD for each frequency bin, is an appropriate realization of the quadrature receiver Figure 5.16, as it was for the narrowband processor in Section 4.4.1. The length of the DFT is matched to the duration of the pulse.

We consider a set of CW pulses with a center frequency of 3 kHz and lengths ranging from 5 ms to 1 s, as listed in column 1 of Table 5.1. Such a set of pulse lengths allows us to minimize the effects of reverberation noise for contacts with various values of doppler shift (Figure 5.15). Assuming that the maximum relative velocity is 30 knots in magnitude and that $c = 3000$ knots, the compression parameter, $\delta = 2v/c$, will fall in the range of $-0.02 \leq \delta \leq 0.02$. The magnitude of the maximum doppler shift is then equal to 60 Hz.

Table 5.1
Parameters for an FFT-Based Processor for CW Pulses

	Ambiguity Contours		*FFT Parameters*	
Pulsewidth (ms)	*Doppler (Hz)*	*Range (ms)*	*Length N*	*Resolution (Hz)*
5.0	176	2.93	No FFT processing	
10.0	88.0	5.86	No FFT processing	
62.5	14.1	36.6	16	16
125	7.04	73.3	32	8
250	3.52	147	64	4
500	1.76	293	128	2
1000	0.88	586	256	1

We wish to obtain an FFT-based implementation of Figure 5.16 for the 62.5 to 1000 ms pulses listed in Table 5.1. (The short pulses will be treated later.) An appropriate value for the sampling frequency at the input to the FFT is obtained by considering the effects of spectral width and doppler shift. The spectrum of the transmitted pulse is equal to

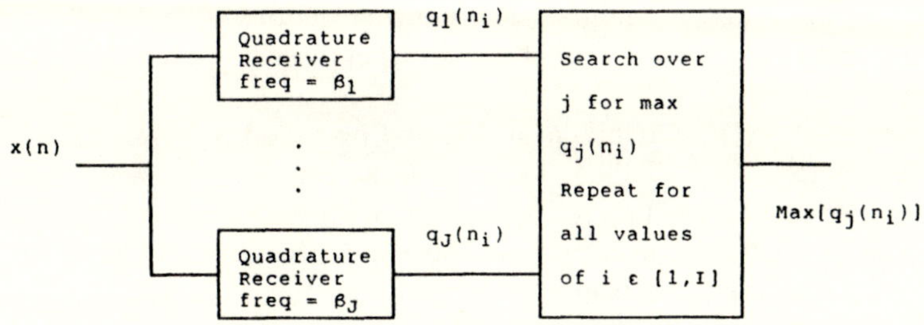

Figure 5.16 Implementation of the multiple alternative hypotheses test for range and doppler shift. After: [10] A.D. Whalen, *Detection of Signals in Noise*, Academic Press, New York, 1971.

$$\frac{\sin^2[\pi(f - f_0)T]}{[\pi(f - f_0)T]^2}$$

Therefore, the width of the spectrum between first zero crossings is $f = 2/T$, where T is the pulse duration in seconds. Because this frequency width varies inversely with T, it achieves a maximum value of 32 Hz for the 62.5 ms pulse (the shortest pulse processed by the FFT). As a result of doppler shift, the received center frequency occurs in the range of $f_0 - 60 \leq f \leq f_0 + 60$. After shifting the center of the spectrum to zero frequency, the highest frequency in this basebanded data is then $60 + 32/2 = 76$ Hz. Selecting a sampling frequency of 256 Hz at the input to the FFT results in a transition region for the low-pass filter of $f_a - f_p = (256 - 76) - 76 = 104$ Hz. We can show that such a transition region is provided by FIR filters of reasonable length.

To establish the sampling frequency for the A/D converter, we note that the highest frequency in the analog data is equal to the sum of the nominal center frequency, the maximum doppler shift, and half the maximum pulse spectral width. The 5 ms pulse has a spectral width of 400 Hz, centered on f_0. Therefore, the highest frequency in the analog signal is 3000 Hz + 60 Hz + 200 Hz = 3260 Hz. If we allow aliasing in the transition region of the analog filter, a sampling frequency of 8192 Hz results in a transition ratio of $f_p/(f_s - f_p) = 3260/4932 = 0.66$. This corresponds to a 6-pole–6-zero elliptic antialiasing filter (Figure 1.3). It is desirable, for the considerations given in Section 1.1.1, to use a shorter analog filter by oversampling the analog signal at 16384 Hz.

We would like to have the resolution of the FFT be at least as fine as the 3-dB doppler-shift intercepts of the ambiguity diagram in Figure 5.6. In addition, we wish to calculate the FFT at a rate consistent with the 3-dB range intercepts. The width of the doppler-shift intercepts is $2(0.44 f_s/N) = 0.88/T$ Hz, and the width of the range intercepts is $2(0.293\, N/f_s) = 0.586T$ s. These ambiguity contours are provided in columns 2 and 3 of Table 5.1 for doppler shift and range, respectively.

From these two columns, it again is apparent that, as the pulse length is decreased, the ability to accurately estimate doppler shift decreases whereas the quality of the range estimates increases.

The length of the FFT is given by Tf_s and provided in column 4 of Table 5.1. We thus have matched the FFT length to the pulse length. The last column in Table 5.1 indicates the resolution (f_s/N) associated with each pulse. It is apparent that this resolution is too coarse to accommodate the 3-dB doppler-shift ambiguity contours. For example, for the 1-s pulse, the FFT bins are centered at 0 Hz, 1 Hz, 2 Hz, and so forth. As discussed in [17], a second set of FFT bins can be generated for each pulse length by shifting the data at the input to the FFT by 0.5 Hz. The bins generated by performing an FFT on these shifted data are then centered at 0.5 Hz, 1.5 Hz, 2.5 Hz, *et cetera*. A second set of FFTs is generated in the same manner for the remaining pulse lengths in Table 5.1. Thus the doppler-shift domain has been sampled with sufficient resolution so that a processor bin is available that provides less than 3-dB mismatch loss for any allowable value of doppler shift.

The range domain is sampled at a rate consistent with the range contours in column 3 of Table 5.1 simply by appropriately selecting the FFT redundancy.

No information concerning doppler-shift can be obtained from the 5 ms or 10 ms pulses, because the 3-dB doppler-shift intercepts are equal to ±88 Hz and ±44 Hz, respectively. We therefore provide filters matched to the spectra of these pulses, instead of performing an FFT. The processing for the case of the CW pulse is shown in Figure 5.17.

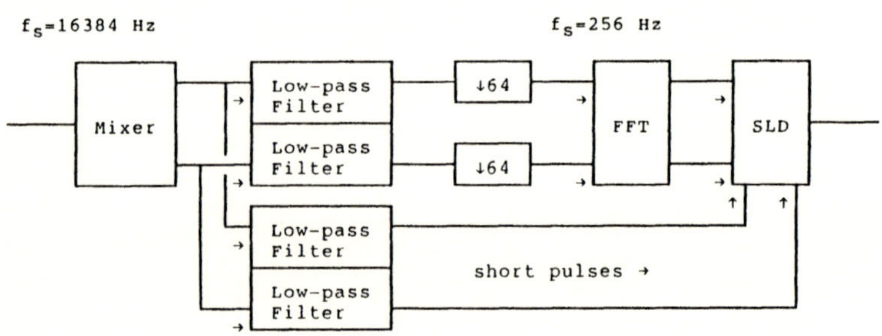

Figure 5.17 Signal processing for the CW pulse.

Processing for the LFM Pulse

We consider an LFM pulse with a duration of $T = 1$ s, a center frequency of $f_0 = 3000$ Hz, and a sweep extent of $W = 200$ Hz. As mentioned in Section 5.2.3, the velocity tolerance can be determined based on the assumption that the pulse is narrowband (so that correlation loss is due primarily to temporal mismatch) or wideband (where a mismatch in slope is the predominant cause of correlation loss). From (5.33), the wideband velocity tolerance is $v = 2614/(WT) = 13.1$ knots, and the tolerance from the narrowband approximation is $v = 435\ W/f_0 = 29$ knots (5.34). We therefore use the wideband assumption for our particular set of LFM pulse parameters.

Because the desired range of velocities is $-30 \leqslant v \leqslant 30$ knots, the use of a replica with the same value of slope as the transmitted pulse will result in a correlation loss that exceeds 3 dB for $|v| > 13$ knots. It therefore is necessary to provide additional replicas with different values of slope. For instance, we could specify three replicas with slopes corresponding to velocities of $v_1 = -26$ knots, $v_2 = 0$ knot, and $v_3 = 26$ knots. The correlation loss due to a mismatch in slope then would be less than 3 dB with one of these replicas for any velocity in the range of interest.

From (5.20), these replicas are

$$r_i(n) = \exp\{j2\pi[f_0 n(1 - \delta_i)/f_s + mn^2(1 - \delta_i)^2/(2f_s^2)]\} \qquad (5.63)$$

where $\delta_i = 2v_i/c$ so that $\delta_1 = 2(-26)/3000 = -0.0173$, $\delta_2 = 0$, and $\delta_3 = 2(26)/3000 = 0.0173$. The correlation loss due to slope mismatch will be less than 3-dB for velocities in the range $-39 \leqslant v \leqslant -13$ knots when replica 1 is used. The corresponding limits are $-13 \leqslant v \leqslant 13$ knots for $r_2(n)$ and $13 \leqslant v \leqslant 39$ knots using $r_3(n)$.

To obtain an FFT-based implementation, we first determine a value for the sampling frequency at the input to the FFT. The maximum doppler shift again is equal to ±60 Hz. From [6], the spectrum of the LFM pulse to a good approximation is contained in a frequency region of 200 Hz centered at f_0. The highest frequency in the basebanded data is then 100 Hz + 60 Hz = 160 Hz. We select a sampling frequency of 512 Hz at the input to the FFT, which results in low-pass filters of reasonable lengths. The replica correlation processing is shown in Figure 5.18, where the replicas are defined in (5.63).

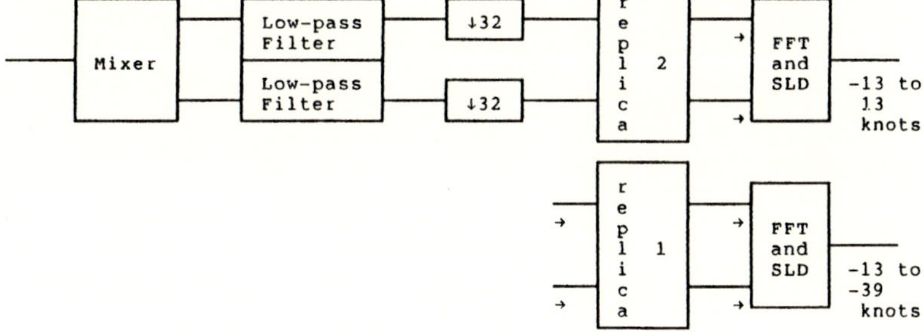

Figure 5.18 FFT-based correlation receiver for the LFM pulse.

As with the CW-pulse processor, the received signal is oversampled at 16384 Hz to reduce the length of the analog antialiasing filter. A simple analog filter provides a reasonably linear phase response in the passband of the filter. The composite phase response of the analog and the low-pass FIR filters therefore is expected to be quite good. (Recall that a symmetric FIR filter has a phase response that is exactly linear.) A nonlinear phase response results in frequency dispersion of the LFM pulse (that is, the delay through the system is frequency dependent). Such dispersion reduces the output of the replica correlator.

A search over the FFT bins of each of the three correlation receivers in Figure 5.18 is used to estimate the doppler shift. The length of each FFT is matched to the pulse length by selecting $N = Tf_s = 512$.

5.3.4 Block Diagram of an Active Sonar Processor

A simplified block diagram of an active sonar processor is provided in Figure 5.19, in which the same array of sensors is used for both the transmitting and the receiving functions. In addition to the usual AGC function used in passive sonar processors, we may want to adjust the gain of the system in response to large returns from surface and bottom scatterers that are received shortly after a transmission. Such a function is often called *time-varying gain* (TVG) [14, p. 314]. After A/D conversion, a receiver beamformer is used to generate a set of beams for processing.

A mixing operation is performed where the mixer frequency is set equal to the transmitted frequency. At this point, it may be desirable to attempt to remove the predominantly low-frequency reverberation noise through a high-pass filter. The lower

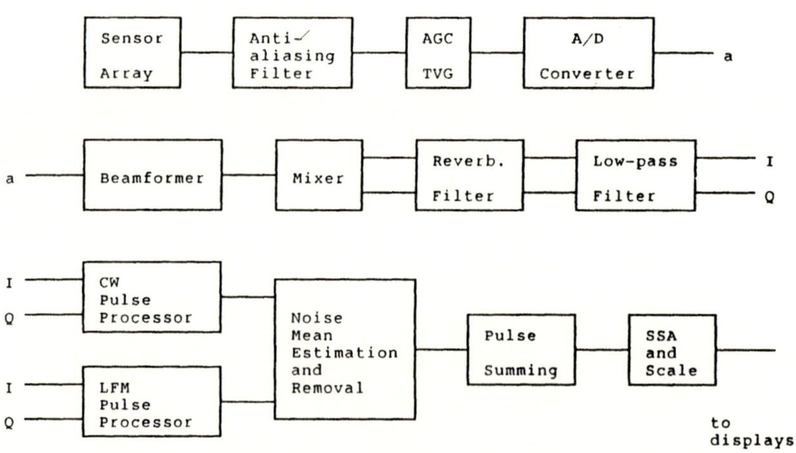

Figure 5.19 Simplified block diagram of an active sonar processor.

frequency of the passband of this filter may have a value of a few Hz. Next, low-pass filtering and decimation is provided to define the width of the frequency band to be processed. Depending on the type of pulse selected, either the CW-pulse or the LFM-pulse processor is implemented next. The FFT length is again a function of pulse length. After detection, a noise-mean-estimation-and-removal algorithm is implemented. This may be a two-dimensional algorithm where normalization is performed over time as well as frequency, as described in [18].

In the event that multiple pulses are transmitted, a summing function is used to sum over the corresponding return pulses. Next, spatial trends in background noise can be removed with the same type of sector-space-averaging algorithm that was discussed for the passive broadband processor in Section 4.3.2. The data then are scaled for display in a variety of possible formats, such as range-bearing display or frequency-bearing display.

5.3.5 APPENDIX

This appendix demonstrates that the time-bandwidth product for the Gaussian pulse is equal to the minimum possible value given in (5.53). This is done for a pulse with and without angle modulation.

No Angle Modulation

We select the following Gaussian pulse:

$$\tilde{s}(t) = \frac{1}{(\pi)^{1/4}\sqrt{\sigma}} \exp\left[-t^2/(2\sigma^2)\right] \tag{5.64}$$

Using [19, eq. (3.3213)], it is straightforward to show that

$$\int_{-\infty}^{\infty} |\tilde{s}(t)|^2 \, dt = 1$$

and, because (5.64) is an even function of t, that

$$\int_{-\infty}^{\infty} t|\tilde{s}(t)|^2 \, dt = 0$$

We calculate the Fourier transform of $\tilde{s}(t)$ with [19, eq. (3.8964)] and obtain

$$S(f) = \sqrt{2}(\pi)^{1/4}\sqrt{\sigma} \exp[-2(\pi f \sigma)^2] \tag{5.65}$$

We next evaluate the quantities, t_d^2 and f_d^2. From (5.49), we have that

$$t_d^2 = \int_{-\infty}^{\infty} t^2 |\tilde{s}(t)|^2 \, \mathrm{d}t$$

After substituting the definition for $\tilde{s}(t)$ (5.64), we obtain

$$\begin{aligned} t_d^2 &= \frac{1}{\sqrt{\pi}\sigma} \int_{-\infty}^{\infty} t^2 \exp(-t^2/\sigma^2) \, \mathrm{d}t \\ &= \sigma^2/2 \end{aligned} \tag{5.66}$$

where the integral in (5.66) can be evaluated using [19, eq. (3.4612)]. Also from (5.49) we have the following expression for f_d^2, the extent of $\tilde{s}(t)$ in the frequency domain:

$$f_d^2 = \int_{-\infty}^{\infty} (2\pi f)^2 |S(f)|^2 \, \mathrm{d}f$$

Substituting for $S(f)$ from (5.65) yields

$$\begin{aligned} f_d^2 &= 8(\pi)^{5/2}\sigma \int_{-\infty}^{\infty} f^2 \exp[-(2\pi f\sigma)^2] \, \mathrm{d}f \\ &= 1/(2\sigma^2) \end{aligned} \tag{5.67}$$

We now calculate $t_d^2 f_d^2$, using (5.66) and (5.67), and thus verify that the minimum value of $1/4$ is achieved for the Gaussian pulse with no angle modulation.

Angle Modulation

We include a phase factor in (5.64) and obtain the following as our pulse of interest:

$$\tilde{s}(t) = \frac{1}{(\pi)^{1/4}\sqrt{\sigma}} \exp[-t^2/(2\sigma^2)] \exp(-\mathrm{j}kt^2) \tag{5.68}$$

which is of the form $b(t)\,\mathrm{e}^{\mathrm{j}\theta(t)}$. We observe that f_d^2 can be expressed by using Parseval's identity as

$$
\begin{aligned}
f_d^2 &= \int_{-\infty}^{\infty} (2\pi f)^2 |S(f)|^2 \, \mathrm{d}f \\
&= \int_{-\infty}^{\infty} |\tilde{s}'(t)|^2 \, \mathrm{d}t
\end{aligned} \tag{5.69}
$$

where $\tilde{s}'(t)$ denotes the derivative of $\tilde{s}(t)$ with respect to time. If $s_1(t)$ and $S_1(f)$, as well as $s_2(t)$ and $S_2(f)$, are Fourier transform pairs, a general form of Parseval's identity is

$$
\int_{-\infty}^{\infty} s_1(t) s_2^*(t) \, \mathrm{d}t = \int_{-\infty}^{\infty} S_1(f) S_2^*(f) \, \mathrm{d}f
$$

Taking the squared magnitude of the time derivative of (5.68) yields

$$
|\tilde{s}'(t)|^2 = \frac{t^2/\sigma^4 + 4k^2 t^2}{\sqrt{\pi}\sigma} \exp(-t^2/\sigma^2) \tag{5.70}
$$

By substituting (5.70) into (5.69) and evaluating the integral, we obtain

$$
f_d^2 = 1/(2\sigma^2) + 2k^2\sigma^2 \tag{5.71}
$$

The expression for t_d^2 for the pulse in (5.68) is the same as for the pulse with no angle modulation, so t_d^2 is given by (5.66). We next evaluate (ωt), which from (5.50) is equal to

$$
(\omega t) = \int_{-\infty}^{\infty} t \, \mathrm{d}\theta(t)/\mathrm{d}t \, b^2(t) \, \mathrm{d}t \tag{5.72}
$$

where from (5.68)

$$
\begin{aligned}
b(t) &= \frac{1}{(\pi)^{1/4}\sqrt{\sigma}} \exp[-t^2/(2\sigma^2)] \\
\theta(t) &= -kt^2
\end{aligned}
$$

We thus can rewrite (5.72) and obtain

$$
\begin{aligned}
(\omega t) &= -\frac{4k}{\sqrt{\pi}\sigma} \int_0^{\infty} t^2 \exp(-t^2/\sigma^2) \, \mathrm{d}t \\
&= -k\sigma^2
\end{aligned} \tag{5.73}
$$

We are now in a position to verify that equality holds in (5.53) for the pulse defined in (5.68). From (5.66) and (5.71), $t_d^2 f_d^2 = (\sigma^2/2)[1/(2\sigma^2) + 2k^2\sigma^2] = (k\sigma^2)^2 + 1/4$. But from (5.73), this is equal to $(\omega t)^2 + 1/4$. Therefore, we have established that the Gaussian pulse with angle modulation has the smallest possible time-bandwidth product.

REFERENCES

1. Urick, R.J., *Principles of Underwater Sound,* McGraw-Hill, New York, 1975.
2. Burdic, W.S., *Underwater Acoustic System Analysis,* Prentice-Hall, Englewood Cliffs, NJ, 1984.
3. Van Trees, H.L., *Detection, Estimation, and Modulation Theory, Part III, Radar/Sonar Signal Processing and Gaussian Signals in Noise,* John Wiley & Sons, New York, 1971.
4. DeFerrari, H.A., and L. Nghiem-Phu, "Scattering Function Measurements for a 7-NM Propagation Range in the Florida Straits," *J. Acoustical Society of America,* Vol. 56, No. 1, July 1974, pp. 47–52.
5. Veenkant, R.L., "Investigation of the Propagation Stability of a Doubly Spread Underwater Acoustic Channel," *IEEE Trans. on Acoustics, Speech, Signal Processing,* Vol. ASSP-25, No. 2, April 1977, pp. 109–116.
6. Glisson, T.H., C.I. Black, and A.P. Sage, "On Sonar Signal Analysis," *IEEE Trans. on Aerospace and Electronic Systems,* Vol. AES-6, No. 1, January 1970, pp. 37–49.
7. Abramowitz, M., and I.A. Stegun, eds., *Handbook of Mathematical Functions with Formulas, Graphs, and Mathematical Tables,* National Bureau of Standards, Applied Mathematics Series 55, U.S. Government Printing Office, Washington, DC, 1972.
8. Kramer, S.A. "Doppler and Acceleration Tolerances of High-Gain, Wideband Linear FM Correlation Sonars," *Proc. IEEE,* Vol. 55, No. 5, May 1967, pp. 627–636.
9. Stewart, J.L., E.C. Westerfield, and M.K. Brandon, "Optimum Frequencies for Active Sonar Detection," *J. Acoustical Society of America,* Vol. 33, No. 9, September 1961, pp. 1216–1222.
10. Whalen, A.D., *Detection of Signals in Noise,* Academic Press, New York, 1971.
11. Van Trees, H.L. "Optimum Signal Design and Processing for Reverberation-Limited Environments," *IEEE Trans. on Military Electronics,* Vol. MIL-19, No. 3-4, July–October 1965, pp. 212–229.
12. Harris B., and S.A. Kramer, "Asymptotic Evaluation of the Ambiguity Functions of High-Gain FM Matched Filter Sonar Systems," *Proc. IEEE,* Vol. 56, No. 12, December 1968, pp. 2149–2157.
13. Baggeroer, A.B., "Sonar Signal Processing," Chapter 6 of *Applications of Digital Signal Processing,* ed. A.V. Oppenheim, Prentice-Hall, Englewood Cliffs, NJ, 1978.
14. Winder, A.A., "Sonar System Technology," *IEEE Trans. on Sonics and Ultrasonics,* Vol. SU-22, No. 5, September 1975, pp. 291–332.
15. Kincaid, T.G., "Optimum Waveforms for Correlation Detection in the Sonar Environment: Noise-Limited Conditions," *J. Acoustical Society of America,* Vol. 43, No. 2, February 1968, pp. 258–268.
16. Kincaid, T.G., "On Optimum Waveforms for Correlation Detection in the Sonar Environment: Reverberation-Limited Conditions," *J. Acoustical Society of America,* Vol. 44, No. 3, September 1968, pp. 787–796.
17. Glisson, T.H., C.I. Black, and A.P. Sage, "On Digital Replica Correlation Algorithms with

Applications to Active Sonar," *IEEE Trans. on Audio Electroacoustics,* Vol. AU-17, No. 3, September 1969, pp. 190–197.

18. Morgan, D.R., "Two-Dimensional Normalization Techniques," *IEEE J. Oceanic Engineering,* Vol. OE-12, No. 1, January 1987, pp. 130–142.

19. Gradshteyn, I.S., and I.M. Ryzhik, *Table of Integrals, Series, and Products,* Academic Press, New York, 1980.

Chapter 6
Conventional Passive Sonar System

A set of typical signal processing requirements is advanced. Using the hardware implementations developed in Chapter 4, we calculate the computational throughput, input-output data bandwidth, and memory necessary to satisfy these requirements. A generic signal processing unit is proposed that has some desirable attributes for sonar signal processing.

6.1 SONAR SYSTEM SPECIFICATION

We consider an example in which the input to the sonar system consists of sets of beams from two arrays: 50 beams from a *low-frequency array* (LFA) sampled at 3000 Hz, and 250 beams from a *medium-frequency array* (MFA) sampled at 12,000 Hz. It is necessary to simultaneously provide four broadband processing bands, six narrowband processing bands, and one modulation processing band for each beam from the LFA. That is, we wish to examine four distinct spectral regions for broadband energy and six distinct spectral regions for narrowband components. Because of the higher sampling frequency and larger number of beams, only one narrowband processing band is required for the MFA. These requirements are summarized in Figure 6.1. The *frequency-azimuth* (FRAZ) display is defined in Section 6.2.2.

It frequently is desirable, in a large sonar processing system such as the one implied by the requirements in Figure 6.1, to arrange for all displays (broadband, narrowband, and modulation) to be updated at the same basic rate. An averaging time of about 10 seconds is assumed to be consistent with the anticipated short-term stability of signals of interest. As discussed in Section 6.1.2, a reasonable choice for such a standard update rate is one that can be expressed as a ratio of a power of 2 and the input sampling frequency. The update rate is chosen to be $2^{15}/3000 \simeq 10.923$ s, for short-time integration, and ten times this value for long-time integration.

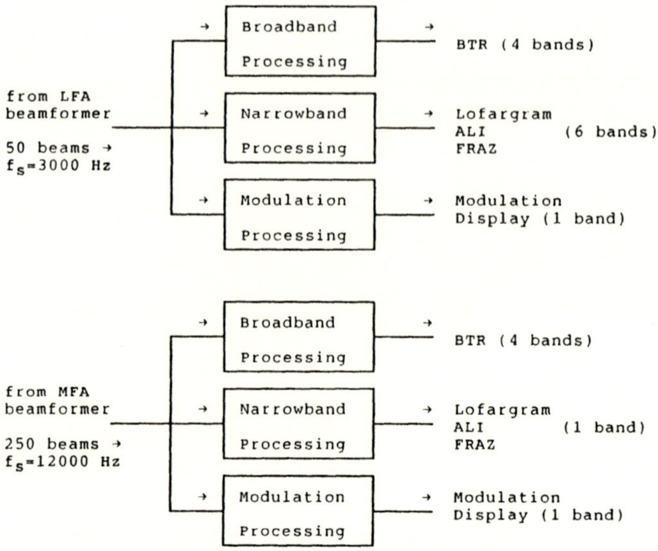

Figure 6.1 Sonar system requirements.

FIR filters will be employed extensively in the sonar system. The lengths of these filters are estimated using (1.5), which is repeated here

$$N = \frac{-10\log_{10}(\delta_1\delta_2) - 13}{14.6\,(f_a - f_p)/f_s} + 1 \tag{6.1}$$

As discussed in Section 1.2.1, $f_a - f_p$ is the transition region and f_s is the input sampling frequency. A plot of filter response is given in Figure 6.2, where f_{so} is the

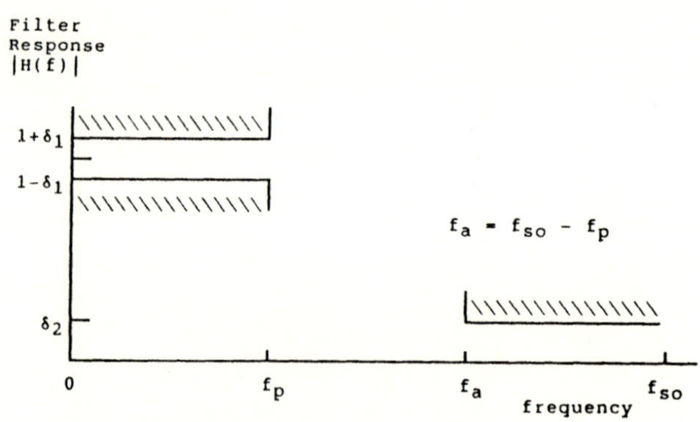

Figure 6.2 Magnitude frequency response for the low-pass filter.

output sampling frequency of the filter. Because aliasing is permitted in the transition region, the stopband edge, f_a, is equal to $f_{so} - f_p$. A passband ripple of 0.5 dB ($\delta_1 = 2.88 \cdot 10^{-2}$) and a stopband attenuation of 70 dB ($\delta_2 = 3.16 \cdot 10^{-4}$) are used for all FIR filters.

6.1.1 Broadband Processing

Consistent with the common display update rate, it is desired that ITA outputs be generated every 10.92 s and that the LTA update rate be 109.2 s. A spectral region with a width of 500 Hz is to be analyzed for each of the four broadband processing bands. The broadband processor in Figure 4.8 is implemented four times in parallel for both the LFA and the MFA. The bandpass filter approximation for $f(n)$, the filter that follows the beamformer, is used here. Thus each implementation of Figure 4.8 examines a different 500 Hz region of the received spectrum, by specifying a separate center frequency in Figure 4.7.

The broadband processing parameters are given in Table 6.1. The filter lengths are determined from (6.1) by using the parameters listed in this table. After the lowpass filter, we can reduce the sampling frequency to f_{so}. The computational rate for

Table 6.1
Broadband Processing Parameters

Array	f_s (Hz)	f_p (Hz)	f_a (Hz)	N	D	f_{so} (Hz)	ITA	LTA
LFA	3000	250	500	32	4	750	8192	81920
MFA	12000	250	500	124	16	750	8192	81920

one beam and one band is provided in Table 6.2 for the LFA, in real operations per s. (An operation is either a multiplication or an addition.)

The computations associated with the MFA filter can be reduced by decomposing the decimation factor of 16 into factors of 8 and 2. In this manner, the 124-stage filter in Table 6.1 can be replaced by a 35-stage filter and decimation by 8, followed by an 18-stage filter and decimation by 2. This requires 71 percent of the computations of the single-filter implementation. Note that the ripple is set equal to 0.25 dB ($\delta_1 = 1.44 \cdot 10^{-2}$), rather than 0.5 dB, in the estimation of the lengths of the two filters, to ensure that the composite filter response satisfies the 0.5 dB specification. The computational rate for one beam and one band of the MFA is given in Table 6.3.

Table 6.2
Computational Rate for One LFA Broadband Processing Band

Algorithm		*Rate* (10^3 op/s)
Mixer	2 mult/sample × 3000 Hz	6.0
Filter	2 × 32 mult/sample × 750 Hz	48.0
	2 × 32 add/sample × 750 Hz	48.0
SLD	2 mult/sample × 750 Hz	1.5
	1 add/sample × 750 Hz	0.75
STA	1 add/sample × 750 Hz	0.75
TOTAL		$1.05 \cdot 10^5$ op/s

Table 6.3
Computational Rate for One MFA Broadband Processing Band

Algorithm		*Rate* (10^3 op/s)
Mixer	2 mult/sample × 12000 Hz	24.0
Filter 1	2 × 35 mult/sample × 1500 Hz	105.0
	2 × 35 add/sample × 1500 Hz	105.0
Filter 2	2 × 18 mult/sample × 750 Hz	27.0
	2 × 18 add/sample × 750 Hz	27.0
SLD	2 mult/sample × 750 Hz	1.5
	1 add/sample × 750 Hz	0.75
STA	1 add/sample × 750 Hz	0.75
TOTAL		$2.91 \cdot 10^5$ op/s

An optimization procedure for decomposing large decimation factors is given in Crochiere and Rabiner [1, Section 5.3]. This decomposition can be optimized in terms of either multiplication rate or coefficient storage requirements. For the 124-stage filter in Table 6.1, the implementation obtained by decomposing $D = 16$ into 8 and 2 requires the lowest computational rate of any two-filter structure.

In both Tables 6.2 and 6.3, the computations required for the SSA, prescale, ITA, and LTA functions in Figure 4.8 are not included because of their comparatively low computational rates.

6.1.2 Narrowband Processing

The STI output rate is 10.92 s and that for the LTI and ALI displays is 109.2 s. It is desired to display 800 frequency bins. Each narrowband processing, or vernier, band is defined by selecting one of six resolutions (ranging in width from 1.46 Hz to 0.046 Hz) and a center frequency about which these bins are located. Time-domain windowing, an FFT length of 2048, and a redundancy of 2 are desired.

Low-Frequency Array

For the LFA, six vernier bands are provided by implementing Figure 4.11 six times. The output of all beams is processed for each vernier band. Table 6.4 provides a list of narrowband processing parameters.

Table 6.4
Narrowband Processing Parameters for the LFA

Resolution (Hz)	f_p (Hz)	f_a (Hz)	N	D	f_{so} (Hz)	FFT rate (FFT/s)	STI
1.46	—	—	—	1	3000	2.93	32
0.732	293	1207	10	2	1500	1.46	16
0.336	146	604	18	4	750	0.732	8
0.183	73.2	302	35	8	375	0.366	4
0.0916	36.6	151	69	16	188	0.183	2
0.0458	18.3	75.4	136	32	93.8	0.0916	1

The utility of the $2^{15}/3000$ display update rate can be deduced from Table 6.4. The update rate is equal to the ratio of the STI value and the FFT rate. FFTs are calculated at a rate equal to $(f_s R)/(ND)$, where R is the redundancy and N is the length of the FFT. The display update rate is thus equal to

$$\text{Update rate} = \frac{ND\text{STI}}{f_s R}$$

where $R = 2$. By setting the update rate equal to $2^k/f_s$, where k is an integer, we can generate output for display at the same rate for all values of resolution in Table 6.4 and still satisfy the requirement that the STI values be integers.

From Table 6.4, the highest FFT rate corresponds to the 1.46-Hz resolution whereas the 0.732-Hz resolution requires the largest number of filter computations. We assume that it is not desirable to have to redistribute algorithms in the signal processor after redefining vernier-band parameters. Therefore, sufficient processing resources should be dedicated to accommodate these extremes in FFT and filter computational requirements.

We again note that the computational rate for the filtering function can be reduced by decomposing large decimation factors. For example, the optimum two-factor decomposition (in terms of both computational rate and coefficient storage) of $D = 32$ in Table 6.4 is into factors of 8 and 4. The associated filters of 26 and 20 stages, respectively, require 91 percent of the computations and 34 percent of the coefficient storage of the single-filter implementation.

The number of real operations per second required to implement one vernier band for one LFA beam is given in Table 6.5. We assume that the two-pass split-window algorithm is used for noise mean estimation and removal. The window length is chosen to be 32 in this algorithm.

Table 6.5
Computational Rate for One LFA Narrowband Processing Band

Algorithm		*Rate* (10^3 op/s)
Mixer	2 mult/sample × 3000 Hz	6.0
Filter	2 × 10 mult/sample × 1500 Hz	30.0
	2 × 10 add/sample × 1500 Hz	30.0
Window	2 × 2048 mult/FFT × 2.93 Hz	12.0
FFT	5 × 2048 $\log_2$ 2048 × 2.93 Hz	330.0
SLD	2 mult × 800 bins × 2.93 Hz	4.7
	1 add × 800 bins × 2.93 Hz	2.4
STI	1 add × 800 bins × 2.93 Hz	2.4
Noise Mean Estimation and Removal	100 ops × 800 bins × 1/10.92 Hz	7.4
TOTAL		$4.25 \cdot 10^5$ op/s

Medium-Frequency Array

The MFA processing parameters are listed in Table 6.6 and the computational rate for one beam is provided in Table 6.7. By choosing a length-2048 FFT and a set of 800 bins for display, the sampling frequency at the input to the FFT is $f_{so} = 5.12\, f_p$. From Example 4.4.1, this results in a higher computational rate for

Table 6.6
Narrowband Processing Parameters for the MFA Using a Length-2048 FFT

Resolution (Hz)	f_p (Hz)	f_a (Hz)	N	D	f_{so} (Hz)	FFT rate (FFT/s)	STI
1.46	586	2414	18	4	3000	2.93	32
0.732	293	1207	35	8	1500	1.46	16
0.336	146	604	69	16	750	0.732	8
0.183	73.2	302	136	32	375	0.366	4
0.0916	36.6	151	270	64	188	0.183	2
0.0458	18.3	75.4	540	128	93.8	0.0916	1

Table 6.7
Computational Rate for One MFA Narrowband Processing Band Using a Length-2048 FFT

Algorithm		*Rate* (10^3 op/s)
Mixer	2 mult/sample × 12000 Hz	24.0
Filter	2 × 18 mult/sample × 3000 Hz	108.0
	2 × 18 add/sample × 3000 Hz	108.0
Window	2 × 2048 mult/FFT × 2.93 Hz	12.0
FFT	5 × 2048 $\log_2$ 2048 × 2.93 Hz	330.0
SLD	2 mult × 800 bins × 2.93 Hz	4.7
	1 add × 800 bins × 2.93 Hz	2.4
STI	1 add × 800 bins × 2.93 Hz	2.4
Noise Mean Estimation and Removal	100 ops × 800 bins × 1/10.92 Hz	7.4
TOTAL		$5.99 \cdot 10^5$ op/s

the FFT function than for the low-pass filter function. This is supported by Table 6.7, where the filter function requires 40 percent of the total computations for filtering and the FFT. The alternative is to use a length-1024 FFT so that $f_{so} = 2.56\ f_p$. The same number of output bins is provided but the filter requirements are more severe because of the shorter-length FFT. This is indicated in Table 6.8.

By replacing the 95-stage filter in Table 6.8 (1.46 Hz resolution) with a 20-stage filter and decimation by 4, followed by a 27-stage filter and decimation by 2, the computational rate is reduced by about 30 percent. We can show that two-filter

Table 6.8
Narrowband Processing Parameters for the MFA Using a Length-1024 FFT

Resolution (Hz)	f_p (Hz)	f_a (Hz)	N	D	f_{so} (Hz)	FFT rate (FFT/s)	STI
1.46	586	914	95	8	1500	2.93	32
0.732	293	457	189	16	750	1.46	16
0.336	146	229	372	32	375	0.732	8
0.183	73.2	115	737	64	188	0.366	4
0.0916	36.6	57.2	1494	128	93.8	0.183	2
0.0458	18.3	28.6	2987	256	46.9	0.0916	1

Table 6.9
Computational Rate for One MFA Narrowband Processing Band Using a Length-1024 FFT

Algorithm		*Rate* (10^3 op/s)
Mixer	2 mult/sample × 12000 Hz	24.0
Filter 1	2 × 20 mult/sample × 3000 Hz	120.0
	2 × 20 add/sample × 3000 Hz	120.0
Filter 2	2 × 27 mult/sample × 1500 Hz	81.0
	2 × 27 add/sample × 1500 Hz	81.0
Window	2 × 1024 mult/FFT × 2.93 Hz	6.0
FFT	5 × 1024 $\log_2$ 1024 × 2.93 Hz	150.0
SLD	2 mult × 800 bins × 2.93 Hz	4.7
	1 add × 800 bins × 2.93 Hz	2.4
STI	1 add × 800 bins × 2.93 Hz	2.4
Noise Mean Estimation and Removal	100 ops × 800 bins × 1/10.92 Hz	7.4
TOTAL		$5.99 \cdot 10^5$ op/s

realizations for the remaining resolutions also require fewer computations than single-filter implementations. Table 6.9 presents the computational rate for implementing a vernier band for one beam using a length-1024 FFT.

As indicated in Table 6.9, the filtering function represents about 73 percent of the total filtering and FFT computations (as opposed to 40 percent with the length-2048 FFT). The overall computation rates are equal for $f_{so} = 5.12 f_p$ (Table 6.7) and $f_{so} = 2.56 f_p$ (Table 6.9). Given specific signal processing hardware, we can select the technique that is implemented more efficiently. No choice is to be made for the LFA because the coarsest resolution (1.46 Hz) requires no filtering.

Before discussing the modulation processing, note that the 0.046-Hz resolution in Table 6.8 requires a filter consisting of 2987 stages. A dramatic reduction in both computation rate and coefficient storage can be realized by decomposing the decimation factor of 256 into 32 and 8, as shown in Figure 6.3. This decomposition requires the fewest computations and fewest number of filter coefficients of any two-filter implementation. The computation rate for the structure in Figure 6.3 is $1.69 \cdot 10^5$ op/s, which is 30 percent of that for the single filter. Only 40 coefficients are required instead of 2987.

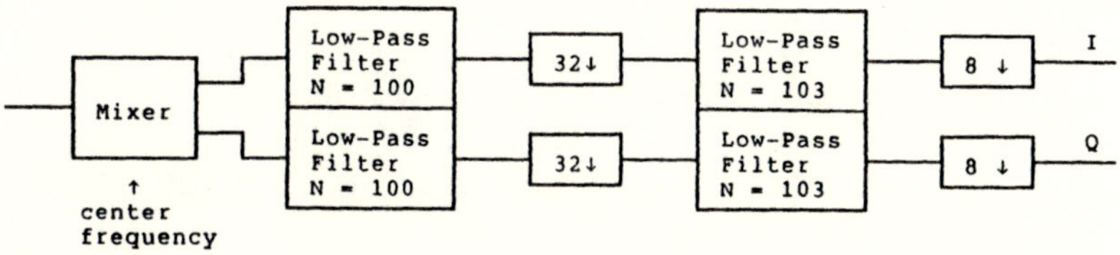

Figure 6.3 Efficient realization of decimation by 256.

6.1.3 Modulation Processing

The display output for modulation processing is updated at the standard interval of 10.92 s. As with narrowband processing, each output consists of 800 bins. A single resolution of 0.183 Hz is provided. Spectral regions with a width of 500 Hz for the LFA and 1000 Hz for the MFA are selected for analysis to determine if modulation is present.

The processing shown in Figure 4.18 is implemented for each beam of each array. As was done for broadband processing (Section 6.1.1), the filter $f(n)$ in Figure 4.18 is approximated with a 500 Hz bandpass filter for the LFA. Because of the broader range of operating frequencies, a bandwidth of 1000 Hz is specified for the MFA. The center frequency is selectable for each array. Table 6.10 provides a list

Table 6.10
Modulation Processing Parameters

	Filter 1					*Filter 2*					
Array	f_p (Hz)	f_a (Hz)	*N*	*D*	f_{so} (Hz)	f_p (Hz)	f_a (Hz)	*N*	*D*	f_{so} (Hz)	STI
LFA	250	500	32	4	750	146.4	228.6	25	2	375	16
MFA	500	1000	63	8	1500	146.4	228.6	48	4	375	16

of processing parameters. The first 800 bins of the FFT correspond to the passband (0 to 146.4 Hz) of the second low-pass filter.

The computation rates for one LFA beam and one MFA beam are given in Tables 6.11 and 6.12, respectively. To obtain the required resolution of 0.183 Hz, an FFT with a length of $N = f_{so}/\text{resolution} = 375/0.183 = 2048$ is used. A redundancy of 8 corresponds to an FFT rate of 375 Hz/(256 samples/FFT) = 1.46 FFT/s. Because the input data to the FFT are real, the real-data algorithm in Figure 1.33 is used.

Table 6.11
Computational Rate for LFA Modulation Processing

Algorithm		*Rate* (10^3 op/s)
Mixer	2 mult/sample × 3000 Hz	6.0
Filter 1	2 × 32 mult/sample × 750 Hz	48.0
	2 × 32 add/sample × 750 Hz	48.0
SLD	2 mult/sample × 750 Hz	1.5
	1 add/sample × 750 Hz	0.75
1P1Z Filter[1]	1 mult/sample × 1500 Hz	0.75
	2 add/sample × 1500 Hz	1.5
Filter 2	25 mult/sample × 375 Hz	9.4
	25 add/sample × 375 Hz	9.4
Window	2048 mult/FFT × 1.46 Hz	3.0
FFT	$(5 \times 1024 \log_2 1024 + 14 \times 1024)$ × 1.46 Hz	95.7
SLD	2 mult × 800 bins × 1.46 Hz	2.4
	1 add × 800 bins × 1.46 Hz	1.2
STI	1 add × 800 bins × 1.46 Hz	1.2
Noise Mean Estimation and Removal	100 ops × 800 bins × 1/10.92 Hz	7.4
TOTAL		$2.36 \cdot 10^5$ op/s

Note: 1P1Z refers to the one-pole–one-zero high-pass filter in Figure 4.18.

6.2 TOTAL COMPUTATIONAL RATE AND INPUT-OUTPUT RATE

Based on the results of Section 6.1, we can calculate the total computational rate and input-output rate for the sonar processing system.

Table 6.12
Computational Rate for MFA Modulation Processing

Algorithm		*Rate* (10^3 op/s)
Mixer	2 mult/sample × 12000 Hz	24.0
Filter 1	2 × 63 mult/sample × 1500 Hz	189.0
	2 × 63 add/sample × 1500 Hz	189.0
SLD	2 mult/sample × 1500 Hz	3.0
	1 add/sample × 1500 Hz	1.5
1P1Z Filter[1]	1 mult/sample × 1500 Hz	1.5
	2 add/sample × 1500 Hz	3.0
Filter 2	48 mult/sample × 375 Hz	18.0
	48 add/sample × 375 Hz	18.0
Window	2048 mult/FFT × 1.46 Hz	3.0
FFT	(5 × 1024 $\log_2$ 1024 + 14 × 1024)	
	× 1.46 Hz	95.7
SLD	2 mult × 800 bins × 1.46 Hz	2.4
	1 add × 800 bins × 1.46 Hz	1.2
STI	1 add × 800 bins × 1.46 Hz	1.2
Noise Mean Estimation and Removal	100 ops × 800 bins × 1/10.92 Hz	7.4
TOTAL		$5.58 \cdot 10^5$ op/s

Note: 1P1Z refers to the one-pole–one-zero high-pass filter in Figure 4.18.

For the LFA, the total rate is

Computational Rate

$$\begin{aligned} &= 50 \text{ beams} \times 4 \text{ bands} \times 0.105 \text{ Mflops} && \text{(Table 6.2)} \\ &\quad + 50 \text{ beams} \times 6 \text{ bands} \times 0.425 \text{ Mflops} && \text{(Table 6.5)} \\ &\quad + 50 \text{ beams} \times 1 \text{ band} \times 0.236 \text{ Mflops} && \text{(Table 6.11)} \\ &= 160 \text{ Mflops} \end{aligned}$$

Similarly, for the MFA,

Computational Rate

$$\begin{aligned} &= 250 \text{ beams} \times 4 \text{ bands} \times 0.291 \text{ Mflops} && \text{(Table 6.3)} \\ &\quad + 250 \text{ beams} \times 1 \text{ band} \times 0.599 \text{ Mflops} && \text{(Table 6.9)} \\ &\quad + 250 \text{ beams} \times 1 \text{ band} \times 0.558 \text{ Mflops} && \text{(Table 6.12)} \\ &= 580 \text{ Mflops} \end{aligned}$$

where *Mflops* denotes 1 million floating point operations per second. The total requirement for both arrays is 740 Mflops. This could increase if the sonar system were used for postmission analysis, where recorded data could be processed at a rate greater than real time.

The value of 740 Mflops represents an estimate of the actual number of real operations. In terms of a realizable system, this number would have to be adjusted by two factors. First, an algorithm in general would not be able to utilize all available operations. We consider the use of a DSP device that performs a multiplication and an addition each instruction cycle. As an example, the mixing function at most is 50 percent efficient because the available additions are not used. This is discussed further in Section 6.3.3. Second, in general it is desirable to provide a "safety margin" in terms of using the capabilities of a processor. That is, if a maximum of $1 \cdot 10^6$ mixing operations can be performed per second, perhaps only 75 percent of these in fact would be used. We assume an average value of 75 percent for algorithm efficiency and an average value 75 percent for loading level. A theoretical capacity of 1320 Mflops therefore is required to provide the $740 \cdot 10^6$ multiplications and additions per second needed to implement the sonar signal processing specified in Section 6.1.

6.2.2 Input-Output Rate

Also of interest is the *input-output* (I/O) rate required for this system. For the purpose of these estimates, we assume that the LFA and MFA beamformers provide 32-bit, or 4-byte floating-point data. Therefore, the input rate to the system is

$$\begin{aligned} \text{I/O rate} &= 50 \text{ beams} \times 3000 \text{ samples/s} \times 4 \text{ bytes/sample} \\ &\quad + 250 \text{ beams} \times 12000 \text{ samples/s} \times 4 \text{ bytes/sample} \\ &= 12.6 \text{ Mbytes/s} \end{aligned}$$

where *Mbyte* denotes megabyte. It is typical of sonar signal processing that the output data rate is considerably lower than the input rate. This is apparent from the output rates provided by Tables 6.13 and 6.14 for the LFA and MFA, respectively. Display data have a word length of 4 bits with the exception of the ALI display. Because this particular display is formed by integrating a number of STI outputs, additional levels of quantization are required. Therefore, 16 bits are used. The total output rate for the sonar processor for both the LFA and the MFA is only 0.044 Mbyte/s.

The FRAZ display is a static display (that is, not presented in a waterfall format), consisting of the output of the 800 processed frequency bins from all beams. It thus offers a presentation of narrowband detections for all values of bearing.

Note that, for simplicity, the LTA, LTI, ALI, and FRAZ displays in Table 6.13 and Table 6.14 are generated each 109.2 s by summing 10 outputs of either

Table 6.13
Display Data Output Rates for the LFA

Function	*Bands*	*Beams*	*Bins*	*Update Rate*	*Bytes*	*Rate* (kbyte/s)
Broadband Processing						
ITA	4 ×	50 ×		1/ 10.92 ×	1/2	0.01
LTA	4 ×	50 ×		1/109.2 ×	1/2	—
Narrowband Processing						
Lofargram						
STI	6 ×	50 ×	800 ×	1/ 10.92 ×	1/2	11.0
LTI	6 ×	50 ×	800 ×	1/109.2 ×	1/2	1.1
ALI	6 ×	50 ×	800 ×	1/109.2 ×	2	4.4
FRAZ	6 ×	50 ×	800 ×	1/109.2 ×	1/2	1.1
Modulation Processing	1 ×	50 ×	800 ×	1/ 10.92 ×	1/2	1.8
TOTAL						0.020 Mbyte/s

Table 6.14
Display Data Output Rates for the MFA

Function	*Bands*	*Beams*	*Bins*	*Update Rate*	*Bytes*	*Rate* (kbyte/s)
Broadband Processing						
ITA	4 ×	250		× 1/ 10.92 ×	1/2	0.05
LTA	4 ×	250		× 1/109.2 ×	1/2	—
Narrowband Processing						
Lofargram						
STI	1 ×	250 ×	800 ×	1/ 10.92 ×	1/2	9.2
LTI	1 ×	250 ×	800 ×	1/109.2 ×	1/2	0.92
ALI	1 ×	250 ×	800 ×	1/109.2 ×	2	3.7
FRAZ	1 ×	250 ×	800 ×	1/109.2 ×	1/2	0.92
Modulation Processing	1 ×	250 ×	800 ×	1/ 10.92 ×	1/2	9.2
TOTAL						0.024 Mbyte/s

the ITA or the STI functions. In practice, exponential integrators often are used to implement these long-time integration functions. The time constant can be set equal to 109.2 s but the output of these exponential integrators would be sampled at a higher rate.

Although the input rate (12.6 Mbytes/s) and the output rate (0.044 Mbyte/s) for the sonar processor are relatively modest, intermediate data rates can be much higher. As an example, we consider the total input rate to the FFTs in the narrowband processing function. From Table 6.5, 2.93 length-2048 FFTs are calculated per second for one band for each LFA beam. The total data rate is

$$4096 \text{ samples/FFT} \times 2.93 \text{ FFT/s} \times 50 \text{ beams} \times 6 \text{ bands} \times 4 \text{ bytes/sample} = 14.4 \text{ Mbytes/s}$$

(The maximum FFT rate in Table 6.4 is used because, as stated earlier, we can select the coarsest resolution for all six vernier bands.) Similarly, from Table 6.9, 2.93 length-1024 FFTs are performed each second per MFA beam, for a data rate of

$$2048 \text{ samples/FFT} \times 2.93 \text{ FFT/s} \times 250 \text{ beams} \times 1 \text{ band} \times 4 \text{ bytes/sample} = 6.0 \text{ Mbytes/s}$$

Therefore, just the transfer of data to the FFTs represents an intermediate data rate of 20.4 Mbytes/s.

6.3 SONAR SIGNAL PROCESSING ARCHITECTURE

A conceptual signal processing architecture is shown in Figure 6.4. Here a high-speed–bus and memory unit is used to transfer and store data. To this bus are attached a number of processors. One approach for partitioning the sonar signal processing problem in Section 6.1 is to assign an entire function (such as narrowband processing) for a set of beams to a processor. This is shown in Figure 6.5(a), where

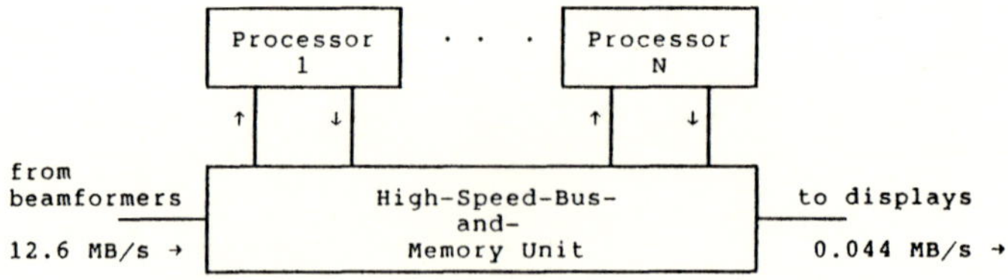

Figure 6.4 Conceptual signal processing architecture (MB = megabyte).

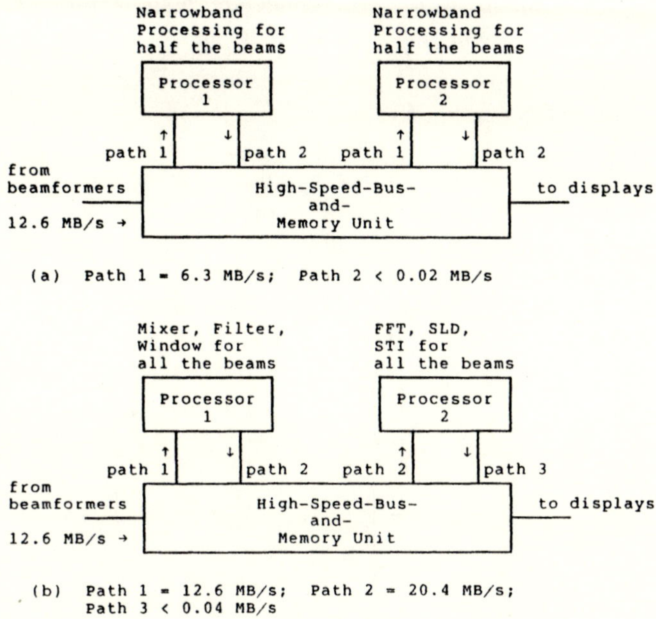

Figure 6.5 Two examples of partitioning the narrowband processing function: (a) path 1 = 6.3 Mbytes/s, path 2 < 0.02 Mbyte/s; (b) path 1 = 12.6 Mbytes/s, path 2 = 20.4 Mbytes/s, path 3 < 0.04 Mbyte/s.

processor 1 executes the narrowband processing for the first half of the beams, and processor 2 performs the same processing for the second half. The input data path for each processor must transfer 6.3 Mbytes/s. The output data path transfers display data at a rate less than 0.02 Mbyte/s. Therefore, the total I/O rate for these two processors is about 12.7 Mbytes/s.

A second partitioning requires that processor 1 calculate the mixing, low-pass filtering, and windowing functions for all beams. The second processor then performs the FFT, SLD, and STI functions. As shown in Figure 6.5(b), the total I/O rate for processors 1 and 2 is now about 53 Mbytes/s.

Although this example is rather extreme, it does suggest the importance of performing as much processing as possible in a processor before returning the results to the high-speed bus. This follows from the nature of the sonar signal processing example in Section 6.1. The input data rate is moderate, the output rate is low, but intermediate rates can be quite high because multiple bands are implemented for the same beamformer data.

In the next three subsections, an architecture is presented that can be used to implement the sonar signal processing problem advanced in Section 6.1. We examine successively smaller components of this architecture: a signal processing unit; a signal processor; a DSP card; and finally, a DSP device.

6.3.1 Overall Sonar Processing System

An approach for implementing the sonar signal processing requirements in Section 6.1 is shown in Figure 6.6. A set of processing units, described in Section 6.3.2, executes the required signal processing algorithms. High-rate data from the beamformers are transferred via high-speed buses located in each processing unit. Processed results for display are transferred over the low-data-rate system controller bus to an archiving medium or display devices. This bus also is used to download programs and execution parameters to each processing unit and to monitor unit status.

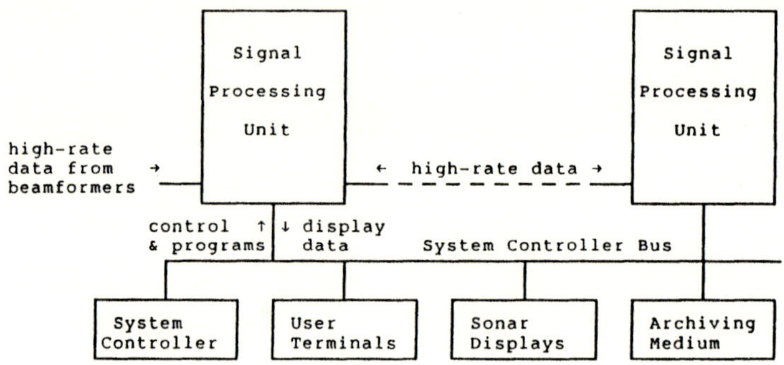

Figure 6.6 Implementation of the sonar signal processing requirements.

6.3.2 Signal Processing Unit

A generic signal processing unit that has some desirable attributes with respect to our sonar processing problem is shown in Figure 6.7. This consists of four signal processors and a high-speed–bus and memory unit. High-rate sonar data are entered via an external interface and stored in memory. These data are then transferred to the signal processors by using the processor interface cards. Here a degree of flexibility is available, in that the same data can be sent to all processors. Each processor then can execute a different set of algorithms. Returning to the narrowband processing example, each processor could implement a different band in parallel on the same data. On the other hand, each signal processor in Figure 6.7 can receive different data and execute the same or a different set of algorithms. This corresponds to the case where each processor would perform the same vernier-band processing on a subset of the total number of beams.

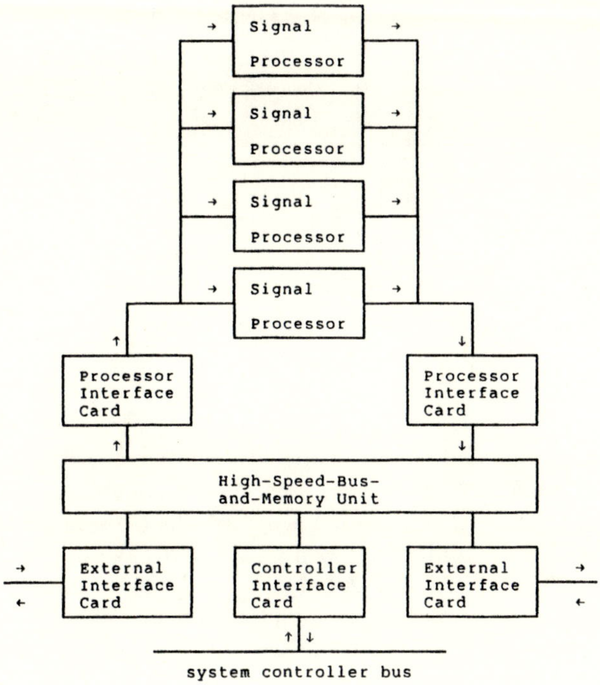

Figure 6.7 Generic signal processing unit.

To minimize the I/O requirements for the high-speed bus, we again attempt to partition the signal processing tasks in such a manner that an entire function is performed in a processor. The low-rate-display output is then placed back on the bus.

To execute the type of processing described in Section 6.1, the appropriate signal processing algorithms can be downloaded from the system controller, via the system controller bus, and transferred to the signal processors. Also downloaded from the controller is a set of commands that orchestrate the movement of data across the various interface cards in the signal processing unit in Figure 6.7. This set of commands is stored in the controller card. Thus, local control is provided in each processing unit. Data memory is present in the high-speed–bus and memory unit and in each signal processor.

6.3.3 Signal Processor

Each signal processor in Figure 6.7 consists of a number of DSP cards and I/O-and-memory cards. This arrangement is illustrated in Figure 6.8. Data are transferred

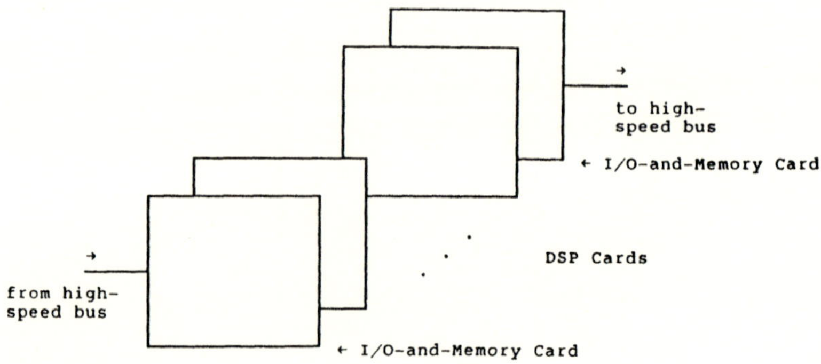

Figure 6.8 Signal processor.

from the high-speed bus and stored in the I/O-and-memory card. Each DSP card contains as a minimum, a DSP device, local data memory, program memory, and associated control logic. It is beyond the scope of this book to describe in detail the design, operation, and capabilities of such a DSP card. We do assume that the DSP device can execute a multiplication and an addition in one instruction cycle and that operations are performed using floating-point arithmetic. The approach taken in the signal processor is that each DSP card in Figure 6.8 executes a particular algorithm, or portion of an algorithm, and transfers the results to other DSP cards for further processing.

Available with the DSP device is a set of signal processing algorithms. Such a list is provided in Table 6.15, where only those algorithms are given that pertain

Table 6.15
Partial Library of DSP Algorithms

Algorithm	*DSP Cards*	*Instruction Cycles per Card*	*Efficiency*
Length-N FFT	K	$(4N \log_2 N)/\mathrm{K}$	62.5%
N-stage FIR Filter	1	$4 + N$ per output sample	$N/(4 + N)$%
N-pole–N-zero IIR Filter	K	$5N/(2\mathrm{K})$ per input sample	80%
Mixer exp $(j2\pi f_m n/f_s)$	1	2 per input	No coefficient update → 50%
	1	3 per input	Coefficient update → 33%
Noise Mean Estimation and Removal	1	100 per input sample	50%
SLD	1	2 per output sample	75% (input is complex)
STI (Sum M Samples)	1	$M + 1$ per output sample	$(M - 1)/(2M + 2)$%
Time-Domain Window	1	$4 + 3N$ per N complex inputs	$N/(4 + 3N)$%

to narrowband processing. Associated with each algorithm is the required number of instruction cycles. Multiple-DSP-card implementations are available for the FFT and the IIR filter. It therefore is possible to balance execution time and number of DSP cards. For instance, an N-pole–N-zero IIR filter requires five instruction cycles per input sample if $N/2$ DSP cards are used. Conversely, $5N/2$ cycles are required for a single-DSP-card implementation.

The FFT is implemented with $(4N \log_2 N)/K$ instruction cycles when using K DSP cards. Thus a single-card implementation executes the FFT in $4N \log_2 N$ cycles. The minimum number of cycles, equal to $4N$, is required when $\log_2 N$ cards are used. In the latter case, a separate DSP card is devoted to each butterfly stage in the FFT.

The last column of Table 6.15 provides the efficiency of the algorithm. A multiplication and an addition are available during each instruction cycle. The efficiency is the ratio of the number of operations performed to the number available. For the FFT, for instance, $(5N \log_2 N)/K$ operations are performed in $(4N \log_2 N)/K$ instruction cycles. A total of $(8N \log_2 N)/K$ operations is available, so the efficiency is 5/8 or 62.5 percent.

A high-level diagram of a typical DSP device is provided in Figure 6.9. Such a device contains a multiplier, an *arithmetic logic unit* (ALU), or adder, and banks of multiplier registers and ALU registers. Data can be read from left memory or right memory. The output of the ALU can be written to either memory or returned to an ALU register. These return paths are used to update partial sums. We can

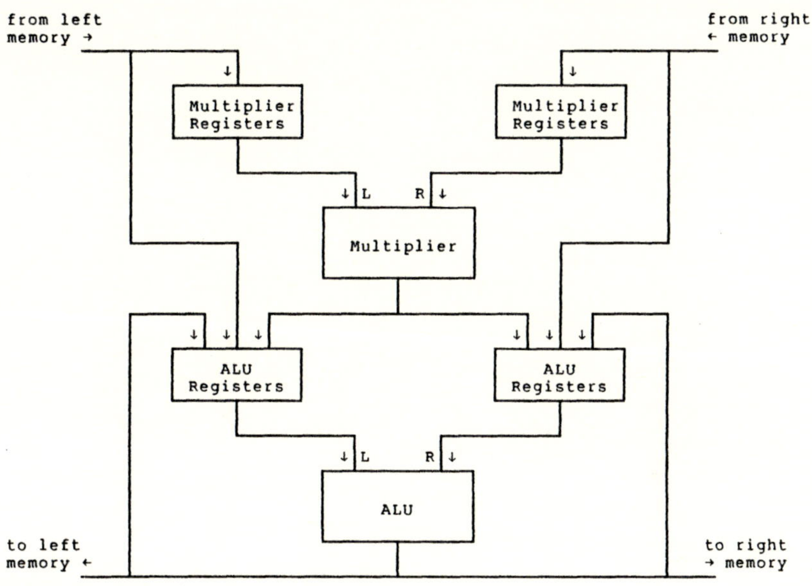

Figure 6.9 Typical digital signal processing device.

perform four steps during a single instruction cycle: a multiplication; an addition; a left-memory access (either read or write); and a right-memory access (either read or write).

An N-stage FIR filter can be implemented by storing the filter impulse response, $h(n)$, in right memory and the sequence to be filtered, $x(n)$, in left memory. Assuming that the algorithm has been initialized, the following steps are executed during one instruction cycle:

1. Move $x(n)$ from left memory to a left-multiplier register.
2. Move $h(n)$ from right memory to a right-multiplier register.
3. Multiply $x(n - 1)$ and $h(n - 1)$.
4. Add $x(n - 2)$ $h(n - 2)$ to the partial sum $ps(n - 3)$.

After initialization, this algorithm can use both the multiplication and the addition that are available each instruction cycle. We see from Table 6.15 that four instruction cycles are required for initialization. Therefore, for long FIR filters, the efficiency approaches 100 percent.

Returning to the signal processor in Figure 6.8, we wish for the purpose of illustration to partition the MFA narrowband processing (Table 6.9) with respect to a set of 10 DSP cards. A block diagram of the processing for 1.46-Hz resolution is provided in Figure 6.10. The clock frequency of the DSP device is 10 MHz. Therefore, the signal processor has a theoretical computational capability of 10 DSP cards $\times$ 2 flop/card/cycle $\times$ $1 \cdot 10^7$ cycles/s or 200 Mflops.

We start by assuming a 2-DSP-card implementation for the length-1024 FFT. From Table 6.15, $(4N \log_2 N)/2$ = 20,480 instruction cycles are required for each FFT calculation. Based on the clock frequency of 10 MHz, an FFT rate of $1 \cdot 10^7$

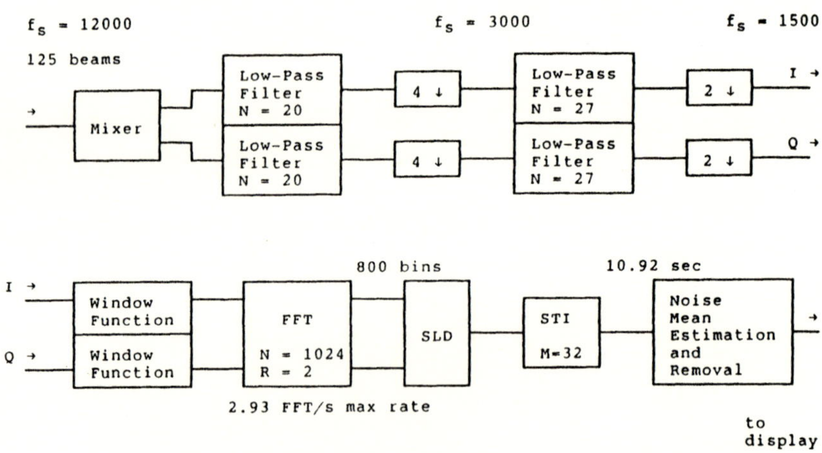

Figure 6.10 MFA narrowband processing for 1.46 Hz resolution.

cycles/s × 1 FFT/20480 cycles = 488 FFT/s can be sustained. The redundancy is equal to 2, so an FFT is calculated each 512 samples for each beam, for an FFT rate of 1500/512 = 2.93 FFTs/s/beam. If 125 beams are processed by this signal processor (one-half the beams from the MFA beamformer), then the required FFT rate is 125 beams × 2.93 FFT/s/beam = 366 FFT/s. This represents a loading of 366/488, or 75 percent, which is deemed acceptable. We next allocate the rest of the processing in Figure 6.10 to the remaining 8 DSP cards. This is done by determining the number of cycles required by one DSP card to execute each of the algorithms in Figure 6.10, using 1 s of data. From Table 6.15, we obtain the following 1-PE loading levels for each of the required algorithms.

Mixer:

$$3 \text{ cycles/sample} \times (12000 \times 125) \text{ samples} = 4.5 \cdot 10^6 \text{ cycles}, \quad \text{or} \quad 45\%$$

Filter 1:

$$(20 + 4) \text{ cycles/sample} \times [(12000 \times 125 \times 2)/4] \text{ samples} = 1.8 \cdot 10^7 \text{ cycles}, \quad \text{or} \quad 180\%$$

Filter 2:

$$(27 + 4) \text{ cycles/sample} \times [(12000 \times 125 \times 2)/8] \text{ samples} = 1.16 \cdot 10^7 \text{ cycles}, \quad \text{or} \quad 116\%$$

Time-domain window:

$$(4 + 3 \times 1024) \text{ cycles/FFT} \times 366.25 \text{ FFTs} = 1.13 \cdot 10^6 \text{ cycles}, \quad \text{or} \quad 11.3\%$$

SLD:

$$2 \text{ cycles/bin} \times 800 \text{ bins/FFT} \times 366.25 \text{ FFTs} = 5.86 \cdot 10^5 \text{ cycles}, \quad \text{or} \quad 5.86\%$$

STI (M = 32):

$$(32 + 1) \text{ cycles/bin} \times 800 \text{ bins/FFT} \times 366.25/32 \text{ FFTs} = 3.02 \cdot 10^5 \text{ cycles}, \quad \text{or} \quad 3.02\%$$

Noise mean estimation and removal:

$$100 \text{ cycles/bin} \times 800 \text{ bins/FFT}$$
$$\times\ 366.25/32 \text{ FFTs} = 9.16 \cdot 10^5 \text{ cycles}, \quad \text{or} \quad 9.16\%$$

Because $1 \cdot 10^7$ cycles are available per card per s, these results indicate that two DSP cards must be devoted to the first low-pass filter and two cards to the second. This is a convenient implementation because one DSP card can be assigned to the in-phase channel and the other to the quadrature channel. The SLD and STI functions can be performed readily in the same DSP card. The partitioning for a 10-DSP-card signal processor is shown in Figure 6.11, with loading levels given in percentages.

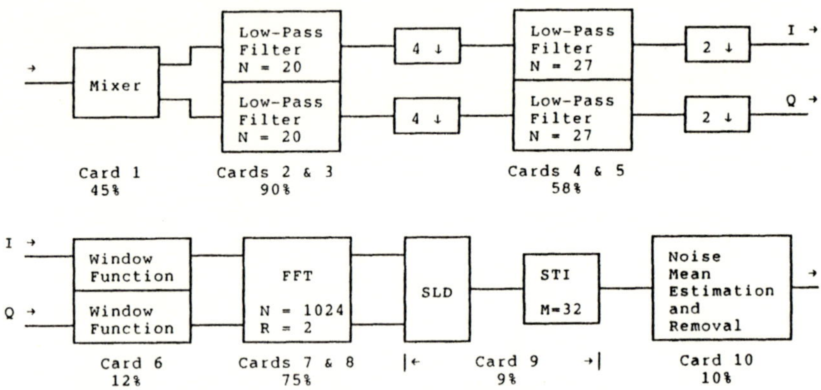

Figure 6.11 Partitioning of MFA narrowband processing relative to 10 DSP cards.

It is of interest to assess the utilization of available operations. This is done in Table 6.16. A utilization value (in units of DSP cards) is obtained for each algorithm by multiplying the loading level, the algorithm efficiency, and the number of DSP cards used for that algorithm. From Table 6.16, the processing capability of 3.74 cards is used by the MFA narrowband processing. Because 10 cards are available in the signal processor, only 37 percent of the available operations are actually utilized.

Referring to the signal processing unit in Figure 6.7, the first signal processor would then perform narrowband analysis for 125 MFA beams and the second processor would execute this processing for the remaining 125 beams. In a similar fashion, the remaining sonar signal processing algorithms in Figure 6.1 can be partitioned or mapped to the signal processors.

Table 6.16
Utilization of Available Operations in the 10-Card Signal Processor

Algorithm	*Loading Level*		*Algorithm Efficiency*		*DSP Cards*		*Utilization* (in DSP cards)
Mixer	0.45	×	0.33	×	1	=	0.15
Filter 1	0.90	×	0.83	×	2	=	1.50
Filter 2	0.58	×	0.87	×	2	=	1.01
Window	0.11	×	0.33	×	1	=	0.036
FFT	0.75	×	0.625	×	2	=	0.94
SLD	(0.06	×	0.75				
STI	+ 0.03	×	0.47)	×	1	=	0.059
Noise Mean Estimation and Removal	0.092	×	0.50	×	1	=	0.046
TOTAL							3.74 cards

Memory Considerations

As indicated in Figure 6.8, an I/O-and-memory card is located at each end of a signal processor. The memory at the input is used to perform what is known as "corner turning." For the example treated in Figure 6.10, data for 125 beams are transferred from the high-speed bus in Figure 6.7 to the signal processor. These data consist of 125 samples at time nT followed by 125 samples at time $(n + 1)T$, and so on, where T is equal to $1/f_s$, or 83.3 μs. This is shown in Figure 6.12. The data are read into memory by columns. It is necessary to form what is effectively the

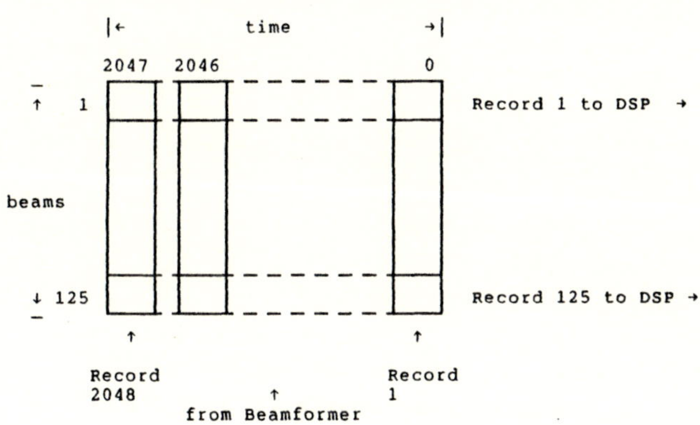

Figure 6.12 Corner-turning memory.

transpose of this matrix by transferring the data to the DSP cards by rows because narrowband processing is performed on a single beam using a set of time samples. It is convenient to store a number of time samples that is equal to a power of 2 for each beam (a value of 2048 is shown in Figure 6.12). For example, the memory requirements for storing 2048 time samples for each beam, assuming 32-bit (4-byte) data and double buffering, is 2 × 125 beams × 2048 samples/beam × 4 bytes/sample = 2.048 Mbytes.

Constant-Geometry FFT

We indicated in Table 6.15 that the FFT algorithm can be implemented using multiple DSP cards to increase the execution time. The radix-2 DIT and DIF butterfly structures that are given in Section 1.3.2 are in-place algorithms. That is, each butterfly operates on a pair of complex numbers in memory locations p and q and returns the results to the same locations. As is shown in Figure 1.32 for the length-16 FFT, these memory locations are a function of the butterfly-stage index. Butterflies in the first stage operate on data stored in locations separated by $N/2$, whereas pairs of data values used by the second stage are separated by $N/4$, and so on. In a multiple-DSP-card implementation, this means that a different program must be stored in each card, in addition to a different set of FFT parameters, W_N^r.

An algorithm that uses the same memory addressing regardless of the butterfly stage is the constant-geometry FFT [2, Chapter 10]. For the mth stage, the DIT butterfly is

$$X_{m+1}(p) = X_m(p/2) + W_N^r X_m[(N + p)/2] \tag{6.2}$$

$$X_{m+1}(p + 1) = X_m(p/2) - W_N^r X_m[(N + p)/2]$$

where $m = 1, \ldots, \log_2 N$ and $p = 0, \ldots, N - 2$ (p even). The constant-geometry DIF butterfly is

$$X_{m+1}(p) = X_m(p/2) + X_m[(N + p)/2] \tag{6.3}$$

$$X_{m+1}(p + 1) = \{X_m(p/2) - X_m[(N + p)/2]\}W_N^r$$

We see from (6.2) and (6.3) that, regardless of the particular butterfly stage, the input addressing is constant, as is the output addressing. This is shown in Figure 6.13 for a length-16 FFT. The penalty that we pay arises because the constant-geometry algorithm is not an in-place algorithm. It requires $2N$ complex registers instead of N registers for the algorithm given in Figure 1.32. Again, the constant-geometry algorithm allows the same program to be loaded into each DSP card in a

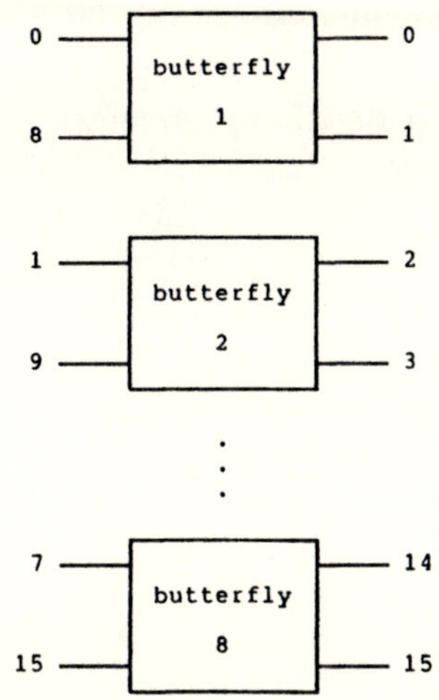

Figure 6.13 Constant-geometry butterfly stage for $N = 16$.

multiple-DSP-card FFT implementation. Versions of this algorithm also are available for higher radices.

6.4 COMPOSITE PASSIVE SONAR DISPLAY

A useful format for displaying the output data generated by passive narrowband processing is the "data cube" format, shown in Figure 6.14 and mentioned in the survey paper by Knight, Pridham, and Kay [3]. The display axes of the cube are time,

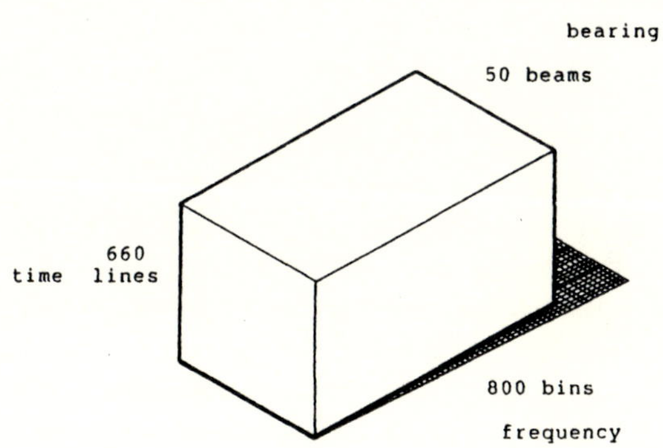

Figure 6.14 Data cube.

frequency, and bearing. By moving through the data cube along each of these three axes, it is possible to observe the following:

1. Bearing *versus* time as a function of frequency; this is the single-frequency BTR face of the data cube.
2. Frequency *versus* time as a function of bearing; this is the lofargram face.
3. Frequency *versus* azimuth as a function of time; this is the FRAZ face.

We wish to form a data cube for a LFA vernier band that corresponds to the last 2 hours of processed data. From Section 6.1.2, the STI update rate is 10.92. That is, 800 frequency bins of 4-bit data are available for each of 50 beams every 10.92 s. About 660 such sets of frequency bins for each beam therefore are generated after 2 hrs.

The storage requirement is equal to 660 STI outputs × 50 beams × 800 bins × 0.5 byte/bin = 13.2 Mbytes. It may be of interest to form data cubes for other LFA vernier bands, as well as for the MFA.

We assume that a sweep rate of 10 frames/s is desired through the data cube along any of the three axes. The lofargram face (frequency *versus* time) imposes the largest I/O requirement of the three faces. This is equal to 10 frames/s × 660 lines × 800 bins × 0.5 byte/bin = 2.7 Mbytes/s.

6.5 CHAPTER SUMMARY

We have shown in this chapter that implicit in the sonar system requirements given in Figure 6.1 is a high degree of parallelism. This parallelism is because (1) the same types of processing (broadband, narrowband, modulation) are performed on large numbers of beams and (2) within each processing type, the data from each beam often are processed by several sets of parameters (bandwidths and center frequencies). We introduced in Figure 6.6 a signal processing architecture that possesses some desirable attributes relative to our system requirements. Successively smaller components of this architecture were examined in some detail; namely, the signal processing unit (Figure 6.7), the signal processor (Figure 6.8), the DSP cards, and finally, the DSP device (Figure 6.9).

Mapping a set of sonar signal processing algorithms onto the architecture shown in Figure 6.6 is facilitated by two features. At the level of the signal processing unit, the same data can be transferred from the high-speed bus and memory unit to the four signal processors. Each processor then can implement different algorithms on these data. Alternatively, each signal processor can receive different data and execute the same or different sets of algorithms. At the level of the signal processor, we can balance execution time and number of DSP cards by (1) using multiple-DSP-card implementations of algorithms, such as the FFT and the IIR filter, and (2) performing more than one algorithm with a single card.

Memory is distributed throughout the signal processing architecture (on the DSP cards, on the I/O-and-memory cards, and in the high-speed bus and memory unit) so that input data, intermediate results, and output data can be stored where they occur. Control is provided in the controller interface card in each signal processing unit, thus simplifying the task of expanding the system by adding additional units. Finally, high-speed data are transferred independently of the system controller bus, which is reserved for control, status data, and low-rate processed data.

REFERENCES

1. Crochiere, R.E., and L.R. Rabiner, *Multirate Digital Signal Processing,* Prentice-Hall, Englewood Cliffs, NJ; 1983.
2. Rabiner, L.R., and B. Gold, *Theory and Application of Digital Signal Processing,* Prentice-Hall, Englewood Cliffs, NJ; 1975.
3. Knight, W.C., R.G. Pridham, and S.M. Kay, "Digital Signal Processing for Sonar," *Proc. IEEE,* Vol. 69, No. 11, November 1981, pp. 1451–1506.

Chapter 7
Matrix Processing for Sonar

This chapter discusses computation-intensive sonar signal processing algorithms, which until recently have been beyond the capabilities of operational systems. The required matrix operations are identified; and numerically stable versions of these operations, such as the QR decomposition and the singular value decomposition, also are discussed.

7.1 COMPUTATION-INTENSIVE SONAR SIGNAL PROCESSING

This section examines high-resolution beamforming and high-resolution spectral estimation algorithms. These algorithms are compared to the conventional processing approaches discussed in Chapter 4. We are particularly interested in the types of matrix operations that a modern sonar signal processor must be able to execute to implement these algorithms.

Note that techniques discussed in Sections 7.1–7.3 typically offer higher resolution and lower leakage (that is, lower response to a signal separated by an appreciable value in angle or frequency) than conventional techniques. They do require a knowledge of the signal and noise environments, however, and often are more sensitive to errors, such as departures from assumed sensor locations.

7.1.1 High-Resolution Beamforming

The Neyman-Pearson broadband processor in Section 3.2.2 includes a set of complex weights applied to the output of the sensors. In contrast to the conventional beamformer, the NP processor incorporates a knowledge of the noise field. The weights from (3.17) are

$$\mathbf{h}^{+}(k) = \mathbf{p}^{+}(k)\,\mathbf{Q}^{-1}(k) \tag{7.1}$$

where we recall from Section 2.3.2 that $\mathbf{p}(k)$ is a vector of phase terms that incorporates the signal arrival angle, $\mathbf{Q}(k)$ is the normalized spatial correlation matrix of the noise, and k is the discrete frequency index. In the event that the noise is spatially uncorrelated, so that $\mathbf{Q}(k) = \mathbf{I}$, the complex weights in (7.1) correspond simply to delays in the time domain (the conventional unweighted beamformer).

The NP processor is given in Figure 3.8 and repeated here as Figure 7.1. In Section 4.3.1, we discussed the effect of the filter $f(n)$ on detection performance. Here we consider the effect of incorporating knowledge of the spatial correlation properties of the signal and noise fields on resolution (that is, the ability to separate two plane-wave signals whose arrival angles are closely spaced).

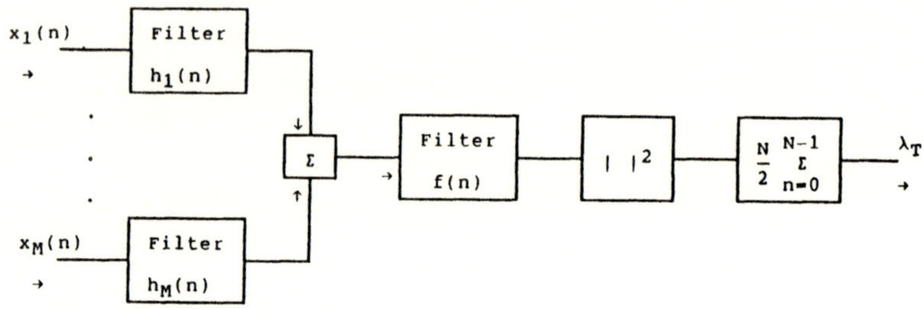

Figure 7.1 Neyman-Pearson broadband processor for an array of sensors.

Maximum Likelihood Method

The power at the output of a beamformer, for the kth frequency, is given by (2.9)

$$P = \mathbf{w}^{+}(k)\mathbf{R}(k)\mathbf{w}(k) \tag{7.2}$$

where again $\mathbf{w}(k)$ is a vector of complex weights and $\mathbf{R}(k)$ is spatial correlation matrix of signal and noise.

We consider an environment consisting of q distinct incoherent plane-wave signals, with unit direction vectors $\mathbf{u}_h$, $h = 1, \ldots, q$, that are impinging on an array of M sensors with position vectors $\mathbf{r}_i$, $i = 1, \ldots, M$. The noise at each sensor is assumed to be statistically independent of the signals and described by the spatial correlation matrix $\sigma_n^2 \mathbf{Q}$. Furthermore, a vector corresponding to a direction of look, $\mathbf{u}_0$, is defined by

$$e(i) = \exp[j(2\pi f/c)\, \mathbf{r}_i \cdot \mathbf{u}_0] \tag{7.3}$$

where $e(i)$ is the ith element of $\mathbf{e}$. The problem addressed by the so-called *maximum likelihood* (ML) method is the following. Find a weight vector $\mathbf{w}_0$ that minimizes the output power in (7.2) subject to the constraint that $\mathbf{w}_0^+\mathbf{e} = 1$. That is, we would like to maintain unity gain in the direction-of-look $\mathbf{u}_0$. The ML method also is known as the *minimum variance* method. (As mentioned in Johnson [1], this minimization problem is not equivalent to the standard maximum likelihood estimation method.)

The spatial correlation matrix is

$$\mathbf{R} = \sigma_n^2\mathbf{Q} + \sum_{h=1}^{q} \sigma_{sh}^2\, \mathbf{p}_h\, \mathbf{p}_h^+ \tag{7.4}$$

where σ_{sh}^2 is the power associated with the hth signal, and the frequency index has been omitted for simplicity. We are dealing with an arbitrary three-dimensional array so for the hth signal, the ith element of $\mathbf{p}_h$ is $p_h(i) = \exp[j(2\pi f/c)\mathbf{r}_i \cdot \mathbf{u}_h]$. Johnson [1] shows that the solution to the constrained optimization problem addressed by the ML method is

$$\mathbf{w}_0 = \frac{\mathbf{R}^{-1}\mathbf{e}}{\mathbf{e}^+\mathbf{R}^{-1}\mathbf{e}} \tag{7.5}$$

which is found by using the Lagrange multiplier technique. It is straightforward to show that $\mathbf{w}_0^+\mathbf{e} = 1$, so the response in the direction of look (7.3) is maintained at unity, as desired. We next calculate the power in this direction, which we denote by $P(\mathbf{e})$. Substituting (7.5) into (7.2) yields

$$P(\mathbf{e}) = \left(\frac{\mathbf{R}^{-1}\mathbf{e}}{\mathbf{e}^+\mathbf{R}^{-1}\mathbf{e}}\right)^+ \mathbf{R} \left(\frac{\mathbf{R}^{-1}\mathbf{e}}{\mathbf{e}^+\mathbf{R}^{-1}\mathbf{e}}\right)$$

$$= \frac{\mathbf{e}^+\mathbf{R}^{-1}\mathbf{e}}{(\mathbf{e}^+\mathbf{R}^{-1}\mathbf{e})^2}$$

which gives the desired result:

$$P(\mathbf{e}) = \frac{1}{\mathbf{e}^+\mathbf{R}^{-1}\mathbf{e}} \tag{7.6}$$

Eigenvector Method

Also discussed in [1] is the *eigenvector* (EV) method. Intuitively, it appears that we could improve the resolution properties of the ML beamformer if the denominator of (7.6) could be made smaller in those directions that correspond to signals and larger in directions corresponding to noise. We assume q distinct incoherent plane-wave signals and spatially white noise, so that $\mathbf{Q} = \mathbf{I}$. Therefore, we may be in a flow noise-limited environment or an isotropic ambient noise-limited environment where the sensor spacing is equal to $\lambda/2$. We can show that the q largest eigenvalues of the $M \times M$ spatial correlation matrix, $\mathbf{R}$, are associated with the signals. The corresponding q eigenvectors span the signal subspace (that is, they are linear combinations of the signal vectors, $\mathbf{p}_h$). Furthermore, the $M - q$ remaining eigenvectors span a noise subspace orthogonal to the signal subspace. We can express $\mathbf{R}$ in terms of its eigenvalues λ_i and eigenvectors $\mathbf{v}_i$, $i = 1, \ldots, M$, by using the spectral theorem for a matrix:

$$\mathbf{R} = \sum_{i=1}^{M} \lambda_i \mathbf{v}_i \mathbf{v}_i^+$$

The inverse of $\mathbf{R}$ has an equally simple form:

$$\mathbf{R}^{-1} = \sum_{i=1}^{M} \frac{1}{\lambda_i} \mathbf{v}_i \mathbf{v}_i^+ \tag{7.7}$$

Next, we form a truncated correlation matrix by including only the $M - q$ eigenvectors that span the noise subspace. This matrix is denoted by $\mathbf{R}_{\mathrm{EV}}$, and from (7.7) $\mathbf{R}_{\mathrm{EV}}^{-1}$ is equal to

$$\mathbf{R}_{\mathrm{EV}}^{-1} = \sum_{i=1}^{M-q} \frac{1}{\lambda_i} \mathbf{v}_i \mathbf{v}_i^+ \tag{7.8}$$

This truncated correlation matrix then is used in (7.5) to find the corresponding optimum weight vector for the EV beamformer:

$$\mathbf{w}_0 = \frac{\mathbf{R}_{\mathrm{EV}}^{-1}\mathbf{e}}{\mathbf{e}^+\mathbf{R}_{\mathrm{EV}}^{-1}\mathbf{e}} \tag{7.9}$$

We now calculate the power in the direction of look by substituting the expression for the truncated correlation matrix (7.8) into (7.6):

$$P(\mathbf{e}) = \left[\mathbf{e}^+ \left(\sum_{i=1}^{M-q} \frac{1}{\lambda_i} \mathbf{v}_i \mathbf{v}_i^+\right) \mathbf{e}\right]^{-1}$$

$$= \left(\sum_{i=1}^{M-q} \frac{1}{\lambda_i} \mathbf{e}^+ \mathbf{v}_i \mathbf{v}_i^+ \mathbf{e}\right)^{-1}$$

which is equivalent to

$$P(\mathbf{e}) = \left(\sum_{i=1}^{M-q} \frac{1}{\lambda_i} |\mathbf{e}^+ \mathbf{v}_i|^2 \right)^{-1} \tag{7.10}$$

Because vectors in the signal subspace are orthogonal to the noise subspace spanned by $\mathbf{v}_i$, $i = 1, \ldots, M - q$, the inner product $\mathbf{e}^+\mathbf{v}_i$ is zero if $\mathbf{e}$ is in the signal subspace. Thus, with the assumption of uncorrelated noise, the denominator of (7.10) is equal to zero for signal directions. The power at the output of the beamformer given by (7.10) therefore is infinite in those directions in the signal subspace.

The *multiple signal classification* (MUSIC) method of Schmidt [2] sets the small-noise eigenvalues equal to unity in the formation of the truncated correlation matrix. That is,

$$\mathbf{R}_{\mathrm{M}}^{-1} = \sum_{i=1}^{M-q} \mathbf{v}_i \mathbf{v}_i^+$$

where the subscript, M, denotes the MUSIC algorithm.

Knowledge of the number of sources present in the data is required by both the EV and the MUSIC methods. In general this is not known, but can be estimated by examining the eigenvalues of $\mathbf{R}$. Statistical procedures for estimating q are described by Wax and Kailath [3] and Williams and Johnson [4], among others.

Example 7.1.1

Consider the case of one signal. From (7.4), the spatial correlation matrix is

$$\mathbf{R} = \sigma_n^2 \mathbf{I} + \sigma_s^2 \mathbf{p}\mathbf{p}^+ \tag{7.11}$$

The largest eigenvalue, λ_M, of $\mathbf{R}$ in (7.11) corresponds to the signal and the associated eigenvector, $\mathbf{v}_M$, is equal to $\mathbf{p}$. To find λ_M, we solve $\mathbf{R}\mathbf{v}_M = \lambda_M \mathbf{v}_M$, which from (7.11) is

$$(\sigma_n^2 \mathbf{I} + \sigma_s^2 \mathbf{p}\mathbf{p}^+)\mathbf{p} = (\sigma_n^2 + M\sigma_s^2)\mathbf{p}$$

Thus $\lambda_M = \sigma_n^2 + M\sigma_s^2$. The remaining $M - 1$ eigenvalues are equal to σ_n^2 and the associated eigenvectors, $\mathbf{v}_i$, $i = 1, \ldots, M - 1$, are orthogonal to $\mathbf{p}$. (We used the fact that $\mathbf{R}$ in (7.11) has one eigenvalue equal to $\sigma_n^2 + M\sigma_s^2$ and $M - 1$ eigenvalues equal to σ_n^2 in the ML estimation of broadband parameters in Section 3.3.2.)

The power at the output of the MV beamformer (7.10) is

$$P(\mathbf{e}) = \left(\sum_{i=1}^{M-1} \frac{1}{\sigma_n^2} |\mathbf{e}^+\mathbf{v}_i|^2 \right)^{-1} \tag{7.12}$$

Evaluating (7.12) for $\mathbf{e} = \mathbf{p}$, that is, steering the beam in the direction of the signal, ideally would result in an infinite peak in the output power because $\mathbf{p}^+\mathbf{v}_i = 0$ for $i = 1, \ldots, M - 1$. The departure from this ideal performance observed in practice is discussed next. (End of example.)

The expression for the output power of the EV beamformer (7.10) was obtained by assuming that (1) the number of signals is known; (2) the signals and noise are not correlated; and (3) the noise is not correlated from sensor to sensor. Departure from the ideal performance predicted by (7.10) results if any of these assumptions is violated. In practice, small-signal eigenvalues may be confused with noise eigenvalues in the statistical determination of an estimate for q. In addition, the noise field in fact may be correlated. Even if the noise field is not correlated, in practice we can only estimate the spatial correlation matrix $\mathbf{R}$ based on a set of K observations of the output of the array. This estimated correlation matrix thus can appear to correspond to an environment in which the noise field is correlated from sensor to sensor as well as with the signal (thus violating assumptions (2) and (3)).

A comparison of the resolution performance of the unweighted conventional beamformer, the ML beamformer, and the EV beamformer is provided by Johnson [1]. A linear array of 10 equally spaced sensors is used, where the spacing is equal to $\lambda/2$. The environment consists of two plane-wave signals, with arrival angles of 5° and −5° from broadside. The noise is not correlated from sensor to sensor, and the signal powers and noise power equal 1 ($\sigma_{s1}^2 = \sigma_{s2}^2 = \sigma_n^2 = 1$). Thus, the theoretical correlation matrix is (7.4)

$$\mathbf{R} = \mathbf{I} + \mathbf{p}_1\mathbf{p}_1^+ + \mathbf{p}_2\mathbf{p}_2^+$$

In [1], $\mathbf{R}$ is estimated by using 50 observations of the frequency-domain data. Because $M = 10$ sensors and $q = 2$ signals, the truncated correlation matrix (7.8) used in the EV method consists of 8 terms. The results for the conventional beamformer are given in Figure 7.2(a). Here the beamformer output power (7.2) is plotted as a function of direction of look, or steering angle, which is varied from −90° to 90°. Figure 7.2(b) provides the output power of the ML beamformer, as defined by (7.6). Finally, the output power of the EV beamformer (7.10) is given in Figure 7.2(c). Figures 7.2(b) and 7.2(c) are obtained by using the weight vectors defined by (7.5) and (7.9), respectively, in which $\mathbf{e}$ is varied to correspond to directions of look from −90° to 90°. Note that the information presented in Figure 7.2 is in the form of a single-frequency version of the bearing-time-record display discussed in Section 4.3.3.

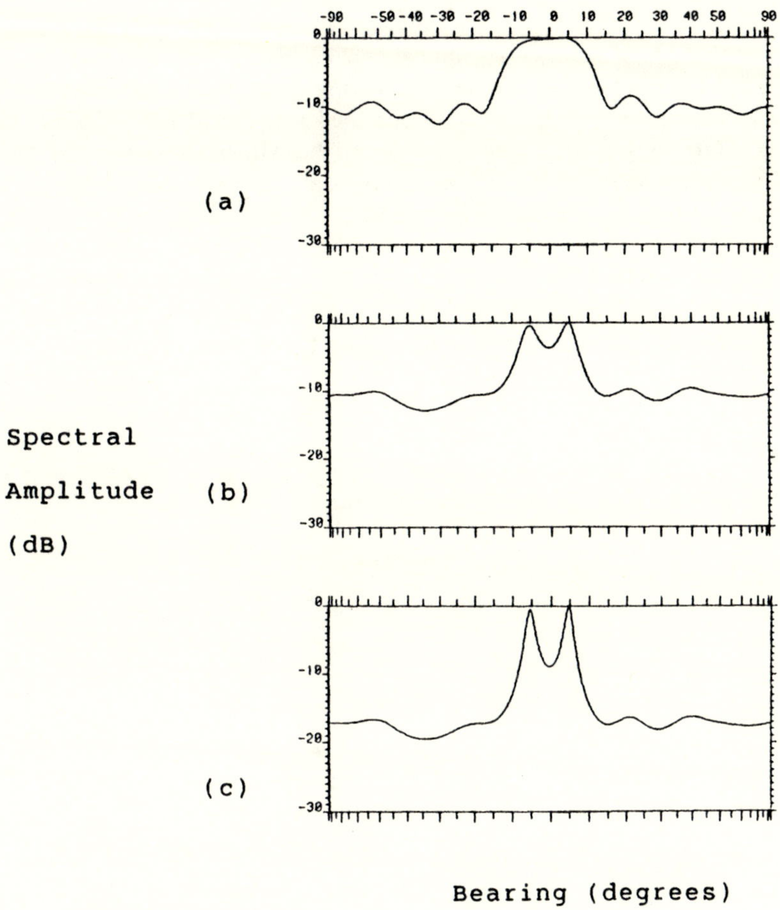

Figure 7.2 Comparison of resolution by using three different sets of weights: (a) unity weights; (b) ML weights; and (c) EV weights.
(*Source:* [1] D.H. Johnson, "The Application of Spectral Estimation Methods to Bearing Estimation Problems," *Proc. IEEE,* Vol. 70, No. 9, September 1982, pp. 1018–1028. © 1982 IEEE. Reprinted with permission.)

That is, Figure 7.2 does not provide plots of beam power patterns generated with a particular weight vector.

It is apparent from Figure 7.2(a) that the conventional beamformer with unity weighting is unable to resolve the signals at $-5°$ and $5°$. This is not surprising because of the broad 3-dB beamwidth given by (2.6)

$$\Delta\psi_{3dB} \approx \sin^{-1}[\sin\psi_o + 0.44c/(Mfd)] - \sin^{-1}[\sin\psi_o - 0.44c/(Mfd)]$$

In our example, $\psi_o = \pm 5$, $M = 10$, and $d = \lambda/2$. Thus 3-dB beamwidth is about 10.1°. Figure 7.2(b) indicates that the ML beamformer is able to resolve the two signals, by including knowledge of the combined signal and noise fields. The EV beamformer offers improved performance, over the ML beamformer, by using the truncated correlation matrix. The peaks in power at −5° and 5° in Figure 7.2(c) are finite, and not infinite as predicted by (7.10), because only an estimate of **R** is available.

It is instructive to compare the resolution performance of the conventional and the ML beamformers for a single signal. As shown in Section 7.4.1, such a comparison actually can be performed by hand. The spatial correlation matrix is (7.11)

$$\mathbf{R} = \sigma_n^2 \mathbf{I} + \sigma_s^2 \mathbf{p}\mathbf{p}^+$$

For the conventional beamformer with unity weights, the weight vector **w** is simply equal to **e**. From (7.2), the power in the direction of look is

$$\begin{aligned} \underline{P}_C(\mathbf{e}) &= \mathbf{e}^+\mathbf{R}\mathbf{e} \\ &= \sigma_n^2 M + \sigma_s^2 |\mathbf{e}^+\mathbf{p}|^2 \end{aligned}$$

where the subscript C denotes the conventional beamformer. We normalize by the output power in the direction of the signal, $P_C(\mathbf{p})$, and obtain

$$\underline{P}_C(\mathbf{e}) = \frac{\sigma_n^2 M + \sigma_s^2 |\mathbf{e}^+\mathbf{p}|^2}{\sigma_n^2 M + \sigma_s^2 M^2} \tag{7.13}$$

where the underline in (7.13) denotes a normalized quantity. The normalized output power for the ML beamformer is derived in Section 7.4.1 and repeated here

$$\underline{P}_{ML}(\mathbf{e}) = \frac{\sigma_n^2 M}{\sigma_n^2 M + \sigma_s^2 (M^2 - |\mathbf{e}^+\mathbf{p}|^2)} \tag{7.14}$$

Equations (7.13) and (7.14) are evaluated for **p** corresponding to a plane wave signal at 0°, $M = 10$ sensors, and $\sigma_n^2 = \sigma_s^2 = 1$. The results, presented in Table 7.1 as functions of the direction of look, indicate the improvement in resolution of the ML beamformer relative to the conventional beamformer. (Note that $|\mathbf{e}^+\mathbf{p}|$ is equal to the beam pattern in (2.4) with $d = \lambda/2$.)

Intuitive Explanation for the Improved Resolution of the ML Beamformer

For a steering angle of 5°, for example, the plane wave at 0° is considered part of the *noise field*. A weight vector for the ML beamformer (7.5) is generated that is

Table 7.1
Normalized Output Power for the Conventional and the ML Beamformers

Direction of Look (degrees)	$\lvert e^{+}p\rvert^2$	*Normalized Output Power* (dB)	
		Conventional	*ML*
0.0	100.0	0.0	0.0
2.5	85.4	−0.6	−3.9
5.0	51.5	−2.5	−7.7
10.0	2.23	−9.5	−10.3
15.0	4.07	−8.9	−10.3

designed to minimize the output power with the constraint of unity response in the 5° direction. Thus this weight vector will effectively form a null in the direction of the signal at 0°. This is shown next.

Beam Power Pattern for the ML Beamformer

We derive the following expression for the ML beam power pattern in Section 7.4.1:

$$|b(\psi_x)|^2 = \frac{|(\mathbf{e}^+\mathbf{x})(\sigma_n^2 + M\sigma_s^2) - (\mathbf{e}^+\mathbf{p})(\mathbf{p}^+\mathbf{x})\sigma_s^2|^2}{[M(\sigma_n^2 + M\sigma_s^2) - |\mathbf{e}^+\mathbf{p}|^2\sigma_s^2]^2} \tag{7.15}$$

where ψ_x is the angle at which the beam power pattern is evaluated. The sensor spacing is again $d = \lambda/2$, so the ith element of $\mathbf{x}$ is $x(i) = \exp[j\pi(i-1)\sin\psi_x]$.

The beam power pattern for the conventional beamformer with unity weights, $M = 10$ sensors, $d = \lambda/2$, and $\psi_o = 5°$, is shown in Figure 7.3. Equation (7.15) for the ML beamformer is plotted in Figure 7.4 with the same set of parameter values as in Figure 7.3. In addition, $\sigma_n^2 = \sigma_s^2 = 1$, and $\mathbf{p}$ corresponds to 0°. Clearly from Figure 7.4, a null has been formed in the direction of what is treated as a plane-wave interference at 0°.

From Section 7.4.1, the weight vector for the ML beamformer is

$$\mathbf{w}_0 = \frac{\mathbf{e}(\sigma_n^2 + M\sigma_s^2) - \mathbf{p}\mathbf{p}^+\mathbf{e}\sigma_s^2}{M(\sigma_n^2 + M\sigma_s^2) - |\mathbf{e}^+\mathbf{p}|^2\sigma_s^2} \tag{7.16}$$

We consider the special case of $\mathbf{e} = \mathbf{p}$, so that the direction of look is equal to the direction of the signal. The weight vector (7.16) reduces to $\mathbf{w}_0 = \mathbf{p}/M$, which is

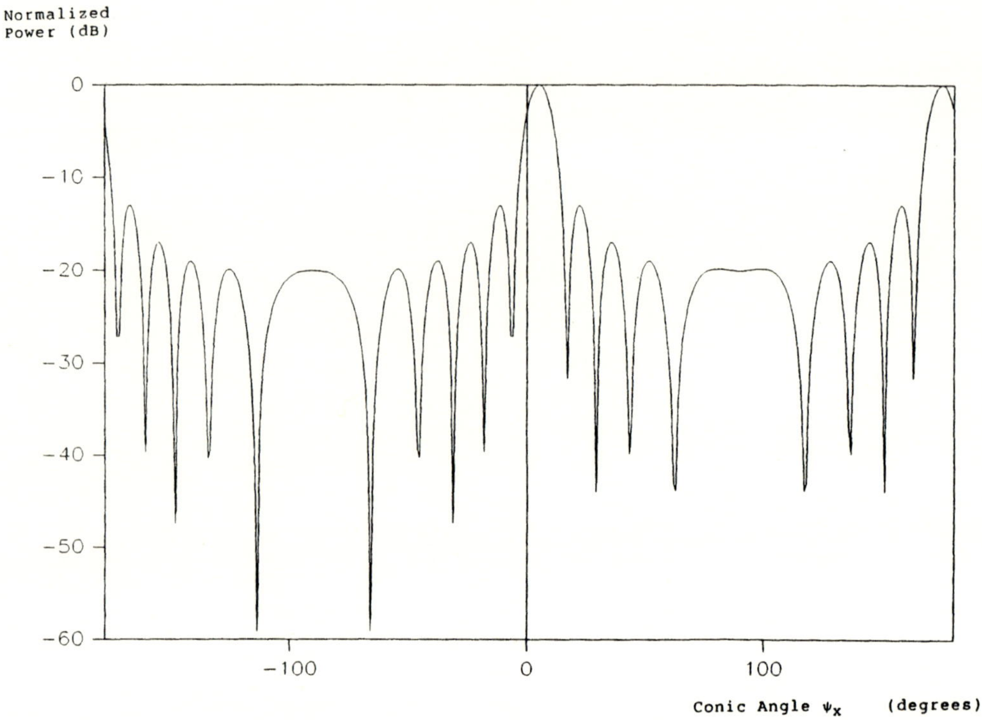

Figure 7.3 Beam pattern for a 10-sensor linear array with unity weights (plane wave at 0°, beam steered to 5°, $d = \lambda/2$ spacing).

just the conventional beamformer. Setting **e** equal to **p** means that there is no longer a plane-wave interference at 0°. Thus the noise environment is spatially uncorrelated, so no special structure can be utilized in the calculation of the weight vector.

To this point, the noise has been assumed to be spatially uncorrelated from sensor to sensor, so that $\mathbf{Q} = \mathbf{I}$. Johnson [5] discusses the effects of nonwhite ambient noise on the performance of the ML beamformer.

Matrix Operations

The ML beamformer requires a matrix-vector multiply, a matrix inversion, and an eigensystem solution. (An eigensystem consists of the eigenvalues and eigenvectors of a matrix.) We can calculate the eigensystem of the $M \times M$ spatial correlation matrix directly. This approach can be numerically ill-conditioned if the ratio of the largest and smallest eigenvalues, $\lambda_{max}/\lambda_{min}$, of **R** is large. A more numerically stable approach is the *singular value decomposition* (SVD) algorithm, which is defined in

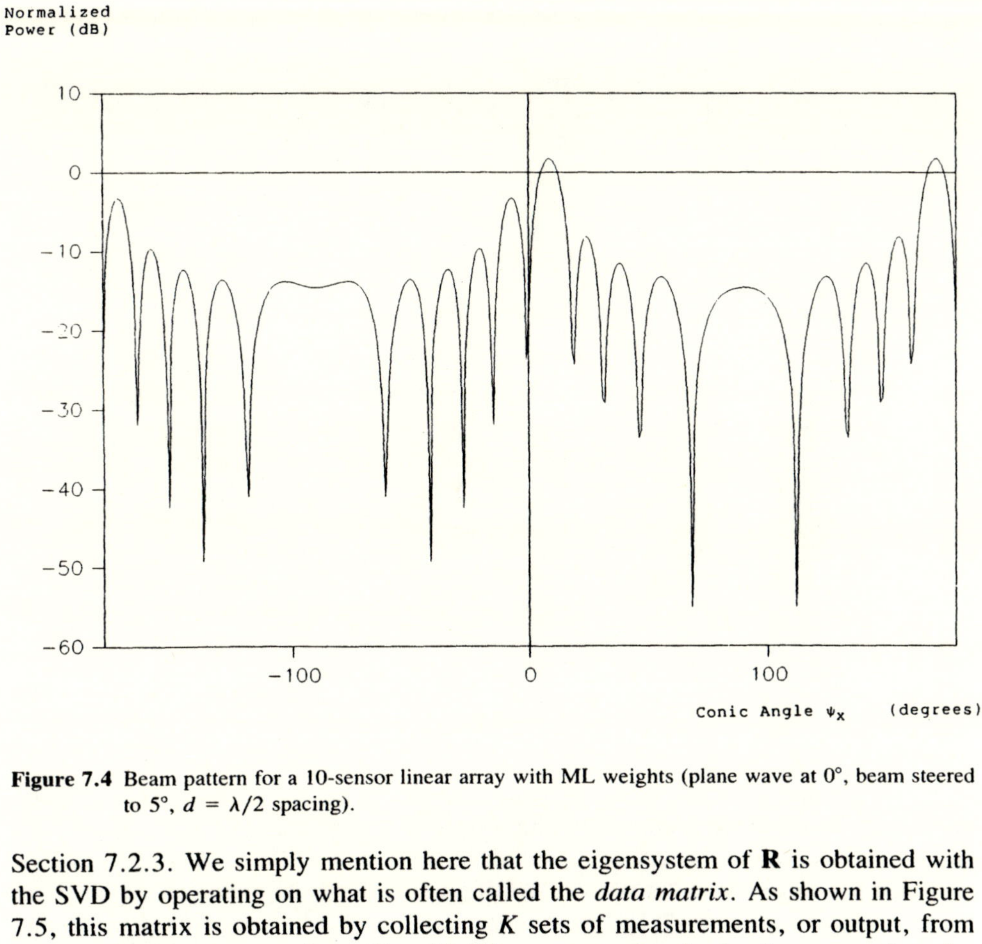

Figure 7.4 Beam pattern for a 10-sensor linear array with ML weights (plane wave at 0°, beam steered to 5°, $d = \lambda/2$ spacing).

Section 7.2.3. We simply mention here that the eigensystem of **R** is obtained with the SVD by operating on what is often called the *data matrix*. As shown in Figure 7.5, this matrix is obtained by collecting K sets of measurements, or output, from each of the M sensors, where $K \geq M$. We are operating in the frequency domain, so the outputs could consist of the contents of the jth DFT bin of each sensor.

Example 7.1.2

We assume $M = 3$ sensors and $K = 4$ measurements. The data matrix is

$$\mathbf{A} = \begin{bmatrix} x_1(1) & x_2(1) & x_3(1) \\ x_1(2) & x_2(2) & x_3(2) \\ x_1(3) & x_2(3) & x_3(3) \\ x_1(4) & x_2(4) & x_3(4) \end{bmatrix} \tag{7.17}$$

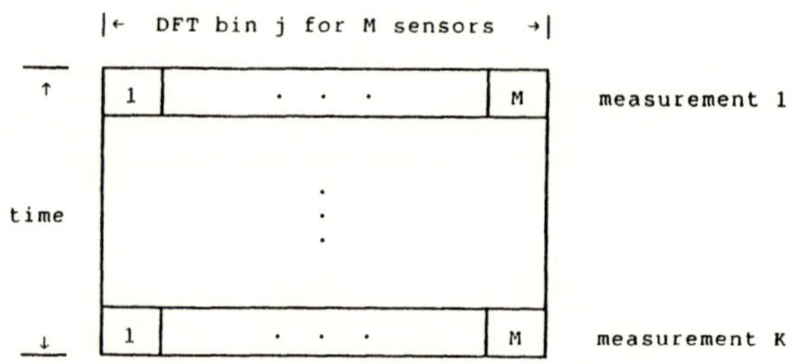

Figure 7.5 Data matrix composed of K measurements from M sensors ($K \geqslant M$).

The subscripts in (7.17) correspond to the sensors whereas the integers in parentheses denote the time at which the data are observed. Calculating $\mathbf{A}^+\mathbf{A}$ yields

$$\mathbf{A}^+\mathbf{A} = \begin{bmatrix} \sum_{k=1}^{4} |x_1(k)|^2 & \sum_{k=1}^{4} x_1^*(k)x_2(k) & \sum_{k=1}^{4} x_1^*(k)x_3(k) \\ \sum_{k=1}^{4} x_1(k)x_2^*(k) & \sum_{k=1}^{4} |x_2(k)|^2 & \sum_{k=1}^{4} x_2^*(k)x_3(k) \\ \sum_{k=1}^{4} x_1(k)x_3^*(k) & \sum_{k=1}^{4} x_2(k)x_3^*(k) & \sum_{k=1}^{4} |x_3(k)|^2 \end{bmatrix} \tag{7.18}$$

We see that $\mathbf{A}^+\mathbf{A}$ in (7.18) is just K times the estimate of the 3×3 correlation matrix, $\hat{\mathbf{R}}$. That is, the spatial correlation matrix that appears in the expression for the ML weight vector (7.5) is estimated by calculating $\hat{\mathbf{R}} = (1/K)\,\mathbf{A}^+\mathbf{A}$. Intuitively, we might expect to encounter numerical difficulties when performing an eigensystem solution of the matrix in (7.18) because the data has been squared, thus increasing the required dynamic range. Operating with the data matrix (7.17), on the other hand, should result in improved numerical stability.

7.1.2 Adaptive-Combiner Beamformer

Where fixed beams are available, we can reduce the effects of plane-wave interferences by adaptively combining the output of these beams. An example of such an adaptive-combiner beamformer is shown in Figure 7.6. Here, a signal beam contains a desired signal, $s(n)$, as well as contributions from a number of interfering sources. This combined sequence is designated $y(n)$ in Figure 7.6. A separate set of "interference beams" is steered to each of the M interferences.

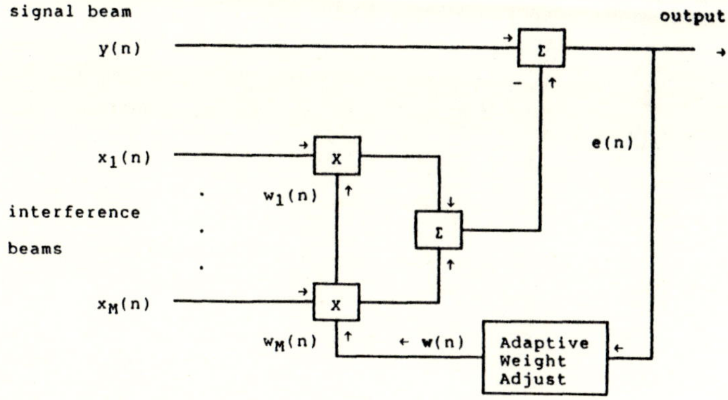

Figure 7.6 Adaptive-combiner beamformer.

As shown in Figure 7.6, an error signal, $e(n)$, is generated by subtracting a weighted sum of the interference beam output from the signal beam output. For simplicity, we assume that the interference beams output is not correlated with the desired signal. Because of this assumption, minimizing $E\{e^2\}$ will remove the interferences from the signal beam. By observing that $e(n)$ is equal to $y(n) - \mathbf{x}^T(n)\mathbf{w}(n)$, the expression that we wish to minimize is

$$E\{(\mathbf{x}^T\mathbf{w} - y)^2\} = \mathbf{w}^T\mathbf{R}_x\mathbf{w} - 2\mathbf{r}_{xy}^T\mathbf{w} + E\{y^2\} \tag{7.19}$$

where the time index, n, has been omitted for convenience. Information concerning the correlation of the interference beams is contained in $\mathbf{R}_x = E\{\mathbf{x}\mathbf{x}^T\}$, and the correlation between the output of the signal beam and the set of interference beams is found in $\mathbf{r}_{xy} = E\{y\mathbf{x}\}$. Typically $\mathbf{R}_x$ and $\mathbf{r}_{xy}$ are estimated from the data.

The weight vector that results from minimizing (7.19) is

$$\mathbf{w}_{\text{opt}} = \mathbf{R}_x^{-1}\mathbf{r}_{xy} \tag{7.20}$$

Expression (7.20) is called the *Wiener-Hopf equation*. The beams steered toward the interfering sources must have nulls in the signal direction to satisfy our requirement that the output of the interference beams is not correlated with the desired signal.

Example 7.1.3

We assume that the output of the signal beam consists of the desired signal as well as contributions from two interfering sources. That is, $y(n) = s(n) + i_1(n) + i_2(n)$, where $s(n)$, $i_1(n)$, and $i_2(n)$ are assumed to be sample sequences of zero mean uncorrelated Gaussian random processes. The variances of $i_1(n)$ and $i_2(n)$ are σ_1^2 and σ_2^2, respectively. A beam is steered to each interfering source so that $x_1(n) = \alpha_1 i_1(n)$ and $x_2(n) = \alpha_2 i_2(n)$, where α_1 and α_2 are constants. Based on this information, we find that

$$\mathbf{R}_x = \begin{bmatrix} \alpha_1^2\sigma_1^2 & 0 \\ 0 & \alpha_2^2\sigma_2^2 \end{bmatrix}$$

$$\mathbf{r}_{xy} = \begin{bmatrix} \alpha_1\sigma_1^2 \\ \alpha_2\sigma_2^2 \end{bmatrix}$$

The solution of the Wiener-Hopf equation (7.20) then is

$$\begin{aligned} \mathbf{w}_{\text{opt}} &= \begin{bmatrix} 1/(\alpha_1^2\sigma_1^2) & 0 \\ 0 & 1/(\alpha_2^2\sigma_2^2) \end{bmatrix} \begin{bmatrix} \alpha_1\sigma_1^2 \\ \alpha_2\sigma_2^2 \end{bmatrix} \\ &= \begin{bmatrix} 1/\alpha_1 \\ 1/\alpha_2 \end{bmatrix} \end{aligned}$$

With this weight vector, the output of the adaptive-combiner beamformer in Figure 7.6 is

$$\begin{aligned} e(n) &= y(n) - \mathbf{x}^T(n)\mathbf{w}_{\text{opt}}(n) \\ &= [s(n) + i_1(n) + i_2(n)] - [\alpha_1 i_1(n) \quad \alpha_2 i_2(n)] \begin{bmatrix} 1/\alpha_1 \\ 1/\alpha_2 \end{bmatrix} \\ &= s(n) \end{aligned}$$

(End of example.)

We have succeeded in recovering the signal in this example because of the assumptions that (1) the interference beam output contains no signal components; and (2) $\mathbf{R}_x$ and $\mathbf{r}_{xy}$ are known. In practice, it often is difficult to satisfy assumption (1). In addition, we usually are required to estimate $\mathbf{R}_x$ and $\mathbf{r}_{xy}$. As a result, the ideal performance obtained in this example in general is not achieved.

A traditional approach to solving for the optimum weight vector (7.20) is the *least mean square* (LMS), or Widrow-Hoff, algorithm [6, Section 4.2]. In this algorithm, the weight vector is updated by

$$\mathbf{w}(j + 1) = \mathbf{w}(j) - 2k_s e(j)\mathbf{x}(j) \tag{7.21}$$

where j is the time index and k_s is a scalar that controls the rate of convergence. Equation (7.21) is satisfactory for stationary data with little eigenvalue "spread"; that is, where the ratio of the largest eigenvalue and the smallest eigenvalue of $\mathbf{R}_x$ is small. The convergence of (7.21) is slow, however, in a nonstationary environment and for large eigenvalue spread. Another approach involves solving (7.20) directly and updating the inverse matrix recursively as discussed in Section 7.2.6.

A thorough treatment of the topic of sonar array processing is provided by Owsley in [7]. Included in this treatment is a discussion of conditions under which high-resolution beamforming can be expected to offer appreciable improvement over conventional beamforming.

7.1.3 High-Resolution Spectral Estimation

As discussed in Section 4.4.1, maximum likelihood estimates of the parameters of a sine wave in white Gaussian noise can be obtained by performing a length-N DFT. This implementation assumes that the frequency of the sine wave is equal to one of the N equally spaced basis frequencies of the DFT. Because of the assumption of white noise, no use is made of the actual time-correlation properties of the background noise. The resolution of the length-N DFT is proportional to $1/N$ and can be improved only by increasing the length of the data vector processed. As in the case of beamforming in Section 7.1.1, we would expect to be able to improve the resolution performance of the estimator if we include knowledge of the signal and noise environments.

Comprehensive treatments of the topic of modern spectral estimation are provided in a tutorial paper by Kay and Marple [8] and in their respective texts, [9 and 10].

Found in [8] is an interesting figure that illustrates the results of applying various spectral estimation techniques to the same length-64 input sequence. The true power spectral density is shown in Figure 7.7(a). As indicated by this psd, the input sequence presents a rather challenging test of the ability of an estimation technique to (1) resolve two closely spaced narrowband lines; (2) detect a weak line in the presence of strong lines; and (3) contend with a mixture of narrowband and wideband processes. (The abbreviations that appear in Figure 7.7 are *autoregressive-moving average* (ARMA) and *energy spectral density* (ESD).) As can be seen from Figure 7.7(b), the periodogram technique using the FFT (Section 4.4) is unable to resolve the lines at $0.20\,f_s$ and $0.21\,f_s$. Because specific assumptions are inherent in each technique concerning the spectral structure of the input data, no single technique in Figure 7.7 is able to accurately approximate the true psd.

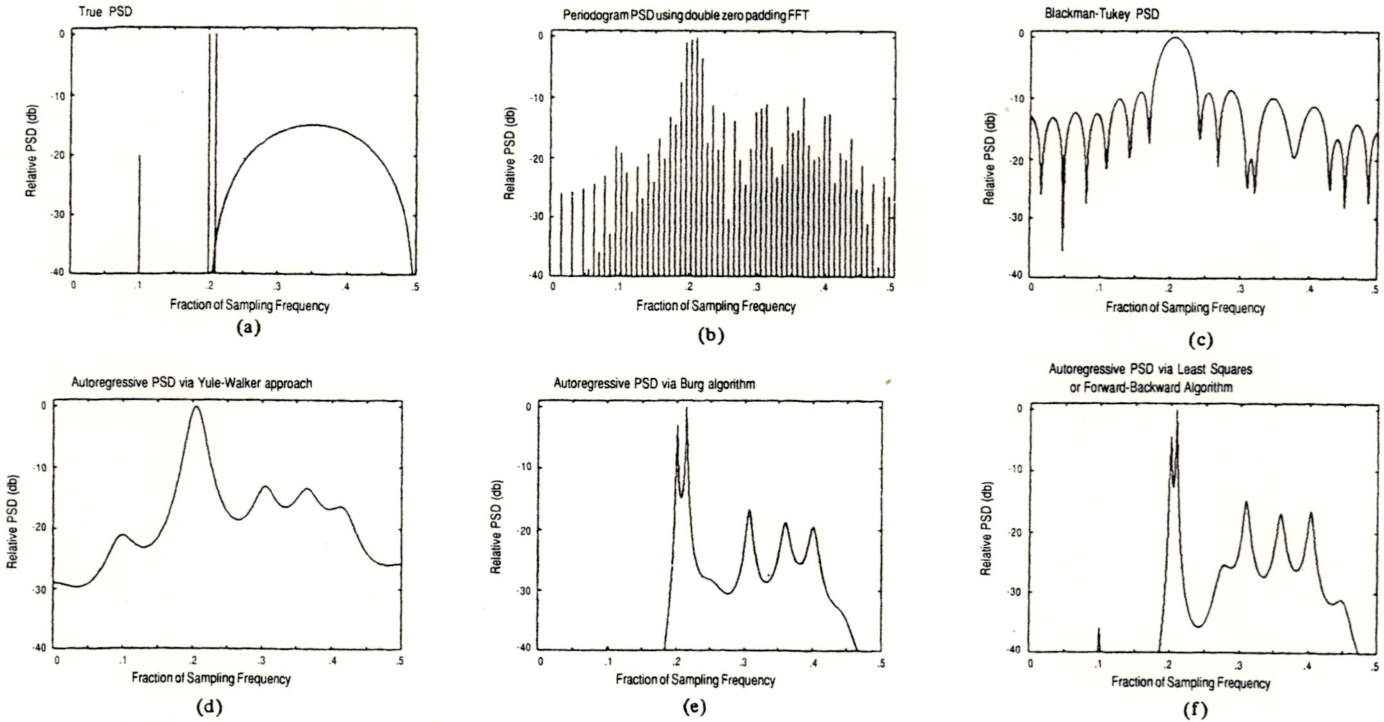

Figure 7.7 The results of applying the same length-64 sample sequence to various spectral estimation techniques.
(*Source:* [8] S.M. Kay and S.L. Marple, Jr., "Spectrum Analysis—A Modern Perspective," *Proc. IEEE*, Vol. 69, No. 11, November 1981, pp. 1380–1419. (© 1981 IEEE. Reprinted with permission.)

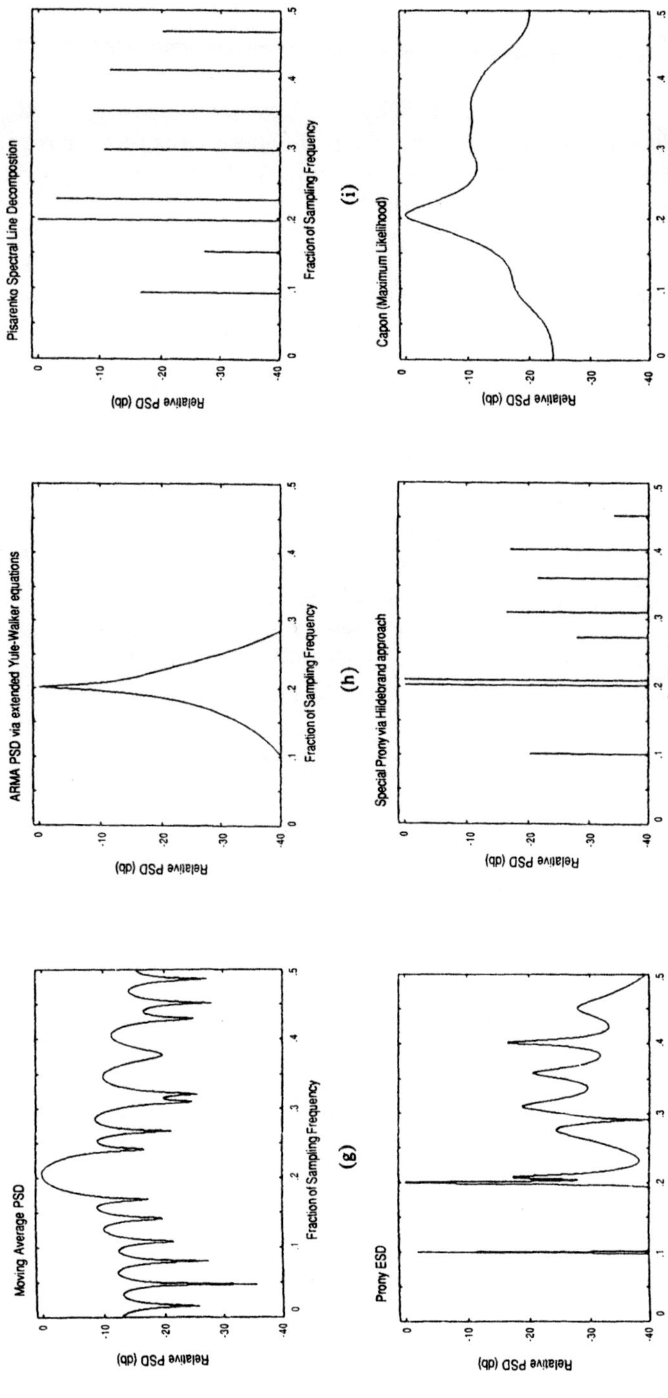

Figure 7.7 (cont'd)

Prony Spectral Estimation Technique

We briefly discuss the Prony technique, which is treated extensively in [8–10]. A length-N complex data sequence, $x(\mathrm{m})$, is assumed. The goal is to approximate this data sequence with a sum of p damped exponentials. Using the notation in [10, Section 11.2], the Prony estimate of $x(m)$ is

$$\hat{x}(m) = \sum_{k=1}^{p} A_k \exp(\mathrm{j}\theta_k) \exp[(\alpha_k + \mathrm{j}2\pi f_k)(m - 1)/f_s] \tag{7.22}$$

where $\hat{x}(m)$ is the estimate provided by the model. We can rewrite (7.22) in the following form:

$$\hat{x}(m) = \sum_{k=1}^{p} h_k z_k^{m-1} \tag{7.23}$$

where $h_k = A_k \exp(\mathrm{j}\theta_k)$ and $z_k = \exp[(\alpha_k + \mathrm{j}2\pi f_k)/f_s]$. Thus the Prony model is considerably more general than that of the DFT, where the signal is assumed to consist of a set of nondamped exponentials equally spaced in frequency.

It is interesting to note that we can fit the data exactly with (7.23) for $N = 2p$; that is, for the case where the sample size is exactly twice the number of damped exponentials in the model. The parameters of the Prony model are found [10, Section 11.4] through the following steps:

1. Solve

$$\sum_{m=0}^{p} x(n - m)\, a(m) = 0$$

for the $a(m)$, where $a(0) = 1$ and $p + 1 \leqslant n \leqslant N$.

2. Find z_k, $k = 1, \ldots, p$, the p complex roots of the polynomial:

$$\sum_{m=0}^{p} a(m)\, z^{p-m}$$

3. Find h_k by solving

$$\sum_{k=1}^{p} z_k^{m-1}\, h_k = x(m)$$

where $1 \leq m \leq p$. The model parameters are recovered by setting

$$A_k = |h_k|$$

$$\theta_k = \tan^{-1}\left[\operatorname{Im}(h_k)/\operatorname{Re}(h_k)\right]$$

$$\alpha_k = f_s \ln|z_k|$$

$$f_k = (f_s/2\pi) \tan^{-1}\left[\operatorname{Im}(z_k)/\operatorname{Re}(z_k)\right]$$

Example 7.1.4

We let $p = 3$ and $N = 6$. Step 1 in the Prony technique then requires that we solve

$$\begin{bmatrix} x(3) & x(2) & x(1) \\ x(4) & x(3) & x(2) \\ x(5) & x(4) & x(3) \end{bmatrix} \begin{bmatrix} a(1) \\ a(2) \\ a(3) \end{bmatrix} = - \begin{bmatrix} x(4) \\ x(5) \\ x(6) \end{bmatrix}$$

for $a(m)$, $m = 1, 2, 3$. Next, we find the roots z_k, $k = 1, 2, 3$ of

$$z^3 + a(1)z^2 + a(2)z^1 + a(3) = 0$$

Finally, the h_k are determined by solving

$$\begin{bmatrix} z_1^0 & z_2^0 & z_3^0 \\ z_1^1 & z_2^1 & z_3^1 \\ z_1^2 & z_2^2 & z_3^2 \end{bmatrix} \begin{bmatrix} h_1 \\ h_2 \\ h_3 \end{bmatrix} = \begin{bmatrix} x(1) \\ x(2) \\ x(3) \end{bmatrix}$$

(End of example.)

In general, the model order and the length of the data do not satisfy $N = 2p$. For the more likely case where $N > 2p$, the least squares Prony technique is applicable. This technique requires the solution of a least squares problem in both steps 1 and 3 of the form:

$$\mathbf{Ay} = \mathbf{b} \tag{7.24}$$

where $\mathbf{A}$ is $m \times n$ and $m > n$. The *normal equations* approach is implemented by premultiplying (7.24) by $\mathbf{A}^+$, yielding

$$\mathbf{A}^+\mathbf{A}\mathbf{y} = \mathbf{A}^+\mathbf{y} \tag{7.25}$$

and inverting $\mathbf{A}^+\mathbf{A}$ (we assume that the rank of $\mathbf{A}$ is equal to the number of columns so that $\mathbf{A}^+\mathbf{A}$ is nonsingular). Consistent with the discussion in Section 7.1.1, we have "squared" the data in (7.25), thus possibly introducing numerical instability. A preferable technique, to be discussed in Section 7.2.2, is QR decomposition, which operates with $\mathbf{A}$ and not $\mathbf{A}^+\mathbf{A}$.

We observe that the ML and the EV methods introduced in Section 7.1.1 also have utility as high-resolution spectral estimators [8–10].

7.1.4 Required Matrix Operations

Based on the brief look at high-resolution techniques provided in the previous sections, we can compile a set of matrix operations that a modern sonar signal processor should be able to execute. Parametric techniques, such as *autoregressive* (AR), autoregressive-moving average, and Prony, in which model parameters are estimated, typically require the least squares solution of an overdetermined set of linear equations. As shown by Marple [10, Chapter 9], the sequential-data autoregressive algorithm can be executed with the LMS technique encountered in Section 7.1.2, or with a recursive least squares method.

Nonparametric techniques, such as the ML (or minimum variance) method and the EV method, require the constrained or unconstrained minimization of a quadratic form. An example of such a quadratic form is the power at the output of the beamformer (7.2), where $P = \mathbf{w}^+\mathbf{R}\mathbf{w}$. The ML method requires the inversion of the Hermitian Toeplitz matrix $\mathbf{R}$. (A matrix is Toeplitz if all the elements of each diagonal are equal.) In addition the EV method requires an eigensystem solution.

7.2 MODERN MATRIX TECHNIQUES

As an introduction to this section, in Table 7.2 we summarize classical and new approaches to solving matrix-based problems. The new approaches typically offer the possibility of improved numerical stability.

For the least squares problem, we wish to solve a set of overdetermined linear equations $\mathbf{A}\mathbf{x} = \mathbf{b}$ so that $\|\mathbf{A}\mathbf{x} - \mathbf{b}\|^2$ is minimized, where $\mathbf{A}$ is $m \times n$, with $m > n$. The notation $\|\cdot\|$ denotes the norm operator, which will be introduced in Section 7.2.1. In the classical approach, the normal equations, $\mathbf{A}^T\mathbf{A}\mathbf{x} = \mathbf{A}^T\mathbf{b}$, are formed and $\mathbf{x}_{LS}$ is found by inverting $\mathbf{A}^T\mathbf{A}$, where $\mathbf{x}_{LS}$ is the least squares solution. (The rank of $\mathbf{A}$ must be equal to n for $\mathbf{A}^T\mathbf{A}$ to be nonsingular.) As mentioned in Section 7.1.1,

Table 7.2
Comparison of Matrix Techniques

Algorithm	*Application*	*Classical Approach*	*Modern Approach*
Least Squares $\mathbf{Ax} = \mathbf{b}$ minimize $\|\mathbf{Ax} - \mathbf{b}\|$	Parametric techniques (AR, ARMA, Prony) Interference cancellation	Normal equations $\mathbf{A}^T\mathbf{Ax} = \mathbf{A}^T\mathbf{b}$	QR Decomposition $\mathbf{QRx} = \mathbf{b}$ SVD $\mathbf{PDQ}^T\mathbf{x} = \mathbf{b}$
Kalman filter	State estimation	Covariance matrix $\mathbf{P}$	Factored matrix $\mathbf{UDU}^T$
Symmetric eigensystem problem	Nonparametric techniques (EV, MUSIC)	$\mathbf{A}^T\mathbf{Ax}_k = \lambda_k\mathbf{x}_k$	SVD QR Jacobi

we are effectively multiplying the data by itself when forming $\mathbf{A}^T\mathbf{A}$. A method that operates with the data matrix directly is QR decomposition, where $\mathbf{A}$ is decomposed into the product of an orthogonal matrix and an upper triangular matrix.

An $n \times n$ matrix, $\mathbf{Q}$, is an orthogonal matrix if the columns of $\mathbf{Q}$ are orthonormal, so that $\mathbf{Q}^T\mathbf{Q} = \mathbf{QQ}^T = \mathbf{I}$. Two columns $\mathbf{v}_i$ and $\mathbf{v}_j$ are orthonormal if their inner product satisfies

$$\mathbf{v}_i^T\mathbf{v}_j = \begin{cases} 1 & \text{for } i = j \\ 0 & \text{for } i \neq j \end{cases}$$

and orthogonal if $\mathbf{v}_i^T\mathbf{v}_i$ is not normalized to unity. If the elements of $\mathbf{Q}$ are complex, $\mathbf{Q}^+\mathbf{Q} = \mathbf{QQ}^+ = \mathbf{I}$, and $\mathbf{Q}$ is called a *unitary matrix*. The elements r_{ij} of an $n \times n$ upper (lower) triangular matrix $\mathbf{R}$ are zero for $i > j$ (for $i < j$).

In the area of state estimation, the classical Kalman filter formulation involves the covariance matrix $\mathbf{P}$, which again is obtained by squaring the data. Techniques with better numerical properties are available, such as the *square root* Kalman filter of Bierman [11] that operates with a decomposed formulation of $\mathbf{P}$. Specifically, $\mathbf{P}$ is written as $\mathbf{P} = \mathbf{UDU}^T$, where $\mathbf{U}$ is an orthogonal matrix and $\mathbf{D}$ is a diagonal matrix.

As discussed in Section 7.1, various high-resolution beamforming and spectral estimation techniques require an eigensystem solution (that is, the calculation of the eigenvalues and eigenvectors of a matrix). The classical approach is to formulate the $M \times M$ correlation matrix $\mathbf{R} = \mathbf{A}^T\mathbf{A}$ and solve the system of equations $\mathbf{A}^T\mathbf{Ax}_k = \lambda_k\mathbf{x}_k$, where $k = 1, \ldots, M$. Instead of forming $\mathbf{R}$, the more numerically stable SVD operates with the data matrix **A** to obtain $\mathbf{PDQ}^T$. The columns of **P** and **Q** are the eigenvectors of $\mathbf{AA}^T$ and $\mathbf{A}^T\mathbf{A}$, respectively, and the elements of the diagonal matrix **D** are the singular values (the positive square roots of the eigenvalues) of $\mathbf{A}^T\mathbf{A}$ (or equivalently, of $\mathbf{AA}^T$).

The majority of the material in Section 7.2 deals with real matrices, although in general the results can be extended to matrices with complex elements.

7.2.1 Numerical Stability

To appreciate the numerical problems that can be encountered with the more classical methods, we first introduce a *condition number* that can be associated with a matrix. We start by considering a length-n vector consisting of real elements. The Euclidean norm of this vector is defined by

$$\|\mathbf{x}\|_2 = (\mathbf{x}^T\mathbf{x})^{1/2} = \left[\sum_{i=1}^{n} x^2(i)\right]^{1/2} \tag{7.26}$$

where the subscript 2 identifies the norm in (7.26) as the Euclidean, or l_2, norm. Because we will deal only with the Euclidean norm, this subscript is omitted in the following development. The norm of **A**, where **A** is an $m \times n$ real matrix with $m \geqslant n$ and rank equal to n, is defined by Golub and Van Loan [12, Section 2.3.1] as

$$\|\mathbf{A}\| = \sup_{\mathbf{x}\neq 0} \frac{\|\mathbf{Ax}\|}{\|\mathbf{x}\|}$$

where the *supremum,* or least upper bound, is designated by sup. The following condition number, $\kappa(\mathbf{A})$, can be associated with **A**:

$$\kappa(\mathbf{A}) = \frac{\mu_{\max}(\mathbf{A})}{\mu_{\min}(\mathbf{A})} \tag{7.27}$$

where $\mu_{\max}(\mathbf{A})$ and $\mu_{\min}(\mathbf{A})$ denote the positive square roots of the maximum and minimum eigenvalues of $\mathbf{A}^T\mathbf{A}$, respectively. (That is, $\mu_{\max}(\mathbf{A})$ and $\mu_{\min}(\mathbf{A})$ are singular values of $\mathbf{A}^T\mathbf{A}$.)

We consider a set of linear equations, $\mathbf{Ax} = \mathbf{b}$, where the least squares solution is denoted by $\mathbf{x}_{\text{LS}}$. It is of interest to determine the effect on $\mathbf{x}_{\text{LS}}$ of small changes in the observation vector, **b**, and the data matrix, **A**. That is we would like to find the constant, C, in the equation

$$\frac{\|\delta \mathbf{x}_{\mathrm{LS}}\|}{\|\mathbf{x}_{\mathrm{LS}}\|} \leq C \max\left\{\frac{\|\delta \mathbf{A}\|}{\|\mathbf{A}\|}, \frac{\|\delta \mathbf{b}\|}{\|\mathbf{b}\|}\right\} \tag{7.28}$$

(The notation $\delta\mathbf{A}$ or $\delta\mathbf{b}$ is used to designate a small change in $\mathbf{A}$ or $\mathbf{b}$, respectively.) Such changes may arise from approximate experimental data or from rounding or truncation errors. Golub and Van Loan [12, Section 5.3] show that C is proportional to

$$\kappa^2(\mathbf{A}) \tag{7.29}$$

when the least squares problem is solved by using the normal equations approach. If $\mathbf{Ax} = \mathbf{b}$ is solved using Householder QR decomposition (Section 7.2.2), C is approximately proportional to

$$\kappa(\mathbf{A}) + \rho_{\mathrm{LS}}\kappa^2(\mathbf{A}) \tag{7.30}$$

where $\rho_{\mathrm{LS}} = \|\mathbf{Ax}_{\mathrm{LS}} - \mathbf{b}\|$ is the residual error of the unperturbed problem. We thus see from (7.28) and (7.30) that, for small residual errors, the sensitivity of $\mathbf{x}_{\mathrm{LS}}$ to changes in $\mathbf{A}$ or $\mathbf{b}$ is bounded by $\kappa(\mathbf{A})$ when Householder QR decomposition is used.

It is apparent that a small condition number is desirable to reduce the sensitivity of the least squares solution to changes in $\mathbf{A}$ or $\mathbf{b}$. The condition number is small if the columns of $\mathbf{A}$ are strongly independent, and it attains a minimum value of unity if the columns of $\mathbf{A}$ are orthonormal (so that $\mathbf{A}^T\mathbf{A} = \mathbf{I}$). Large values of $\kappa(\mathbf{A})$ occur if the columns of $\mathbf{A}$ are nearly dependent. Based on (7.29) and (7.30), we would expect that the solution using QR decomposition will possess better numerical stability than the solution obtained with the normal equations approach, given that the residual error is small.

Example 7.2.1

Consider a least squares problem where

$$\mathbf{A} = \begin{bmatrix} 1 & 1 \\ 0 & \varepsilon \\ \varepsilon & 0 \end{bmatrix}, \quad \mathbf{b} = \begin{bmatrix} 0 \\ -1 \\ 1.5 \end{bmatrix}$$

Using the normal equations approach, we find $\mathbf{x}_{\mathrm{LS}}$ by solving $(\mathbf{A}^T\mathbf{A})^{-1}\mathbf{A}^T\mathbf{b}$. In this example,

$$\mathbf{A}^T\mathbf{A} = \begin{bmatrix} 1 + \varepsilon^2 & 1 \\ 1 & 1 + \varepsilon^2 \end{bmatrix} \tag{7.31}$$

The eigenvalues of $\mathbf{A}^T\mathbf{A}$ are $\lambda_{max} = 2 + \varepsilon^2$ and $\lambda_{min} = \varepsilon^2$. From (7.27), the condition number of $\mathbf{A}$ is $\kappa(\mathbf{A}) = (\sqrt{2 + \varepsilon^2})/\varepsilon \approx \sqrt{2}/\varepsilon$ for $\varepsilon \ll 1$. If the machine precision (that is, the smallest number that can be stored by the computer) is such that $1 + \varepsilon^2$ is represented by 1, $\mathbf{A}^T\mathbf{A}$ in (7.31) is singular, and this least squares problem cannot be solved. We show in Example 7.2.3 that it may be possible to solve this problem using QR decomposition because $\mathbf{A}^T\mathbf{A}$ is not formed. (End of example.)

7.2.2 QR Decomposition

We are interested in solving the set of linear equations

$$\mathbf{A}\mathbf{x} = \mathbf{b} \tag{7.32}$$

by minimizing $\|\mathbf{A}\mathbf{x} - \mathbf{b}\|^2$, where $\mathbf{A}$ is an $m \times n$ real matrix with $m \geq n$ and the rank of $\mathbf{A}$ is equal to n. The matrix $\mathbf{A}$ is decomposed into the following matrices:

$$\mathbf{Q} = [\mathbf{Q}_1\ \mathbf{Q}_2] \quad \text{and} \quad \begin{bmatrix} \mathbf{R} \\ \mathbf{0} \end{bmatrix} \tag{7.33}$$

where the orthogonal matrix $\mathbf{Q}$ is $m \times m$, $\mathbf{Q}_1$ is $m \times n$, and $\mathbf{Q}_2$ is $m \times (m - n)$. $\mathbf{R}$ is an $n \times n$ upper triangular matrix, and $\mathbf{0}$ is the $(m - n) \times n$ null matrix. (In a null matrix, each element is equal to zero.) The decomposition of $\mathbf{A}$ is shown in Figure 7.8.

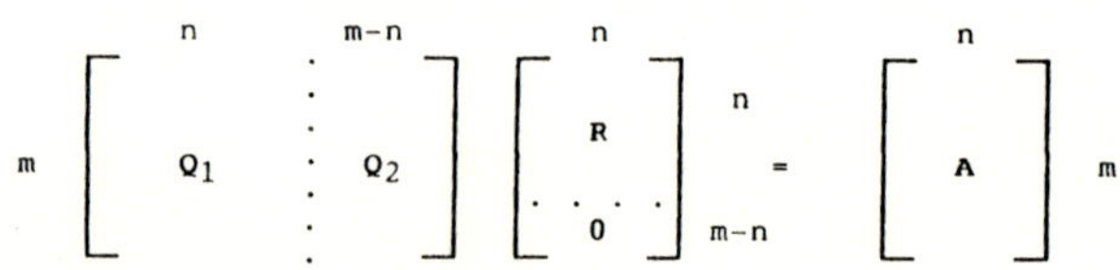

Figure 7.8 QR decomposition of the $m \times n$ matrix $\mathbf{A}$.

Substituting (7.33) into (7.32) yields

$$[\mathbf{Q}_1\ \mathbf{Q}_2] \begin{bmatrix} \mathbf{R} \\ \mathbf{0} \end{bmatrix} \mathbf{x} = \mathbf{b}$$

The least squares solution is found by solving

$$\mathbf{R}\mathbf{x} = \mathbf{Q}_1^T\mathbf{b} \tag{7.34}$$

where the fact that $\mathbf{Q}$ is an orthogonal matrix has been used. Because $\mathbf{R}$ is upper triangular, we can obtain $\mathbf{x}$ in (7.34) by back substitution, which is defined in (7.47).

QR Decomposition by Givens Rotations

We first discuss how we might decompose $\mathbf{A}$ into the factors shown in (7.33) by using Givens rotations. We start with the first two rows of $\mathbf{A}$ and premultiply (7.32) by the rotation matrix $\mathbf{G}_1$, which is shown in Figure 7.9(a). The first two rows of $\mathbf{G}_1\mathbf{A}$ are

$$\begin{bmatrix} c & s \\ -s & c \end{bmatrix} \begin{bmatrix} a_{11} & a_{12} & a_{13} & \cdots \\ a_{21} & a_{22} & a_{23} & \cdots \end{bmatrix}$$

or equivalently

$$\begin{bmatrix} ca_{11} + sa_{21} & ca_{12} + sa_{22} & ca_{13} + sa_{23} & \cdots \\ -sa_{11} + ca_{21} & -sa_{12} + ca_{22} & -sa_{13} + ca_{23} & \cdots \end{bmatrix} \tag{7.35}$$

$$\mathbf{G}_1 = \left[\begin{array}{cc:c} c & s & \\ -s & c & \mathbf{0} \\ \hdashline & & \\ \mathbf{0} & & \mathbf{I}_{m-2} \end{array}\right]$$

(a) First Rotation

$$\mathbf{G}_2 = \left[\begin{array}{ccc:c} c & 0 & s & \\ 0 & 1 & 0 & \mathbf{0} \\ -s & 0 & c & \\ \hdashline & & & \\ & \mathbf{0} & & \mathbf{I}_{m-3} \end{array}\right]$$

(b) Second Rotation

Figure 7.9 First two rotation matrices of the Givens QR decomposition algorithm.

The intent is to transform **A** into $[\mathbf{R}^T \ \mathbf{0}^T]^T$. We perform the first step of this process by selecting c and s so that $-sa_{11} + ca_{21} = 0$ in (7.35). In addition, we want the submatrix

$$\begin{bmatrix} c & s \\ -s & c \end{bmatrix}$$

to be orthogonal, which means that $c^2 + s^2 = 1$. With this pair of constraints, we find that

$$\begin{aligned} c &= \frac{a_{11}}{(a_{11}^2 + a_{21}^2)^{1/2}} \\ s &= \frac{a_{21}}{(a_{11}^2 + a_{21}^2)^{1/2}} \end{aligned} \tag{7.36}$$

The quantities c and s in (7.36) are called *rotation coefficients*. The second pass of the algorithm places a zero in the (3, 1) location of $\mathbf{G}_1\mathbf{A}$ by premultiplying (7.35) by $\mathbf{G}_2$, which is defined in Figure 7.9(b). Here the rotation coefficients are

$$c = \frac{\hat{a}_{11}}{(\hat{a}_{11}^2 + a_{31}^2)^{1/2}}$$

$$s = \frac{a_{31}}{(\hat{a}_{11}^2 + a_{31}^2)^{1/2}}$$

where $\hat{a}_{11}$ is the (1, 1) element of $\mathbf{G}_1\mathbf{A}$. This procedure is continued until **A** has been transformed into $[\mathbf{R}^T \ \mathbf{0}^T]^T$. The product of the matrices $\ldots \mathbf{G}_3\mathbf{G}_2\mathbf{G}_1$ is equal to $\mathbf{Q}^T$. For a 4×3 matrix, six passes are required. That is,

$$\mathbf{Q}^T\mathbf{A} = \begin{bmatrix} X & X & X \\ 1 & X & X \\ 2 & 4 & X \\ 3 & 5 & 6 \end{bmatrix} \tag{7.37}$$

where the first pass places a zero in location 1 in (7.37) by operating on rows 1 and 2, the second pass produces a zero in location 2 by operating on rows 1 and 3, and so on.

Example 7.2.2

We illustrate the Givens-rotations procedure with

$$\mathbf{A} = \begin{bmatrix} 5 & 2 \\ 3 & 4 \\ 1 & 3 \end{bmatrix}, \quad \mathbf{b} = \begin{bmatrix} 1 \\ -1 \\ 3 \end{bmatrix}$$

The three rotation matrices are

$$\begin{aligned} \mathbf{G}_1 &= \begin{bmatrix} 0.8574929 & 0.5144958 & 0.0000000 \\ -0.5144958 & 0.8574929 & 0.0000000 \\ 0.0000000 & 0.0000000 & 1.0000000 \end{bmatrix} \\ \mathbf{G}_2 &= \begin{bmatrix} 0.9856106 & 0.0000000 & 0.1690319 \\ 0.0000000 & 1.0000000 & 0.0000000 \\ -0.1690319 & 0.0000000 & 0.9856106 \end{bmatrix} \\ \mathbf{G}_3 &= \begin{bmatrix} 1.0000000 & 0.0000000 & 0.0000000 \\ 0.0000000 & 0.7192679 & 0.6947328 \\ 0.0000000 & -0.6947328 & 0.7192679 \end{bmatrix} \end{aligned} \tag{7.38}$$

The product, $\mathbf{G}_3\mathbf{G}_2\mathbf{G}_1$, of the three matrices in (7.38) is equal to $\mathbf{Q}^T$, where

$$\mathbf{Q}^T = \begin{bmatrix} 0.8451541 & 0.5070925 & 0.1690319 \\ -0.4707575 & 0.5563489 & 0.6847360 \\ \cdots & \cdots & \cdots \\ 0.2531838 & -0.6582804 & 0.7089181 \end{bmatrix} = \begin{bmatrix} \mathbf{Q}_1^T \\ \cdots \\ \mathbf{Q}_2^T \end{bmatrix} \tag{7.39}$$

The least squares solution is obtained by solving $\mathbf{Rx} = \mathbf{Q}_1^T\mathbf{b}$ by back substitution, where $\mathbf{R} = \mathbf{Q}_1^T\mathbf{A}$. Using (7.39), we find that

$$\begin{bmatrix} 5.9160799 & 4.2257739 \\ 0.0000000 & 3.3380886 \end{bmatrix} \begin{bmatrix} x_1 \\ x_2 \end{bmatrix} = \begin{bmatrix} 0.8451573 \\ 1.0271016 \end{bmatrix} \tag{7.40}$$

The least squares solution (7.40) is $x_1 = -0.0769221$ and $x_2 = 0.3076915$. (End of example.)

Example 7.2.3

Return to Example 7.2.1 but now solve the least squares problem using *QR* decomposition. The exact answer is

$$x_1 = \frac{5 + 3\varepsilon^2}{\varepsilon(4 + 2\varepsilon^2)}, \quad x_2 = -\frac{5 + 2\varepsilon^2}{\varepsilon(4 + 2\varepsilon^2)} \tag{7.41}$$

We assume that the machine precision is sufficient to store $1 + \varepsilon$ but not $1 + \varepsilon^2$. Using the modified Gram-Schmidt *QR* decomposition algorithm in Section 7.4.3 yields the following results:

$$\mathbf{Q}_1 = \begin{bmatrix} 1 & 0 \\ 0 & 1/\sqrt{2} \\ \varepsilon & -1/\sqrt{2} \end{bmatrix} \qquad \mathbf{R} = \begin{bmatrix} 1 & 1 \\ 0 & \sqrt{2}\,\varepsilon \end{bmatrix}$$

$$\mathbf{Q}_1^T\mathbf{b} = \begin{bmatrix} 3\varepsilon/2 \\ -5/(2\sqrt{2}) \end{bmatrix}$$

We solve $\mathbf{Rx} = \mathbf{Q}_1^T\mathbf{b}$ by back substitution and obtain

$$x_1 = \frac{5}{4\varepsilon}, \quad x_2 = -\frac{5}{4\varepsilon} \tag{7.42}$$

Therefore, using *QR* decomposition allows us to solve a nearby least squares problem where the normal equations approach failed because $\mathbf{A}^T\mathbf{A}$ (7.31) is stored as a singular matrix. (A nearby problem is one with data that is slightly perturbed from the original problem.) We note that (7.42) is equal to the exact solution (7.41), but represented with finite precision.

Residual Error

In this section, we demonstrate that the residual error associated with the least squares solution can be obtained via *QR* decomposition. As mentioned earlier, the residual error is equal to

$$\rho_{LS} = \|\mathbf{A}\mathbf{x}_{LS} - \mathbf{b}\| \tag{7.43}$$

Because $\mathbf{Q}$ is an orthogonal matrix, we can rewrite (7.43) as

$$\rho_{LS} = \|\mathbf{Q}^T(\mathbf{A}\mathbf{x}_{LS} - \mathbf{b})\| \tag{7.44}$$

Expression (7.44) can be verified using the definition of the Euclidean norm (7.26):

$$\begin{aligned}\rho_{LS}^2 &= [\mathbf{Q}^T(\mathbf{Ax}_{LS} - \mathbf{b})]^T[\mathbf{Q}^T(\mathbf{Ax}_{LS} - \mathbf{b})] \\ &= (\mathbf{Ax}_{LS} - \mathbf{b})^T\mathbf{QQ}^T(\mathbf{Ax}_{LS} - \mathbf{b}) \\ &= (\mathbf{Ax}_{LS} - \mathbf{b})^T(\mathbf{Ax}_{LS} - \mathbf{b}) \\ &= \|\mathbf{Ax}_{LS} - \mathbf{b}\|^2\end{aligned}$$

By expressing $\mathbf{Q}$ as $[\mathbf{Q}_1\ \mathbf{Q}_2]$, we can rewrite (7.44) in the form

$$\begin{aligned}\rho_{LS}^2 &= \left\| \begin{bmatrix}\mathbf{Q}_1^T \\ \mathbf{Q}_2^T\end{bmatrix} (\mathbf{Ax}_{LS} - \mathbf{b}) \right\|^2 \\ &= \left\| \begin{bmatrix}\mathbf{R} \\ \mathbf{0}\end{bmatrix} \mathbf{x}_{LS} - \begin{bmatrix}\mathbf{Q}_1^T \\ \mathbf{Q}_2^T\end{bmatrix} \mathbf{b} \right\|^2\end{aligned}$$

which is equivalent to

$$\rho_{LS}^2 = \|\mathbf{Rx}_{LS} - \mathbf{Q}_1^T\mathbf{b}\|^2 + \|\mathbf{Q}_2^T\mathbf{b}\|^2 \tag{7.45}$$

We see that the first term in (7.45) is zero because $\mathbf{x}_{LS} = \mathbf{R}^{-1}\mathbf{Q}_1^T\mathbf{b}$ is the solution to the least squares problem. The second term in (7.45) is then the residual error, where

$$\rho_{LS} = \|\mathbf{Q}_2^T\mathbf{b}\| \tag{7.46}$$

Returning to Example 7.2.2, we find that the residual error using (7.46) is

$$[0.2531838 \quad -0.6582804 \quad 0.7089181][1 \quad -1 \quad 3]^T = 3.0382185$$

Back Substitution

The back substitution algorithm is used to evaluate (7.34), namely $\mathbf{Rx} = \mathbf{Q}_1^T\mathbf{b} = \mathbf{c}$, which results from solving the least squares problem using *QR* decomposition. For a set of n equations, the algorithm is

$$\begin{aligned}&x(n) = c(n)/r_{nn} \\ &\text{For } k = n-1, \ldots, 1, \\ &x(k) = \frac{1}{r_{kk}}\left[c(k) - \sum_{i=k+1}^{n} r_{ki}x(i)\right]\end{aligned} \tag{7.47}$$

This algorithm is illustrated in Figure 7.10 for $n = 5$. Back substitution requires on the order of $n^2/2$ multiplications and $n^2/2$ additions.

$$\begin{bmatrix} r_{11} & r_{12} & r_{13} & r_{14} & r_{15} \\ & r_{22} & r_{23} & r_{24} & r_{25} \\ & & r_{33} & r_{34} & r_{35} \\ & 0 & & r_{44} & r_{45} \\ & & & & r_{55} \end{bmatrix} \begin{bmatrix} x(1) \\ x(2) \\ x(3) \\ x(4) \\ x(5) \end{bmatrix} = \begin{bmatrix} c(1) \\ c(2) \\ c(3) \\ c(4) \\ c(5) \end{bmatrix}$$

$$\begin{aligned}
x(5) &= c(5)/r_{55} \\
x(4) &= [c(4) - r_{45}\,x(5)]/r_{44} \\
x(3) &= [c(3) - r_{34}\,x(4) - r_{35}\,x(5)]/r_{33} \\
x(2) &= [c(2) - r_{23}\,x(3) - r_{24}\,x(4) - r_{25}\,x(5)]/r_{22} \\
x(1) &= [c(1) - r_{12}\,x(2) - r_{13}\,x(3) - r_{14}\,x(4) - r_{15}(5)]/r_{11}
\end{aligned}$$

Figure 7.10 Back substitution algorithm for $n = 5$.

Other QR Decomposition Algorithms

The Givens method presented in this section requires the calculation of square roots (7.36). A square root–free Givens method, by Gentleman [13], is provided in Section 7.4.2. Using Gentleman's approach may be preferable because the calculation of the square root frequently requires many instruction cycles. In addition, Sections 7.4.3 and 7.4.4 provide *QR* algorithms using the modified Gram-Schmidt approach and Householder transformations, respectively. We observe that the residual error is generated explicitly by the Givens and the Householder methods. Of the three *QR* decomposition algorithms provided in Section 7.4, only the Gram-Schmidt approach explicitly generates $\mathbf{Q}_1$. A comparison of the numerical properties of these three *QR* decomposition algorithms is beyond the scope of this book, but can be found in [12, Chapter 5].

7.2.3 Singular Value Decomposition

As mentioned earlier (Section 7.2), an $m \times n$ matrix $\mathbf{A}$ can be decomposed into $\mathbf{PDQ}^T$ as shown in Figure 7.11, where (1) the columns of the $m \times n$ $\mathbf{P}$ matrix are orthonormal; (2) $\mathbf{Q}$ is an $n \times n$ orthogonal matrix; and (3) $\mathbf{D}$ is an $n \times n$ diagonal matrix. The columns of $\mathbf{P}$ are the eigenvectors of $\mathbf{AA}^T$ whereas those of $\mathbf{Q}$ are the eigenvectors of $\mathbf{A}^T\mathbf{A}$. The elements of $\mathbf{D}$, called *singular values*, are the positive square roots of the eigenvalues of $\mathbf{A}^T\mathbf{A}$ or, equivalently, of $\mathbf{AA}^T$.

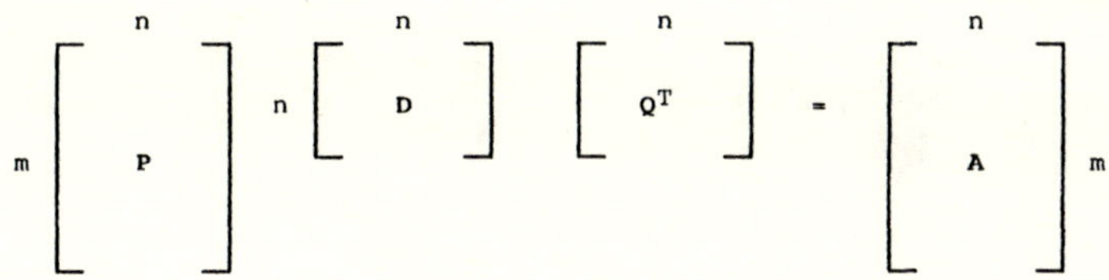

Figure 7.11 Decomposition of **A** into $\mathbf{PDQ}^T$ via the singular value decomposition.

The *SVD* can be calculated using a one-sided orthogonalization algorithm by Nash [14]. We find an orthogonal matrix, **Q**, such that the product **AQ** has orthogonal columns. The kth singular value, d_k, is equal to the Euclidean norm of the kth column of **AQ**. Finally, the kth column of **P** is set equal to the kth column of **AQ**, scaled by $1/d_k$. This algorithm is presented in Section 7.4.5.

One execution of the SVD algorithm, called a *sweep,* consists of $n(n - 1)/2$ rotations on pairs of columns of **A**. Each rotation forces a pair of columns to be orthogonal. Because subsequent rotations can destroy the previously established orthogonality between a pair of columns, a number of such sweeps is required to obtain convergence. (Convergence can be determined by checking the orthogonality of the columns of **AQ**.) Brent and Luk [15] indicate that the number of sweeps is no larger than 10 for $n \leq 1000$. Additional SVD algorithms can be found in [12, Section 8.3].

We encountered the spectral theorem in Section 7.1.1, by which an $n \times n$ symmetric real matrix can be expressed in the form

$$\begin{aligned} \mathbf{A} &= \sum_{i=1}^{n} \lambda_i \mathbf{v}_i \mathbf{v}_i^T \\ &= \mathbf{V}\mathbf{\Lambda}\mathbf{V}^T \end{aligned} \tag{7.48}$$

where **V** is an $n \times n$ orthogonal matrix. The kth column of **V** is the kth eigenvector of **A**, and the elements of the $n \times n$ diagonal matrix $\mathbf{\Lambda}$ are the eigenvalues of **A**. The transformation in (7.48) can be accomplished using Jacobi rotations, *QR* decomposition, or the *SVD* [12, Chapter 8].

7.2.4 LU Decomposition

By using this algorithm, we can solve $\mathbf{Ax} = \mathbf{b}$ by decomposing the nonsingular $n \times n$ matrix **A** into a lower triangular matrix, **L**, and an upper triangular matrix, **U**. The solution to $\mathbf{Ax} = \mathbf{b}$ is found by solving $\mathbf{Lc} = \mathbf{b}$ using forward elimination and then solving $\mathbf{Ux} = \mathbf{c}$ using back substitution. The *LU* decomposition algorithm often is used for inverting **A** by effectively inverting **L** and **U** so the $\mathbf{A}^{-1} = \mathbf{U}^{-1}\mathbf{L}^{-1}$ [16].

A form of LU decomposition that successively solves for a row of $\mathbf{U}$ and then a column of $\mathbf{L}$ is, for $i = 1, \ldots, n$,

$$
\begin{aligned}
u_{ik} &= a_{ik} - \sum_{j=1}^{i-1} l_{ij}u_{jk}, \quad k = i, \ldots, n \\
l_{ki} &= \frac{1}{u_{ii}}\left(a_{ki} - \sum_{j=1}^{i-1} l_{kj}u_{ji}\right), \quad k = i + 1, \ldots, n
\end{aligned}
\tag{7.49}
$$

An example of LU decomposition is provided in Section 7.4.6. As can be seen from (7.49), u_{ii} must be nonzero at the ith step for the LU decomposition algorithm to proceed. If this is not the case, we must exchange the order of the rows (or columns) of $\mathbf{A}$ so that u_{ii} is nonzero. This is sometimes referred to as *pivoting*. A simple example of a nonsingular matrix that requires pivoting to implement LU decomposition is

$$
\mathbf{A} = \begin{bmatrix} 0 & 1 & 2 \\ 2 & 5 & 1 \\ 1 & 4 & 3 \end{bmatrix}
$$

LU decomposition can proceed without pivoting if all ascending principal minors of $\mathbf{A}$ are nonzero. The n ascending principal minors, p_i, are defined by [17]

$$
p_1 = a_{11}, \quad p_2 = \begin{vmatrix} a_{11} & a_{12} \\ a_{21} & a_{22} \end{vmatrix}, \quad p_3 = \begin{vmatrix} a_{11} & a_{12} & a_{13} \\ a_{21} & a_{22} & a_{23} \\ a_{31} & a_{32} & a_{33} \end{vmatrix}, \ldots, \quad p_n = |\mathbf{A}|
$$

where $|\ \ |$ denotes the determinant. Classes of matrices that satisfy this condition include the following:

1. **P** matrix. Each principal minor is positive. This class of matrices is discussed in [18] and has found some applicability in nonlinear circuit theory (see for instance [19, 20]).
2. Strictly column-sum dominant matrix. This class of matrices, described in [21], possesses the property that, for each column, the element on the main diagonal is larger than the sum of the magnitudes of the remaining elements of that column. Specifically,

$$
a_{jj} > \sum_{\substack{i=1 \\ i \neq j}}^{n} |a_{ij}|, \quad j = 1, \ldots, n \tag{7.50}
$$

A strictly row-sum dominant matrix is defined in a manner that is analogous to (7.50). If **A** is symmetric, this class of matrices is referred to as *strictly diagonally dominant*.

3. Positive definite matrix. This class is a special case of the **P** matrix in which **A** is symmetric.

Although the presence of nonzero ascending principal minors guarantees that (7.49) can proceed without the need for pivoting, *LU* decomposition may be ill-conditioned due to small u_{ii}. Therefore, pivoting still may be required.

Cholesky Factorization

If **A** is positive definite, *LU* decomposition generates the product $\mathbf{LL}^T$, or equivalently, $\mathbf{U}^T\mathbf{U}$. This is called *Cholesky factorization,* which was encountered in Section 3.2.1 in the development of the correlation receiver. The algorithm for Cholesky factorization is presented in Section 7.4.7. Because **A** is positive definite, the u_{ii} are positive in (7.49).

7.2.5 Computational Requirements

Table 7.3 provides the numbers of operations (multiplications and additions) required to implement the various matrix operations in Section 7.2. The matrices are assumed to be real. Only operations of order n^3 are included in this table. Results are presented

Table 7.3
Computational Requirements for Various Matrix Algorithms

Algorithm	*Dimensions*	*Computations*
Matrix-Matrix Multiplication	$n \times n$	$2n^3$
Matrix Inverse	$n \times n$	$2n^3$
LU Decomposition	$n \times n$	$2n^3/3$
Cholesky Factorization	$n \times n$	$n^3/3$
QR Decomposition:		
Givens	$m \times n$	$3n^2(m - n/3)$
Givens (square root–free)	$m \times n$	$5n^2(m/2 - n/6)$
Modified Gram-Schmidt	$m \times n$	$2mn^2$
Householder	$m \times n$	$2n^2(m - n/3)$
SVD (Nash [14])		
(one sweep)	$m \times n$	$3n^2(2m + n)$

for an $n \times n$ matrix for matrix-matrix multiplication, matrix inversion, *LU* decomposition, and Cholesky factorization. Computational requirements for *QR* decomposition and the *SVD* (for one sweep) correspond to an $m \times n$ matrix.

Parallel Implementation of Matrix Operations

A large body of recent literature describes parallel implementations of the matrix algorithms listed in Table 7.3. These algorithms exhibit a high degree of regularity, so it is possible to perform portions of the algorithms in parallel using a number of processors. The goal is to maximize the percentage of time that a set of processors is engaged in performing parallel operations and to minimize the time required for transferring data and control. Whereas a single processor requires on the order of n^3 instruction cycles to execute an algorithm in Table 7.3, this can be reduced to n^2 by using n processors, each executing a portion of the algorithm.

Moreno and Lang, in Chapter 8 of this book, treat the parallel implementation of matrix algorithms and present a technique for mapping algorithms onto a set of processors. *LU* decomposition is an example of their technique.

7.2.6 Adding a New Set of Measurements

As mentioned in Section 7.1, the spatial correlation matrices and time correlation matrices that appear in high-resolution techniques in practice are estimated from the data. It is of interest then to augment these matrices with the latest vector of measurements. These measurements often are generated at a fixed rate and may consist of the output of a particular *DFT* bin for each of M sensors (or beams) or N time-domain outputs from a single sensor (or beam). This section discusses techniques that allow the inclusion of such a vector of measurements in (1) the inverse of the correlation matrix; (2) the Cholesky factorization; and (3) the *QR* decomposition. Typically the number of operations required to update the problem is of order n^2. This is in contrast to an operations count of order n^3 associated with recalculating the particular algorithm each time a new vector of measurements becomes available.

Rank-One Update of the Correlation Matrix

We wish to update $\mathbf{R}^{-1}(k-1)$, where $\mathbf{R}(k-1)$ is an $M \times M$ Hermitian matrix generated at time $k-1$, by including a vector of measurements, $\mathbf{x}$, that is obtained at time k. This can be done by using exponential time averaging so that

$$\mathbf{R}(k) = \mu\mathbf{R}(k-1) + (1-\mu)\mathbf{x}(k)\mathbf{x}^{+}(k) \tag{7.51}$$

where μ is a smoothing factor ($0 < \mu < 1$). By applying the matrix identity

$$(\mathbf{A} + \mathbf{BCD})^{-1} = \mathbf{A}^{-1} - \mathbf{A}^{-1}\mathbf{B}(\mathbf{C}^{-1} + \mathbf{DA}^{-1}\mathbf{B})^{-1}\mathbf{DA}^{-1}$$

to (7.51), we obtain the following expression for the inverse of $\mathbf{R}$ at time k:

$$\mathbf{R}^{-1}(k) = \frac{1}{\mu}\,\mathbf{R}^{-1}(k-1) - \frac{1-\mu}{\mu}\left[\frac{\mathbf{R}^{-1}(k-1)\mathbf{x}(k)\mathbf{x}^{+}(k)\mathbf{R}^{-1}(k-1)}{\mu + (1-\mu)\mathbf{x}^{+}(k)\mathbf{R}^{-1}(k-1)\mathbf{x}(k)}\right] \tag{7.52}$$

The updates to $\mathbf{R}(k)$ and $\mathbf{R}^{-1}(k)$ provided in (7.51) and (7.52), respectively, are called *rank-one updates* because the rank of the outer product, $\mathbf{x}(k)\mathbf{x}^{+}(k)$, is equal to one.

Rank-One Update of Cholesky Factorization

Following Owsley [22], we can update $\mathbf{R}$ by operating with the Cholesky factors of $\mathbf{R}$, where at time $k - 1$

$$\mathbf{R}(k-1) = \mathbf{U}^{+}(k-1)\mathbf{U}(k-1)$$

We augment $\mathbf{U}(k-1)$ with the measurement vector obtained at time k and form the $(M+1) \times M$ matrix:

$$\mathbf{B} = \begin{bmatrix} \sqrt{1-\mu}\,\mathbf{x}^{+}(k) \\ \sqrt{\mu}\,\mathbf{U}(k-1) \end{bmatrix} \tag{7.53}$$

A Givens QR decomposition is performed on $\mathbf{B}$ in (7.53) so that

$$\mathbf{Q}^{+}\mathbf{B} = \begin{bmatrix} \hat{\mathbf{U}} \\ \mathbf{0}^{T} \end{bmatrix} \tag{7.54}$$

where $\hat{\mathbf{U}}$ is again upper triangular, and $\mathbf{0}$ is the length-M null vector. We show next that $\hat{\mathbf{U}}$ is the Cholesky factor of $\mathbf{R}(k)$. From (7.54), we have that $\hat{\mathbf{U}}^{+}\hat{\mathbf{U}} = (\mathbf{Q}^{+}\mathbf{B})^{+}(\mathbf{Q}^{+}\mathbf{B}) = \mathbf{B}^{+}(\mathbf{QQ}^{+})\mathbf{B} = \mathbf{B}^{+}\mathbf{B}$. This is equivalent to

$$[\sqrt{1-\mu}\;\;\mathbf{x}(k)\;\;\sqrt{\mu}\,\mathbf{U}^{+}(k-1)]\begin{bmatrix} \sqrt{1-\mu}\,\mathbf{x}^{+}(k) \\ \sqrt{\mu}\,\mathbf{U}(k-1) \end{bmatrix} = (1-\mu)\mathbf{x}(k)\mathbf{x}^{+}(k) + \mu\mathbf{U}^{+}(k-1)\mathbf{U}(k-1)$$

which is the desired rank-one update of $\mathbf{R}$ given in (7.51).

Updating the QR Decomposition

The *QR* decomposition of the $m \times n$ data matrix, **A**, also can be updated with measurements taken at time k. We assume that we have formed

$$\mathbf{A}(k-1) = \mathbf{Q}(k-1)\begin{bmatrix} \mathbf{R}(k-1) \\ \mathbf{0}_1 \end{bmatrix}$$

where as in Section 7.2.2, $\mathbf{Q}(k-1)$ is an orthogonal $m \times m$ matrix, $\mathbf{R}(k-1)$ is an $n \times n$ upper triangular matrix, and $\mathbf{0}_1$ is the $(m-n) \times n$ null matrix. We wish to update $\mathbf{Q}(k-1)$ with $\mathbf{x}(k)$. The augmented data matrix is

$$\mathbf{A}(k) = \begin{bmatrix} \mathbf{x}^+(k) \\ \mathbf{A}(k-1) \end{bmatrix} \tag{7.55}$$

As shown in [12, p. 596], we can premultiply (7.55) with the matrix

$$\begin{bmatrix} 1 & \mathbf{0}_2^T \\ \mathbf{0}_2 & \mathbf{Q}^+(k-1) \end{bmatrix}$$

where $\mathbf{0}_2$ is the length-m null vector. The result is

$$\mathbf{H} = \begin{bmatrix} \mathbf{x}^+(k) \\ \mathbf{Q}^+(k-1)\mathbf{A}(k-1) \end{bmatrix} \tag{7.56}$$

Expression (7.56) is equivalent to

$$\mathbf{H} = \begin{bmatrix} \mathbf{x}^+(k) \\ \mathbf{R}(k-1) \\ \mathbf{0}_1 \end{bmatrix} \tag{7.57}$$

The matrix, **H**, in (7.56) and (7.57) is upper Hessenberg (that is, $h_{ij} = 0$ for $i > j + 1$) and has the structure shown in Figure 7.12. We determine a set of Givens rotation matrices, $\mathbf{G}_n, \ldots, \mathbf{G}_1$, so that

$$\mathbf{G}_n \ldots \mathbf{G}_1\mathbf{H} = \mathbf{G}_n \ldots \mathbf{G}_1 \begin{bmatrix} 1 & \mathbf{0}_2^T \\ \mathbf{0}_2 & \mathbf{Q}^+(k-1) \end{bmatrix} \mathbf{A}(k) = \begin{bmatrix} \mathbf{R}(k) \\ \mathbf{0}_3 \end{bmatrix}$$

where $\mathbf{0}_3$ is the $(m-n+1) \times n$ null matrix. The updated **Q** matrix is then equal to

$$\mathbf{Q}(k) = \begin{bmatrix} 1 & \mathbf{0}_2^T \\ \mathbf{0}_2 & \mathbf{Q}(k-1) \end{bmatrix} \mathbf{G}_1^+ \ldots \mathbf{G}_n^+ \tag{7.58}$$

$$
\begin{bmatrix}
x_1 & x_2 & x_3 & \cdots & & & x_n \\
r_{11} & r_{12} & r_{13} & & & & r_{1n} \\
 & r_{22} & r_{23} & & & & r_{2n} \\
 & & r_{33} & & & & r_{3n} \\
 & & & r_{44} & \cdot & & \\
 & & & & \cdot & \cdot & \\
 & & & & & & r_{nn} \\
0 & \cdot & \cdot & \cdot & \cdot & & 0 \\
 & & & & & & \\
0 & & & & & & 0
\end{bmatrix}
$$

Figure 7.12 Structure of the upper Hessenberg matrix in (7.57).

We note from (7.58) that updating $\mathbf{Q}(k - 1)$ requires n Givens rotations. (These rotations insert zeros along the $(i + 1, i)$ diagonal of $\mathbf{H}$ in Figure 7.12.) This is in contrast to the general *QR* decomposition, which requires on the order of $mn/2$ rotations.

Updating the Symmetric Eigensystem Problem

We close this section by noting that algorithms also are available for updating the eigensystem of a matrix. Specifically, we assume that the eigensystem of $\mathbf{R}(k - 1)$ is available. It is then possible to obtain the eigensystem of $\mathbf{R}(k)$ in (7.51) using an algorithm given by Bunch, Nielsen, and Sorensen [23]. The question of updating the *SVD* is treated in [24].

7.3 HIGH-RESOLUTION BEAMFORMER

A block diagram of a high-resolution beamformer is provided in Figure 7.13. The input consists of a linear array of M equally spaced sensors. The mixing and low-pass filtering functions permit the selection of a segment of the spectrum for further processing. It is desired that a set of N_B beams be formed for each of L frequency bins. The $M \times M$ spatial correlation matrix is calculated for each frequency at time k. These L matrices are the block-diagonal components of the complex power spectral density matrix given in Table 7.4.

One of the algorithms discussed in Section 7.2.6 is used to update the inverses of the $\mathbf{R}_i(k)$, $i = 1, \ldots, L$. These updates are performed at the *FFT* rate; that is, each time a vector of L frequency outputs becomes available from each of the M sensors. A weight vector is generated for each of the N_B beam directions, using (7.5)

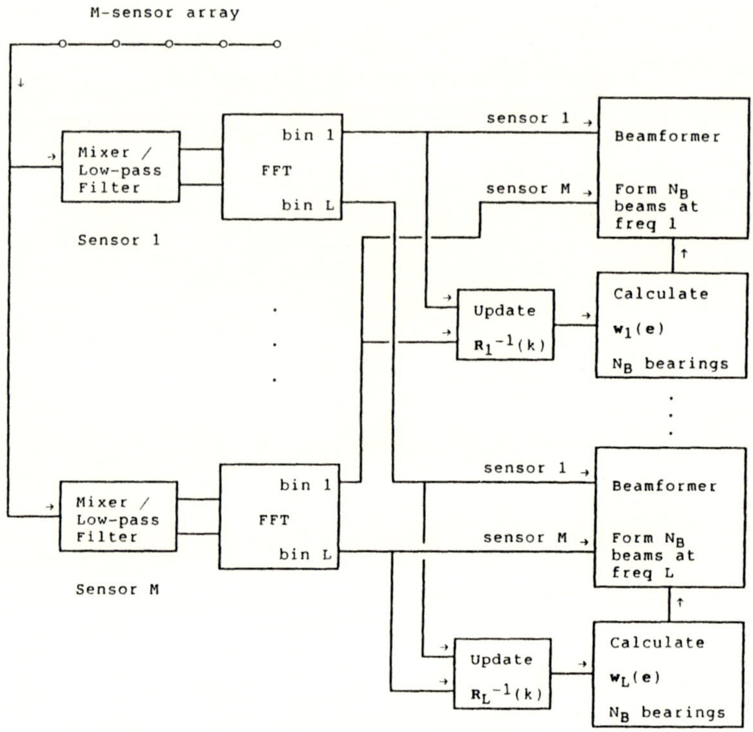

Figure 7.13 Block diagram of a high-resolution beamformer.

Table 7.4
Structure of the Cross-Power Spectral Density Matrix at Time k (each block is $M \times M$)

$$\mathbf{R}(k) = \begin{bmatrix} [\mathbf{R}_1(k)] & & \\ & [\mathbf{R}_2(k)] & \\ & & [\mathbf{R}_L(k)] \end{bmatrix}$$

for instance, if the *ML* beamformer is implemented. That is, the *i*th weight-generation block in Figure 7.13 calculates

$$\mathbf{w}_i(k, \mathbf{e}) = \frac{\mathbf{R}_i^{-1}(k)\mathbf{e}}{\mathbf{e}^+\mathbf{R}_i^{-1}(k)\mathbf{e}}$$

for N_B different steering directions (or equivalently, N_B different vectors $\mathbf{e}$). The beamformer block implements the inner product

$$\mathbf{w}_i^+(k, \mathbf{e})\mathbf{x}_i(k) \tag{7.59}$$

at time k, where $\mathbf{x}_i(k)$ is the length-M vector of *FFT* output for the *i*th frequency. The beamformer calculation (7.59) is performed at the *FFT* rate.

7.4 APPENDICES

Calculations that are required in the high-resolution beamformer section, as well as various matrix algorithms, are provided in the following collection of appendices.

7.4.1 Maximum-Likelihood Beamformer Calculations

In this section, we calculate expressions for (1) the normalized output power; and (2) the beam power pattern for the *ML* beamformer in Section 7.1.1. This is done for a linear array of M equally spaced sensors, where the spacing is equal to $\lambda/2$. The noise field is assumed to be uncorrelated from sensor to sensor.

Normalized Output Power

We derive equation (7.14), the normalized output power of the *ML* beamformer in the direction corresponding to $\mathbf{e}$, where $e(i) = \exp[j\pi(i - 1)\sin\psi_0]$. We assume a single plane-wave signal with conic angle ψ, so that $p(i) = \exp[j\pi(i - 1)\sin\psi]$. The spatial correlation matrix (7.11) can be written in the form

$$\mathbf{R} = \sigma_n^2\left(\mathbf{I} + \frac{\sigma_s^2}{\sigma_n^2}\mathbf{p}\mathbf{p}^+\right) \tag{7.60}$$

The power in the direction corresponding to $\mathbf{e}$ is (7.6)

$$P_{\mathrm{ML}}(\mathbf{e}) = (\mathbf{e}^+\mathbf{R}^{-1}\mathbf{e})^{-1} \tag{7.61}$$

The inverse of **R** in (7.60) can be found by applying the matrix identity

$$(\mathbf{A} + \mathbf{BCD})^{-1} = \mathbf{A}^{-1} - \mathbf{A}^{-1}\mathbf{B}(\mathbf{C}^{-1} + \mathbf{DA}^{-1}\mathbf{B})^{-1}\mathbf{DA}^{-1}$$

where $\mathbf{A} = \mathbf{I}$, $\mathbf{B} = \mathbf{p}$, $\mathbf{C} = \sigma_s^2/\sigma_n^2$, and $\mathbf{D} = \mathbf{p}^+$. The inverse is

$$\begin{aligned} \mathbf{R}^{-1} &= \frac{1}{\sigma_n^2}\left[\mathbf{I} - \mathbf{p}\left(\frac{\sigma_n^2}{\sigma_s^2} + \mathbf{p}^+\mathbf{p}\right)^{-1}\mathbf{p}^+\right] \\ &= \frac{1}{\sigma_n^2}\left(\mathbf{I} - \frac{\sigma_s^2\mathbf{pp}^+}{\sigma_n^2 + M\sigma_s^2}\right) \end{aligned} \tag{7.62}$$

Substituting (7.62) into (7.61) gives

$$P_{\mathrm{ML}}(\mathbf{e}) = \sigma_n^2\left(M - \frac{\sigma_s^2\mathbf{e}^+\mathbf{pp}^+\mathbf{e}}{\sigma_n^2 + M\sigma_s^2}\right)^{-1}$$

After simplifying, we obtain

$$P_{\mathrm{ML}}(\mathbf{e}) = \sigma_n^2\left[\frac{\sigma_n^2 M + \sigma_s^2(M^2 - |\mathbf{e}^+\mathbf{p}|^2)}{\sigma_n^2 + M\sigma_s^2}\right]^{-1}$$

By observing that the power in the direction of the signal is

$$P_{\mathrm{ML}}(\mathbf{p}) = \sigma_n^2\left(\frac{\sigma_n^2 M}{\sigma_n^2 + M\sigma_s^2}\right)^{-1}$$

we find that the normalized output power is

$$P_{\mathrm{ML}}(\mathbf{e}) = \frac{\sigma_n^2 M}{\sigma_n^2 M + \sigma_s^2(M^2 - |\mathbf{e}^+\mathbf{p}|^2)}$$

which appears as (7.14) in Section 7.1.1.

Beam Power Pattern

The optimum weight vector is (7.5)

$$\mathbf{w}_0 = \frac{\mathbf{R}^{-1}\mathbf{e}}{\mathbf{e}^+\mathbf{R}^{-1}\mathbf{e}}$$

After substituting for $\mathbf{R}^{-1}$ from (7.62) and simplifying, we obtain

$$\mathbf{w}_0 = \frac{\mathbf{e}(\sigma_n^2 + M\sigma_s^2) - \mathbf{p}\mathbf{p}^+\mathbf{e}\sigma_s^2}{M(\sigma_n^2 + M\sigma_s^2) - |\mathbf{e}^+\mathbf{p}|^2\sigma_s^2} \tag{7.63}$$

As was done in Section 2.2 for the conventional beamformer, the beam pattern is calculated by assuming that a complex sinusoid, $\exp(j2\pi fn/f_s)$, is propagating through the medium as a plane wave with a conic angle of ψ_x relative to the array. The beam pattern then is determined by obtaining the output of the ML beamformer as ψ_x is varied. For our linear array of M equally spaced sensors, the vector of sensor outputs, $\mathbf{s}$, is defined by

$$s_i(n) = \exp(j2\pi fn/f_s)\exp[j(2\pi f/c)(i-1)d\sin\psi_x] \tag{7.64}$$

for $i = 1, \ldots, M$. Because of the assumption that $d = \lambda/2$, (7.64) is equivalent to

$$s_i(n) = \exp(j2\pi fn/f_s)\exp[j\pi(i-1)\sin\psi_x] \tag{7.65}$$

The output of the ML beamformer is $y(n) = \mathbf{w}_0^+\mathbf{s}$. Using (7.63) and (7.65), $y(n)$ can be expressed in the form

$$y(n) = \exp(j2\pi fn/f_s)\left[\frac{\mathbf{e}^+(\sigma_n^2 + M\sigma_s^2) - \mathbf{e}^+\mathbf{p}\mathbf{p}^+\sigma_s^2}{M(\sigma_n^2 + M\sigma_s^2) - |\mathbf{e}^+\mathbf{p}|^2\sigma_s^2}\right]\mathbf{x} \tag{7.66}$$

where $x_i(n) = \exp[j\pi(i-1)\sin\psi_x]$, $i = 1, \ldots, M$. Consistent with the terminology in Section 2.2, the product of the factor in brackets and $\mathbf{x}$ in (7.66) is $b(\psi_x)$, the complex beam pattern. The beam power pattern then is

$$|b(\psi_x)|^2 = \frac{|(\mathbf{e}^+\mathbf{x})(\sigma_n^2 + M\sigma_s^2) - (\mathbf{e}^+\mathbf{p})(\mathbf{p}^+\mathbf{x})\sigma_s^2|^2}{[M(\sigma_n^2 + M\sigma_s^2) - |\mathbf{e}^+\mathbf{p}|^2\sigma_s^2]^2}$$

which is repeated in Section 7.1.1 as (7.15).

7.4.2 QR Decomposition via Givens Rotations

We consider a set of linear equations, $\mathbf{A}\mathbf{x} = \mathbf{b}$. The $m \times n$ matrix, $\mathbf{A}$, is decomposed into the product of an $m \times n$ matrix, $\mathbf{Q}_1$, with orthonormal columns, and an $n \times n$ upper triangular matrix, $\mathbf{R}$, where $m \geq n$. The rank of $\mathbf{A}$ is equal to n. A square root–free algorithm, by Gentleman [13], is implemented.

Initialize $d_j = 1, j = 1, \ldots, m$. For $i = 1, \ldots, n$, where i is the row index:

$$d_i = d_i(a_{ii})^2$$

For $j = i + 1, \ldots, n + 1$, where $a_{i,n+1} = b_i$:

$$a_{ij} = a_{ij}/a_{ii}$$

For $j = i + 1, \ldots, m$:

$$g_1 = d_i + d_j(a_{ji})^2$$

$$g_2 = d_i d_j/g_1$$

$$c = d_i/g_1$$

$$s = d_j a_{ji}/g_1$$

$$d_i = g_1$$

$$d_j = g_2$$

For $k = i + 1, \ldots, n + 1$ (i.e., include b):

$$\hat{a}_{jk} = a_{jk} - a_{ji}a_{ik}$$

$$\hat{a}_{ik} = ca_{ik} + sa_{jk}$$

$$a_{jk} = \hat{a}_{jk}$$

$$a_{ik} = \hat{a}_{ik}$$

where a_{ij} is the (i, j)th element of $\mathbf{A}$.

The matrix $\mathbf{Q} = [\mathbf{Q}_1\ \mathbf{Q}_2]$ is not generated explicitly. The $\mathbf{R}$ matrix overwrites the $\mathbf{A}$ matrix and $\mathbf{Q}^T\mathbf{b}$ overwrites $\mathbf{b}$.

Example 7.4.1

We repeat the problem defined in Example 7.2.2:

$$\mathbf{A} = \begin{bmatrix} 5 & 2 \\ 3 & 4 \\ 1 & 3 \end{bmatrix}, \quad \mathbf{b} = \begin{bmatrix} 1 \\ -1 \\ 3 \end{bmatrix}$$

Initialize $d_1 = d_2 = d_3 = 1$.

1. Perform the first pass ($i = 1$):

$$d_1 = d_1(a_{11})^2 = 25$$

2. Perform a Givens rotation to insert a zero in location (2, 1):

$$\begin{aligned}
a_{12} &= a_{12}/a_{11} = 0.4, \quad b_1 = b_1/a_{11} = 0.2 \\
g_1 &= d_1 + d_2(a_{21})^2 = 25 + 1 \times 3^2 = 34 \\
g_2 &= d_1 d_2/g_1 = 25/34 = 0.7352941 \\
c &= d_1/g_1 = 25/34 = 0.7352941 \\
s &= d_2 a_{21}/g_1 = 3/34 = 0.0882353 \\
d_1 &= 34, \quad d_2 = 0.7352941 \\
\hat{a}_{22} &= a_{22} - a_{21}a_{12} = 4 - 3 \times 0.4 = 2.8 \\
\hat{a}_{12} &= ca_{12} + sa_{22} = 0.7352941 \times 0.4 + 0.0882353 \times 4 \\
&= 0.6470588 \\
\hat{b}_2 &= b_2 - a_{21}b_1 = -1 - 3 \times 0.2 = -1.6 \\
\hat{b}_1 &= cb_1 + sb_2 = 0.7352941 \times 0.2 + 0.0882353 \times (-1) \\
&= 0.0588235 \\
a_{22} &= 2.8, \quad a_{12} = 0.6470588 \\
b_1 &= 0.0588235, \quad b_2 = -1.6
\end{aligned}$$

3. Perform a Givens rotation to insert a zero in location (3, 1):

$$\begin{aligned}
g_1 &= d_1 + d_3(a_{31})^2 = 34 + 1 \times 1^2 = 35 \\
g_2 &= d_1 d_3/g_1 = 34/35 = 0.9714286 \\
c &= d_1/g_1 = 34/35 = 0.9714286 \\
s &= d_3 a_{31}/g_1 = 1/35 = 0.0285714 \\
d_1 &= 35, \quad d_3 = 0.9714286
\end{aligned}$$

$$\hat{a}_{32} = a_{32} - a_{31}a_{12} = 3 - 1 \times 0.6470588 = 2.3529412$$

$$\hat{a}_{12} = ca_{12} + sa_{32} = 0.9714286 \times 0.6470588 + 0.0285714 \times 3$$
$$= 0.7142856$$

$$\hat{b}_3 = b_3 - a_{31}b_1 = 3 - 1 \times 0.0588235 = 2.9411765$$

$$\hat{b}_1 = cb_1 + sb_3 = 0.9714286 \times 0.0588235 + 0.0285714 \times 3$$
$$= 0.1428570$$

$$a_{32} = 2.3529412, \quad a_{12} = 0.7142856 = r_{12}$$

$$b_1 = 0.1428570, \quad b_3 = 2.9411765$$

4. Perform the second pass ($i = 2$):

$$d_2 = d_2(a_{22})^2 = 0.7352941 \times 2.8^2 = 5.7647057$$

5. Perform a Givens rotation to insert a zero in location (3, 2):

$$b_2 = b_2/a_{22} = -1.6/2.8 = -0.5714286$$

$$g_1 = d_2 + d_3(a_{32})^2 = 5.7647057 + 0.9714286 \times (2.3529412)^2$$
$$= 11.142857$$

$$g_2 = d_2d_3/g_1 = 5.7647057 \times 0.9714286/11.142857$$
$$= 0.5025641$$

$$c = d_2/g_1 = 5.7647057/11.142857 = 0.5173454$$

$$s = d_3a_{32}/g_1 = 0.9714286 \times 2.3529412/11.142857$$
$$= 0.2051282$$

$$d_2 = 11.142857, \quad d_3 = 0.5025641$$

$$\hat{b}_3 = b_3 - a_{32}b_2 = 2.9411765 - 2.3529412 \times (-0.5714286)$$
$$= 4.2857144$$

$$\hat{b}_2 = cb_2 + sb_3 = 0.5173454 \times (-0.5714286) + 0.2051282 \times 2.9411765$$
$$= 0.3076923$$

$$b_2 = 0.3076923, \quad b_3 = 4.2857144$$

The results are

$$\mathbf{R} = \begin{bmatrix} 1 & 0.7142856 \\ 0 & 1 \end{bmatrix}, \quad \mathbf{Q}^T\mathbf{b} = \begin{bmatrix} 0.1428570 \\ 0.3076923 \\ 4.2857144 \end{bmatrix}$$

The residual error is given by $\rho_{LS} = \sqrt{d_3}\, b_3 = 3.0382181$. These results are identical, to six significant figures, to those obtained in Sections 7.4.3 and 7.4.4, after normalizing by the diagonal elements of **R**.

7.4.3 QR Decomposition via Modified Gram-Schmidt

We consider a set of linear equations, $\mathbf{Ax} = \mathbf{b}$. The $m \times n$ matrix, **A**, is decomposed into the product of an $m \times n$ matrix, $\mathbf{Q}_1$, with orthonormal columns, and an $n \times n$ upper triangular matrix, **R**, where $m \geq n$. The rank of **A** is equal to n. The algorithm is as follows. For $i = 1, \ldots, n$, where i is the column index:

$$r_{ii} = \|\mathbf{a}_i\|$$

$$\mathbf{a}_i = \mathbf{a}_i / \|\mathbf{a}_i\|$$

$$c_i = [\mathbf{a}_i, \mathbf{b}]$$

For $k = i + 1, \ldots, n$:

$$r_{ik} = [\mathbf{a}_k, \mathbf{a}_i]$$

$$\mathbf{a}_k = \mathbf{a}_k - [\mathbf{a}_k, \mathbf{a}_i]\, \mathbf{a}_i$$

where $\mathbf{a}_i$ is the ith column of **A**, $\|\mathbf{a}_i\| = [\mathbf{a}_i, \mathbf{a}_i]^{1/2}$ is the norm of the ith column, $[\mathbf{a}_k, \mathbf{a}_i]$ is the dot product of the ith and kth columns, and c_i is the ith element of the vector **c**. On completion, the matrix $\mathbf{Q}_1$ is stored in **A** and the n-length vector **c** contains $\mathbf{Q}_1^T\mathbf{b}$.

Example 7.4.2

We repeat the problem defined in Example 7.2.2:

$$\mathbf{A} = \begin{bmatrix} 5 & 2 \\ 3 & 4 \\ 1 & 3 \end{bmatrix}, \quad \mathbf{b} = \begin{bmatrix} 1 \\ -1 \\ 3 \end{bmatrix}$$

1. Calculate column 1 of **Q** ($i = 1$):

$$r_{11} = \|\mathbf{a}_1\| = (5^2 + 3^2 + 1^2)^{1/2} = 5.9160798$$

$$\mathbf{a}_1 = \begin{bmatrix} 5 \\ 3 \\ 1 \end{bmatrix} 1/5.9160798 = \begin{bmatrix} 0.8451543 \\ 0.5070926 \\ 0.1690309 \end{bmatrix} = \mathbf{q}_1$$

where $\mathbf{q}_i$ denotes the ith column of $\mathbf{Q}_1$.

$$c_1 = [\mathbf{a}_1, \mathbf{b}] = [0.8451543\ 0.5070926\ 0.1690309] \begin{bmatrix} 1 \\ -1 \\ 3 \end{bmatrix}$$

$$= 0.8451544$$

For $k = 2$:

$$r_{12} = [\mathbf{a}_2, \mathbf{a}_1] = [2\ 4\ 3] \begin{bmatrix} 0.8451543 \\ 0.5070926 \\ 0.1690309 \end{bmatrix} = 4.2257717$$

$$\mathbf{a}_2 = \begin{bmatrix} 2 \\ 4 \\ 3 \end{bmatrix} - \begin{bmatrix} 0.8451543 \\ 0.5070926 \\ 0.1690309 \end{bmatrix} 4.2257717 = \begin{bmatrix} -1.5714291 \\ 1.8571424 \\ 2.2857140 \end{bmatrix}$$

2. Calculate column 2 of $\mathbf{Q}_1$ $(i = 2)$:

$$\begin{aligned} r_{22} &= \|\mathbf{a}_2\| \\ &= [(-1.5714291)^2 + 1.8571424^2 + 2.2857140^2]^{1/2} \\ &= 3.3380916 \end{aligned}$$

$$\mathbf{a}_2 = \begin{bmatrix} -1.5714291 \\ 1.8571424 \\ 2.2857140 \end{bmatrix} 1/3.3380916 = \begin{bmatrix} -0.4707567 \\ 0.5563485 \\ 0.6847367 \end{bmatrix} = \mathbf{q}_2$$

$$c_2 = [\mathbf{a}_2, \mathbf{b}] = [-0.4707567\ 0.5563485\ 0.6847367] \begin{bmatrix} 1 \\ -1 \\ 3 \end{bmatrix}$$

$$= 1.0271049$$

The results are

$$\mathbf{Q}_1 = \begin{bmatrix} 0.8451543 & -0.4707567 \\ 0.5070926 & 0.5563485 \\ 0.1690309 & 0.6847367 \end{bmatrix}$$

$$\mathbf{R} = \begin{bmatrix} 5.9160798 & 4.2257717 \\ 0 & 3.3380916 \end{bmatrix}$$

$$\mathbf{c} = \mathbf{Q}_1^T\mathbf{b} = \begin{bmatrix} 0.8451544 \\ 1.0271049 \end{bmatrix}$$

7.4.4 QR Decomposition via Householder Transformations

Consider a set of linear equations, $\mathbf{Ax} = \mathbf{b}$. The $m \times n$ matrix, $\mathbf{A}$, is decomposed into the product of an $m \times n$ matrix, $\mathbf{Q}_1$, with orthogonal columns, and an $n \times n$ upper triangular matrix, $\mathbf{R}$, where $m \geqslant n$. The rank of $\mathbf{A}$ is equal to n. The algorithm is as follows. For $i = 1, \ldots, n$, where i is the column index:

$$r_{ii} = \|\mathbf{a}_i\|$$

$$\hat{\mathbf{a}}_i = [r_{ii}, 0, \ldots, 0]^T, \quad (m - i) \text{ zeros}$$

$$\mathbf{u}_i = (\mathbf{a}_i - \hat{\mathbf{a}}_i)/\|\mathbf{a}_i - \hat{\mathbf{a}}_i\|$$

$$\mathbf{b} = \mathbf{b} - 2[\mathbf{b}, \mathbf{u}_i]\mathbf{u}_i$$

For $k = i + 1, \ldots, n$:

$$\mathbf{a}_k = \mathbf{a}_k - 2[\mathbf{a}_k, \mathbf{u}_i]\mathbf{u}_i$$

$$r_{ik} = i\text{th element of } \mathbf{a}_k$$

where $\mathbf{a}_i$ is the ith column of $\mathbf{A}$, $\|\mathbf{a}_i\| = [\mathbf{a}_i, \mathbf{a}_i]^{1/2}$, the norm of the ith column, and $[\mathbf{a}_k, \mathbf{a}_i]$ is the dot product of the ith and kth columns. The matrix $\mathbf{Q}$ is not explicitly generated and $\mathbf{Q}_1^T\mathbf{b}$ overwrites $\mathbf{b}$. We note that for each value of i, the upper $(i - 1)$ elements of each vector are excluded from the algorithm.

Example 7.4.3

We repeat the problem defined in Example 7.2.2:

$$\mathbf{A} = \begin{bmatrix} 5 & 2 \\ 3 & 4 \\ 1 & 3 \end{bmatrix}, \quad \mathbf{b} = \begin{bmatrix} 1 \\ -1 \\ 3 \end{bmatrix}$$

1. Calculate column 1 of **R** ($i = 1$):

$$r_{11} = \|\mathbf{a}_1\| = (5^2 + 3^2 + 1^2)^{1/2} = 5.9160798$$

$$\hat{\mathbf{a}}_1 = [5.9160798\ 0\ 0]^T$$

$$\|\mathbf{a}_1 - \hat{\mathbf{a}}_1\| = [(5 - 5.9160798)^2 + 3^2 + 1^2]^{1/2}$$

$$= 3.2922943$$

$$\mathbf{u}_1 = \begin{bmatrix} 5 - 5.9160798 \\ 3 \\ 1 \end{bmatrix} 1/3.2922943 = \begin{bmatrix} -0.2782497 \\ 0.9112187 \\ 0.3037396 \end{bmatrix}$$

$$[\mathbf{b}, \mathbf{u}_1] = [1\ -1\ 3] \begin{bmatrix} -0.2782497 \\ 0.9112187 \\ 0.3037396 \end{bmatrix} = -0.2782496$$

$$\mathbf{b} = \mathbf{b} - 2[\mathbf{b}, \mathbf{u}_1]\mathbf{u}_1$$

$$= \begin{bmatrix} 1 \\ -1 \\ 3 \end{bmatrix} - 2\,(-0.2782496) \begin{bmatrix} -0.2782497 \\ 0.9112187 \\ 0.3037396 \end{bmatrix}$$

$$\mathbf{b} = \begin{bmatrix} 0.8451543 \\ -0.4929075 \\ 3.1690308 \end{bmatrix}$$

For $k = 2$:

$$[\mathbf{a}_2, \mathbf{u}_1] = [2\ 4\ 3] \begin{bmatrix} -0.2782497 \\ 0.9112187 \\ 0.3037396 \end{bmatrix} = 3.9995942$$

$$\mathbf{a}_2 = \begin{bmatrix} 2 \\ 4 \\ 3 \end{bmatrix} - 2\,(3.9995942) \begin{bmatrix} -0.2782497 \\ 0.9112187 \\ 0.3037396 \end{bmatrix}$$

$$\mathbf{a}_2 = \begin{bmatrix} 4.2257718 \\ -3.2890101 \\ 0.5703297 \end{bmatrix}$$

$$r_{12} = 4.2257718$$

2. Calculate column 2 of R ($i = 2$), recalling that the first element of each vector is not included in the calculations:

$$r_{22} = \|\mathbf{a}_2\| = [(-3.2890101)^2 + 0.5703297^2]^{1/2}$$
$$= 3.3380928$$

$$\hat{\mathbf{a}}_2 = [3.3380928\ 0]^T$$

$$\|\mathbf{a}_2 - \hat{\mathbf{a}}_2\| = [(-3.2890101 - 3.3380928)^2 + 0.5703297^2]^{1/2}$$
$$= 6.6515989$$

$$\mathbf{u}_2 = \begin{bmatrix} -3.2890101 - 3.3380928 \\ 0.5703297 \end{bmatrix} 1/6.6515989$$

$$\mathbf{u}_2 = \begin{bmatrix} -0.9963173 \\ 0.0857432 \end{bmatrix}$$

$$[\mathbf{b}, \mathbf{u}_2] = [-0.4929075 \quad 3.1690308] \begin{bmatrix} -0.9963173 \\ 0.0857432 \end{bmatrix}$$

$$= 0.7628151$$

The results are

$$\mathbf{R} = \begin{bmatrix} 5.9160798 & 4.2257718 \\ 0 & 3.3380928 \end{bmatrix}, \quad \mathbf{b} = \begin{bmatrix} 0.8451543 \\ 1.0271043 \\ 3.0382184 \end{bmatrix}$$

The third element of **b** is the residual error, ρ_{LS}.

7.4.5 Singular Value Decomposition

We consider a set of linear equations $\mathbf{Ax} = \mathbf{b}$ and obtain the decomposition, $\mathbf{AQ} = \mathbf{PD}$, as per Nash [14]. Here the $n \times n$ matrix, $\mathbf{Q}$, consists of the eigenvectors of

$\mathbf{A}^T\mathbf{A}$, the $m \times n$ matrix, $\mathbf{P}$, has as columns the eigenvectors of $\mathbf{A}\mathbf{A}^T$, and the $n \times n$ diagonal matrix, $\mathbf{D}$, has as its elements the square roots of the eigenvalues of $\mathbf{A}^T\mathbf{A}$ or $\mathbf{A}\mathbf{A}^T$. The algorithm is the following. For $j = 1, \ldots n - 1$, where j is the column index, and for $k = j + 1, \ldots, n$:

$$e = [\mathbf{a}_j, \mathbf{a}_k]$$

$$f = [\mathbf{a}_j, \mathbf{a}_j]$$

$$g = [\mathbf{a}_k, \mathbf{a}_k]$$

$$f = f - g$$

$$h = (4e^2 + f^2)^{1/2}$$

$$c = [(h + f)/(2h)]^{1/2}$$

$$s = e/(ch)$$

Note: if $f < 0$, set $c = 0$, $s = 1$. For $i = 1, \ldots, m$ (apply rotation to $\mathbf{A}$):

$$\hat{a}_{ij} = a_{ij}$$

$$a_{ij} = c\hat{a}_{ij} + sa_{ik}$$

$$a_{ik} = -s\hat{a}_{ij} + ca_{ik}$$

For $i = 1, \ldots, n$ (apply rotation to $\mathbf{Q}$). ($\mathbf{Q}$ is initialized as the $n \times n$ identity matrix.)

$$\hat{q}_{ij} = q_{ij}$$

$$q_{ij} = c\hat{q}_{ij} + sq_{ik}$$

$$q_{ik} = -s\hat{q}_{ij} + cq_{ik}$$

Calculate singular values. For $j = 1, \ldots, n$:

$$d_{jj} = [\mathbf{a}_j, \mathbf{a}_j]^{1/2}$$

For $i = 1, \ldots, m$:

$$p_{ij} = a_{ij}/d_{jj}$$

where a_{ij}, q_{ij}, and p_{ij} denote the (i, j)th element of the matrices **A**, **Q**, and **P**, respectively; $\mathbf{a}_i$ is the ith column of **A**; d_{ii} is the ith element of the diagonal matrix **D**; and $[\mathbf{a}_k, \mathbf{a}_i]$ is the dot product of the ith and kth columns.

Example 7.4.4

We use the same data matrix as in Example 7.2.2 and initialize **Q** as the 2×2 identity matrix.

$$\mathbf{A} = \begin{bmatrix} 5 & 2 \\ 3 & 4 \\ 1 & 3 \end{bmatrix}$$

1. Calculate:

$$e = [\mathbf{a}_1, \mathbf{a}_2] = 25$$

$$f = [\mathbf{a}_1, \mathbf{a}_1] = 35$$

$$g = [\mathbf{a}_2, \mathbf{a}_2] = 29$$

2. Calculate rotation coefficients c and s:

$$f = f - g = 6$$

$$h = (4e^2 + f^2)^{1/2} = 50.358713$$

$$c = [(h + f)/2h]^{1/2} = 0.7480459$$

$$s = e/ch = 0.6636470$$

3. Apply rotation to **A**:

$$a_{11} = 5 \times 0.7480459 + 2 \times 0.6636470 = 5.0675235$$

$$a_{12} = -5 \times 0.6636470 + 2 \times 0.7480459 = -1.8221432$$

$$a_{21} = 3 \times 0.7480459 + 4 \times 0.6636470 = 4.8987257$$

$$a_{22} = -3 \times 0.6636470 + 4 \times 0.7480459 = 1.0012426$$

$$a_{31} = 1 \times 0.7480459 + 3 \times 0.6636470 = 2.7389869$$

$$a_{32} = -1 \times 0.6636470 + 3 \times 0.7480459 = 1.5804907$$

4. Apply rotation to $\mathbf{Q}$:

$$q_{11} = 1 \times 0.7480459 = 0.7480459$$

$$q_{12} = -1 \times 0.6636470 = -0.6636470$$

$$q_{21} = 1 \times 0.6636470 = 0.6636470$$

$$q_{22} = 1 \times 0.7480459 = 0.7480459$$

5. Calculate singular values (elements of $\mathbf{D}$):

$$d_{11} = [\mathbf{a}_1, \mathbf{a}_1]^{1/2} = 7.5617033$$

$$d_{22} = [\mathbf{a}_2, \mathbf{a}_2]^{1/2} = 2.6116362$$

6. Form $\mathbf{P}$:

$$p_{ji} = a_{ji}/d_{ii}; \quad j = 1, 2, 3; \quad i = 1, 2$$

The results are

$$\mathbf{P} = \begin{bmatrix} 0.6701564 & -0.6977018 \\ 0.6478336 & 0.3833775 \\ 0.3622183 & 0.6051726 \end{bmatrix}$$

$$\mathbf{Q} = \begin{bmatrix} 0.7480459 & -0.6636470 \\ 0.6636470 & 0.7480459 \end{bmatrix}$$

$$\mathbf{D} = \begin{bmatrix} 7.5617033 & 0 \\ 0 & 2.6116362 \end{bmatrix}$$

7.4.6 An Example of LU Decomposition

This section provides an example of the Lu decomposition algorithm given in (7.49).

Example 7.4.5

We wish to decompose the matrix

$$\mathbf{A} = \begin{bmatrix} 3 & 1 & 1 \\ -1 & 4 & 2 \\ 1 & 2 & 5 \end{bmatrix}$$

into $\mathbf{L}$ and $\mathbf{U}$ by using (7.49).

For $i = 1$:

$$u_{11} = a_{11} = 3.0000000$$

$$u_{12} = a_{12} = 1.0000000$$

$$u_{13} = a_{13} = 1.0000000$$

$$l_{21} = a_{21}/u_{11} = -0.3333333$$

$$l_{31} = a_{31}/u_{11} = 0.3333333$$

For $i = 2$:

$$u_{22} = a_{22} - \sum_{j=1}^{1} l_{2j}u_{j2} = a_{22} - l_{21}u_{12}$$

$$= 4.3333333$$

$$u_{23} = a_{23} - \sum_{j=1}^{1} l_{2j}u_{j3} = a_{23} - l_{21}u_{13}$$

$$= 2.3333333$$

$$l_{32} = \frac{1}{u_{22}}\left(a_{32} - \sum_{j=1}^{1} l_{3j}u_{j2}\right)$$

$$= (a_{32} - l_{31}u_{12})/u_{22} = 0.3846154$$

For $i = 3$:

$$u_{33} = a_{33} - \sum_{j=1}^{2} l_{3j}u_{j3}$$

$$= a_{33} - l_{31}u_{13} - l_{32}u_{23} = 3.7692308$$

Thus, the result is

$$\mathbf{L} = \begin{bmatrix} 1.0000000 & 0.0000000 & 0.0000000 \\ -0.3333333 & 1.0000000 & 0.0000000 \\ 0.3333333 & 0.3846154 & 1.0000000 \end{bmatrix}$$

$$\mathbf{U} = \begin{bmatrix} 3.0000000 & 1.0000000 & 1.0000000 \\ 0.0000000 & 4.3333333 & 2.3333333 \\ 0.0000000 & 0.0000000 & 3.7692308 \end{bmatrix}$$

7.4.7 Cholesky Factorization

A positive definite $n \times n$ matrix, $\mathbf{A}$, is decomposed into the product of a lower triangular matrix, $\mathbf{L}$, and the transpose of that matrix, $\mathbf{L}^T$. The algorithm is the following. For $i = 1, \ldots, n$:

$$l_{ii} = \left(a_{ii} - \sum_{j=1}^{i-1} l_{ij}^2\right)^{1/2}$$

For $j = i + 1, \ldots, n$:

$$l_{ji} = \frac{1}{l_{ii}}\left(a_{ji} - \sum_{k=1}^{i-1} l_{jk}l_{ik}\right)$$

where a_{ij} and l_{ij} denote the (i, j)th element of $\mathbf{A}$ and $\mathbf{L}$, respectively.

Example 7.4.6

We factor the positive definite matrix

$$\mathbf{A} = \begin{bmatrix} 3 & -1 & 1 \\ -1 & 4 & 2 \\ 1 & 2 & 5 \end{bmatrix}$$

1. Calculate column 1:

$$l_{11} = 1.7320508$$

$$l_{21} = -1/l_{11} = -0.5773503$$

$$l_{31} = 1/l_{11} = 0.5773503$$

2. Calculate column 2:

$$l_{22} = [4 - (-0.5773503)^2]^{1/2} = 1.9148542$$

$$l_{32} = [2 - 0.5773503 \times (-0.5773503)]/1.9148542 = 1.2185436$$

3. Calculate column 3:

$$l_{33} = [5 - (0.5773503)^2 - (1.2185436)^2]^{1/2} = 1.7837652$$

Thus, the result is

$$\mathbf{L} = \begin{bmatrix} 1.7320508 & 0.0000000 & 0.0000000 \\ -0.5773503 & 1.9148542 & 0.0000000 \\ 0.5773503 & 1.2185436 & 1.7837652 \end{bmatrix}$$

REFERENCES

1. Johnson, D.H., "The Application of Spectral Estimation Methods to Bearing Estimation Problems," *Proc. IEEE*, Vol. 70, No. 9, September 1982, pp. 1018–1028.
2. Schmidt, R.O., "Multiple Emitter Location and Signal Parameter Estimation," *IEEE Trans. on Antennas and Propagation*, Vol. AP-34, No. 3, March 1986, pp. 276–280.
3. Wax, M., and T. Kailath, "Detection of Signals by Information Theoretic Criteria," *IEEE Trans. on Acoustics, Speech, and Signal Processing*, Vol. ASSP-33, No. 2, April 1985, pp. 387–392.
4. Williams, D.B., and D.H. Johnson, "Modifying the Sphericity Test for Improved Source Detection with Narrowband Passive Arrays," *Proc. Int. Conf. Acoustics, Speech, and Signal Processing*, Dallas, April 6–9, 1987, pp. 2272–2275.
5. Johnson, D.H. "Properties of Eigenanalysis Methods for Bearing Estimation Algorithms," *Proc. Int. Conf. Acoustics, Speech, and Signal Processing*, Tampa, FL, March 26–29, 1985, pp. 552–555.
6. Monzingo, R.A., and T.W. Miller, *Introduction to Adaptive Arrays*, John Wiley & Sons, New York, 1980.
7. Owsley, N.L., "Sonar Array Processing," in S. Haykin, ed., *Array Signal Processing*, Prentice-Hall, Englewood Cliffs, NJ, 1985.
8. Kay, S.M., and S.L. Marple, Jr., "Spectrum Analysis—A Modern Perspective," *Proc. IEEE*, Vol. 69, No. 11, November 1981, pp. 1380–1419.
9. Kay, S.M., *Modern Spectral Estimation*, Prentice-Hall, Englewood Cliffs, NJ, 1987.
10. Marple, S.L., Jr., *Digital Spectral Analysis with Applications*, Prentice-Hall, Englewood Cliffs, NJ, 1987.
11. Bierman, G.J., *Factorization Methods for Discrete Sequential Estimation*, Academic Press, New York, 1977.
12. Golub, G.H., and C.F. Van Loan, *Matrix Computations*, 2d Ed., Johns Hopkins Press, Baltimore, 1989.
13. Gentleman, W.M., "Least Squares Computations by Givens Transformations without Square Roots," *J. Inst. Mathematics Applications*, Vol. 12, No. 3, December 1973, pp. 329–336.
14. Nash, J.C., "A One-Sided Transformation Method for the Singular Value Decomposition and Algebraic Eigenproblem," *Computer J.*, Vol. 18, No. 1, February 1975, pp. 74–76.
15. Brent, R.P., and F.T. Luk, "The Solution of Singular-Value and Symmetric Eigenvalue Problems on Multiprocessor Arrays," *SIAM J. Sci. Stat. Comput.*, Vol. 6, No. 1, January 1985, pp. 69–84.
16. Dongorra, J.J., J.R. Bunch, C.B. Moler, and G.W. Stewart, *LINPACK User's Guide*, SIAM Publications, Philadelphia, 1979.
17. Balabanian, N., T.A. Bickart, and S. Seshu, *Electrical Network Theory*, John Wiley & Sons, New York, 1969.
18. Fiedler, M., and V. Pták, "Some Generalizations of Positive Definitiveness and Monotonicity," *Numer. Mathematik*, Vol. 9, No. 2, December 1966, pp. 163–172.
19. Sandberg, I.W., and A.N. Willson, Jr., "Some Network-Theoretic Properties of Nonlinear dc

Transistor Networks," *Bell System Technical J.*, Vol. 48, No. 5, May–June 1969, pp. 1293–1311.

20. Nielsen, R.O., and A.N. Willson, Jr., "A Fundamental Result Concerning the Topology of Transistor Circuits with Multiple Equilibria," *Proc. IEEE*, Vol. 68, No. 2, February 1980, pp. 196–208.
21. Varga, R.S., *Matrix Iterative Analysis*, Prentice-Hall, Englewood Cliffs, NJ, 1962.
22. Owsley, N.L., "Systolic Array Adaptive Beamforming," *NUSC Technical Report* No. 7981, Naval Underwater Systems Center, New London, CT, September 21, 1987.
23. Bunch, J.R., C.P. Nielsen, and D.C. Sorensen, "Rank-One Modification of the Symmetric Eigenproblem," *Numer. Mathematik*, Vol. 31, No. 1, 1978, pp. 31–48.
24. Bunch, J.R., and C.P. Nielsen, "Updating the Singular Value Decomposition," *Numer. Mathematik*, Vol. 31, No. 2, 1978, pp. 111–129.

Chapter 8
Mapping Matrix Algorithms onto Processor Arrays*

Jaime H. Moreno
Department of Electrical Engineering
University of Concepción
Concepción, Chile

Tomás Lang
Computer Science Department
University of California at Los Angeles
Los Angeles, California

As discussed in the previous chapter, computation-intensive sonar signal processing algorithms (such as high-resolution beamforming and high-resolution spectral estimation) are based on matrix operations. The real-time implementation of these algorithms is becoming feasible due to the advent of parallel processors and the development of techniques that map the algorithms onto application-specific parallel architectures. This chapter describes a method for mapping matrix algorithms onto processor arrays in which processing elements (cells) have large local memory and communicate only with their nearest neighbors. We refer to such structures as *local-access mesh arrays*. The *mapping technique* is illustrated by using LU decomposition as an example.

8.1 PROCESSOR ARRAYS FOR MATRIX ALGORITHMS

Matrix computations, characterized by having matrix operands or results, are a suitable (albeit computation-intensive) mathematical tool for sonar signal processing and other real-time applications. As stated in Chapter 7, the required matrix operations

*This chapter is based on "Matrix Computations on Systolic-Type Meshes: An Introduction to the Multimesh Graph Method," by J.H. Moreno and T. Lang, which appeared in *Computer*, Vol. 23, No. 4, April 1990, pp. 32–51. [1].

in sonar processing include matrix-vector multiplication, matrix-matrix multiplication, matrix inversion, solution of linear systems of equations, and eigensystems solution. These operations require high computing rates to achieve acceptable execution speed and to meet the time constraints of sonar applications. Thus, parallel algorithms and architectures are needed.

8.1.1 Application-Specific Processor Arrays

Figure 8.1 shows several classes of parallel architectures that have been used for matrix computations. Vector computers, for example, exploit parallelism through vector instructions, extracted from sequential programs by vectorizing compilers. Array computers use a similar type of parallelism. Multiprocessor systems, conversely, exploit parallelism at several levels, including vector operations, concurrent execution of several loop iterations, and block methods.

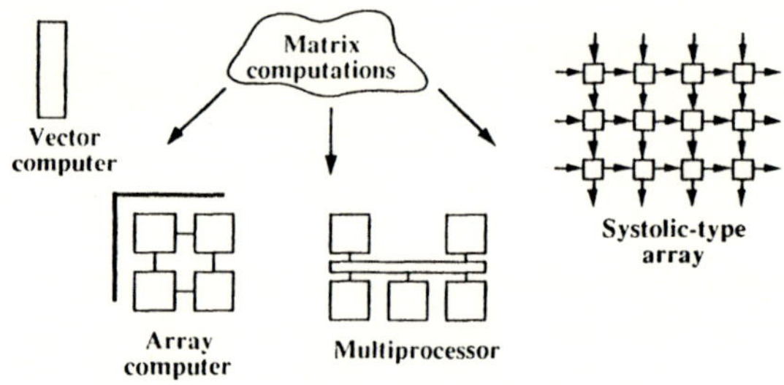

Figure 8.1 Parallel architectures for matrix computations.
Source: [1] J.H. Moreno and T. Lang, "Matrix Computations on Systolic-Type Meshes: An Introduction to the Multimesh Graph Method," *IEEE Computer*, Vol. 23, No. 4, April 1990, pp. 32–51. (© 1990 IEEE. Reprinted with permission.)

Application-specific architectures and, in particular, *systolic-type arrays* [2], have been proposed for matrix computations. These structures can match the fine granularity of parallelism available in the computations and have low overhead in communication and synchronization. Systolic-type arrays (see Figure 8.1) are characterized by synchronous operation of cells, communication only between nearest neighbors, and data flowing in rhythmic fashion through cells. Two classes of systolic-type arrays can be identified: *systolic* and *pseudosystolic*. Systolic cells do not have local memory and require a bandwidth that matches the computing rate; by contrast, pseudosystolic cells have a small-sized local memory, which is used to reduce the bandwidth requirements [1].

In addition to systolic and pseudosystolic cells, another option consists of having cells with large local memory (proportional to the size of the problems to be solved in the array). These cells are suited for an asynchronous (nonsystolic) mode of operation, because a large portion of the data for a problem is stored and used in each cell, with little need for communicating with other cells in the array. Moreover, the array operates in a loosely coupled mode with input data preloaded into the cells' large memories, synchronization achieved through messages or queues, and the cells' bandwidth much lower than the computation rate. We refer to this type of architecture as *local-access arrays* [1, 3], and it shares with systolic-type arrays the characteristic of communicating only with nearest neighbors.

Depending on their range of applicability, there are two kinds of application-specific processor arrays: *algorithm specific* and *class specific*. Algorithm-specific arrays can execute only one particular algorithm. As Figure 8.2(a) illustrates, these structures may be the most appropriate solution for some applications because they offer the possibility of matching an array to a given algorithm and fulfilling specific implementation requirements (such as speed, size, and power consumption). If the application consists of the matrix operation and other computations, the array should be combined with other modules to perform the complete task; thereby composing a heterogeneous system. In addition, such an array is usually connected to a host that performs input-output and control functions.

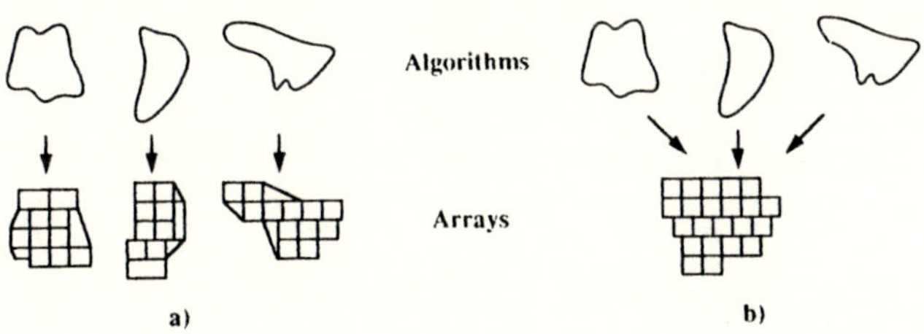

Figure 8.2 Types of application-specific arrays for matrix computations: (a) algorithm specific; (b) class specific.
Source: [1] J.H. Moreno and T. Lang, "Matrix Computations on Systolic-Type Meshes: An Introduction to the Multimesh Graph Method," *IEEE Computer*, Vol. 23, No. 4, April 1990, pp. 32–51. (© 1990 IEEE. Reprinted with permission.)

By contrast with algorithm-specific implementations, in some situations, the set of matrix computations is varied or perhaps not yet defined. In such cases, a better solution is a class-specific array; that is, an array that can be adapted (programmed) for a class of matrix algorithms, as Figure 8.2(b) illustrates. Programming a class-specific array consists of determining:

- the assignment of data and operations to cells;
- the schedule of operations and data transfers;
- the I/O schedule.

There is a large body of literature about algorithm-specific systolic arrays [4–8], and many methods have been proposed for the realization of those structures [9]. As examples, Figure 8.3 depicts algorithm-specific systolic arrays for matrix multiplication [10], LU decomposition [5], and the Faddeev algorithm [11]. The use of systolic and pseudosystolic cells appears in [1, 3], which also gives a graph-based method to design arrays.

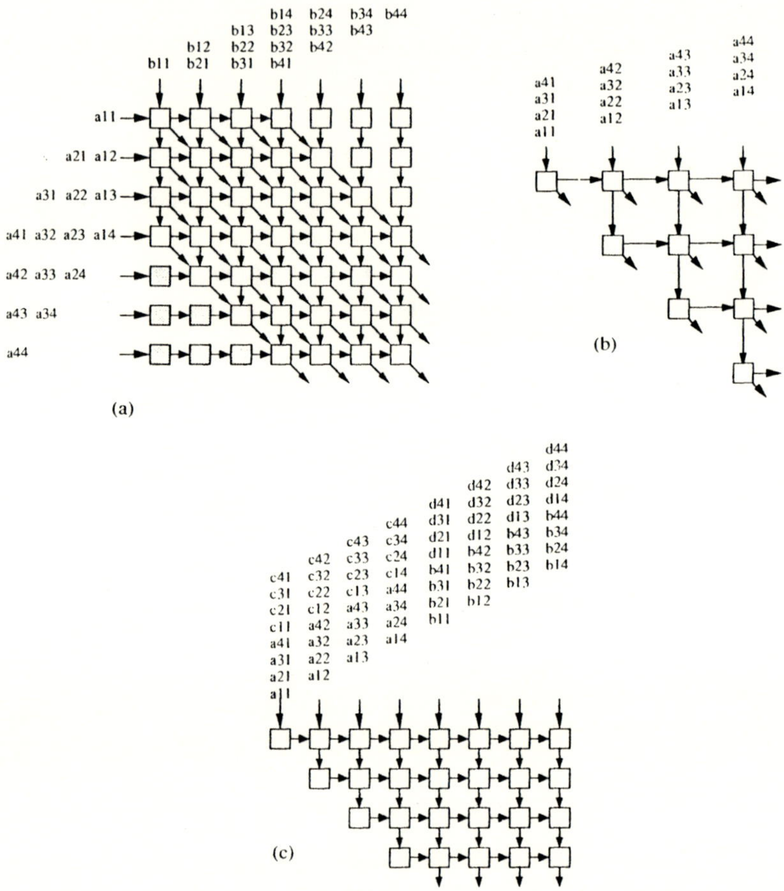

Figure 8.3 Examples of algorithm-specific systolic arrays: (a) matrix multiplication; (b) LU decomposition; (c) Faddeev algorithm.

The remainder of this chapter will focus on mapping matrix algorithms with large data onto class-specific local-access mesh arrays. These arrays are hardware networks of *processing elements* (PEs) with the following basic characteristics:

- linear or two-dimensional structures (a linear array is a mesh of dimension K by 1);
- large local memory (proportional to the size of the problems) and a *register file* (rf) (two outputs, one input) per cell;
- external I/O from host only at the boundaries of the array;
- unidirectional orthogonal links between cells; that is, data flow from cell to cell in one direction only, without data counterflow;
- nearest-neighbor communication only; that is, no capability for broadcasting or routing data through cells without using the data.

Arrays with these characteristics are of interest because they can be implemented by using readily available (off-the-shelf) components (existing processors, memories, and VLSI devices). We only consider arrays with unidirectional links because, as shown in [3], matrix algorithms such as those appearing in sonar processing do not need bidirectional communication.

8.2 THE MULTIMESH GRAPH METHOD

We now present a technique by which matrix algorithms are mapped onto local-access arrays of PEs following a transformational paradigm. That is, the description of an algorithm is successively transformed into a form suitable for decomposition and execution in the cells of an array. This technique is known as the *multimesh dependency graph* (MMG) method.

8.2.1 Stages in a Mapping Technique

We identify the following two stages in the application of a transformational mapping technique: algorithm regularization and mapping onto arrays (see Figure 8.4). Regularization is the derivation of a representation of an algorithm from an initial admissible form. This regularized form must provide a description of the parallelism in the algorithm, in a manner suitable for mapping onto arrays. Moreover, the regularized representation also must be in a form suitable for manipulation in the re-

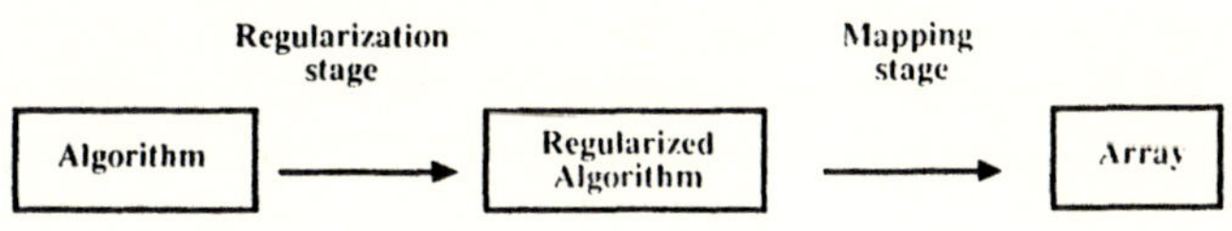

Figure 8.4 The stages in a mapping method.
Source: [1] J.H. Moreno and T. Lang, "Matrix Computations on Systolic-Type Meshes: An Introduction to the Multimesh Graph Method," *IEEE Computer*, Vol. 23, No. 4, April 1990, pp. 32–51. (© 1990 IEEE. Reprinted with permission.)

maining steps of a method. The second stage, mapping onto arrays, determines the allocation of data and operations to cells, the flow of data, the I/O, and the control. For these purposes, the mapping stage uses the regularized description, the topology and structure of the target array, and the characteristics of the cells.

The features of a design method can be stated in terms of the framework described above. In the regularization stage, the important factors are

- the class of algorithms to which the method can be applied; that is, the generality of the initial admissible form of the algorithms;
- the completeness and simplicity of the transformations used to produce a regularized description;
- the effectiveness of the regularized description in conveying the properties of an algorithm in a form suitable for mapping onto arrays.

In the mapping stage the factors correspond to the capabilities and simplicity of the transformations used to map the regularized form onto an array. In particular, they include the capabilities of the transformations:

- to incorporate constraints and restrictions of the target cells, such as limited size of the rf and limited bandwidth of the memory;
- to incorporate different attributes of the functional units, such as pipelining, nonconventional arithmetic, and specialized functionality;
- to determine load balancing, synchronization, and overhead;
- to optimize specific performance measures.

An additional factor, applicable to both stages, is suitability for automation.

8.2.2 Outline of the Multimesh Graph Method

The MMG method is based on transformations on data-dependency graphs. Figure 8.5 shows the basic steps in the two stages. The regularization stage produces the MMG, a three-dimensional graph that incorporates the basic characteristics of a mesh array; namely, unidirectional nearest-neighbor mesh connections and external I/O only at the boundaries. Moreover, the MMG retains all other characteristics of the algorithm, so that the second stage can incorporate requirements and restrictions while mapping the MMG onto an array. Both stages consist of a sequence of simple transformations. As shown in the figure, the method can be used to design algorithm-specific arrays for fixed-size data as well as for partitioned problems [12] and to map algorithms onto class-specific arrays. This chapter concentrates on the latter capabilities; the application of the method to algorithm-specific and fixed-size problems has been described in [1, 3].

We address the case where the matrix size is much larger (its size may not even be predefined) than the array, so that the computation must be partitioned into subproblems. The partitioning approach suitable for local-access cells decomposes

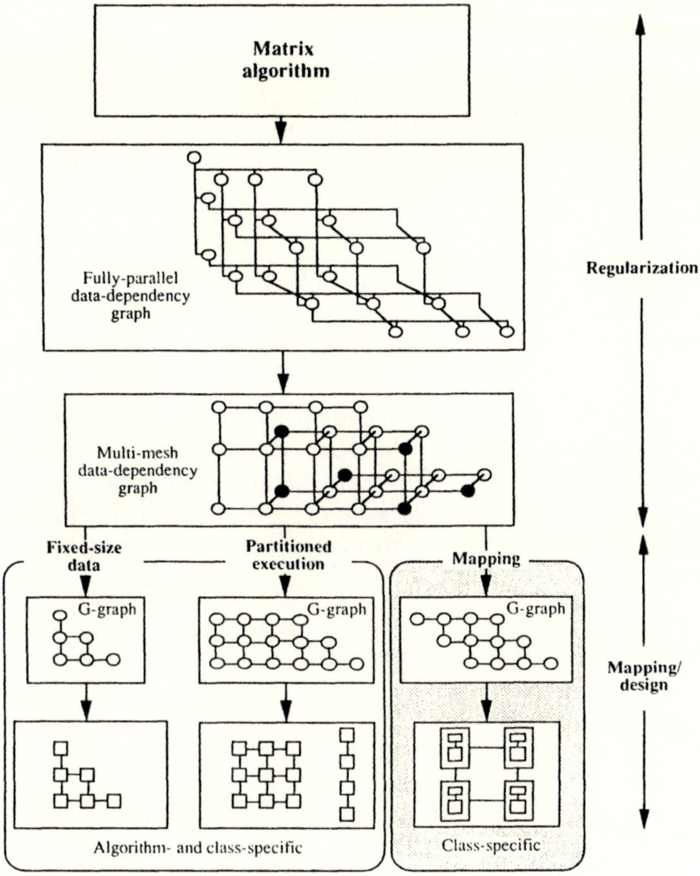

Figure 8.5 The MMG design method.

the entire algorithm into a number of subproblems equal to the number of cells in the target array, and it maps each subproblem onto one cell. As a result, the dependencies among the subproblems should match the interconnection structure of the array, and the large local memory is used to store all the data of the corresponding subproblem. Figure 8.6 shows this technique. The scheme requires a careful selection of subproblems to achieve good load balancing. This type of partitioning is known as *coalescing* [12]; it also has been called *locally sequential globally parallel partitioning* [5, p. 376].

In what follows, we discuss different approaches to coalescing, the allocation of data to cells and memory, and the schedule of primitive operations. We show that the explicit representation of parallelism in the algorithms possible with the MMG

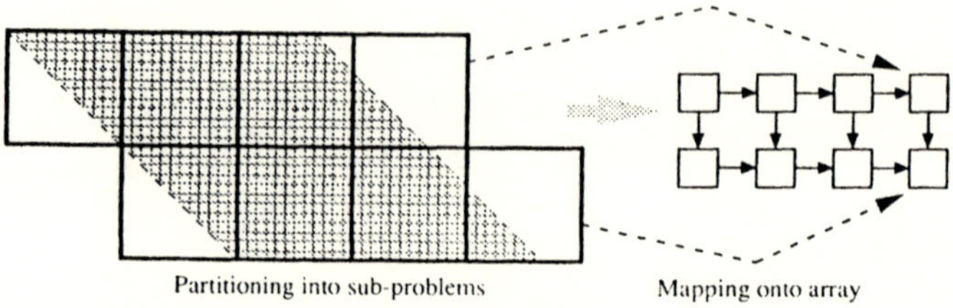

Figure 8.6 Partitioning by coalescing for local-access arrays.
Source: [1] J.H. Moreno and T. Lang, "Matrix Computations on Systolic-Type Meshes: An Introduction to the Multimesh Graph Method," *IEEE Computer*, Vol. 23, No. 4, April 1990, pp. 32–51. (© 1990 IEEE. Reprinted with permission.)

allows the analysis of varied issues including load balancing, synchronization, and overhead in computations and data transfers. Moreover, the regularized representation permits deriving a simple, efficient schedule of operations and data transfer, as well as the control of flow in the execution. These aspects are illustrated by mapping the LU decomposition algorithm onto a linear array and a two-dimensional structure.

8.2.3 Class of Admissible Algorithms

A matrix algorithm in the class suitable for the MMG method is described recursively by an outermost loop and a loop body that contains scalar, vector, and matrix operators, as well as other matrix algorithms (see Figure 8.7). A sequence of these algorithms also is a matrix algorithm. Analysis of matrix computations has shown

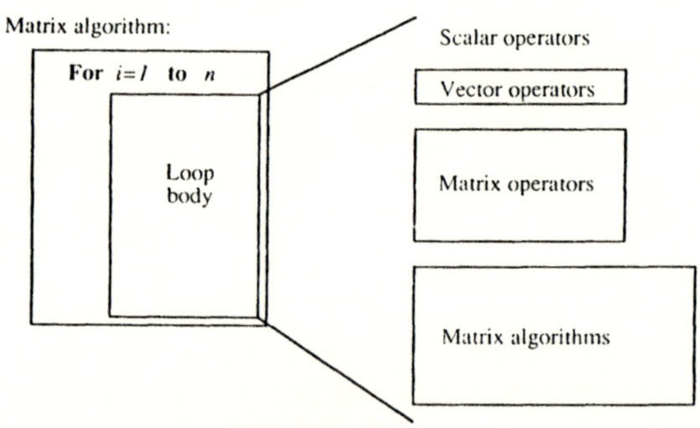

Figure 8.7 Canonical form of a matrix algorithm.
Source: [1] J.H. Moreno and T. Lang, "Matrix Computations on Systolic-Type Meshes: An Introduction to the Multimesh Graph Method," *IEEE Computer*, Vol. 23, No. 4, April 1990, pp. 32–51. (© 1990 IEEE. Reprinted with permission.)

them to consist of primitive operations with up to three operands and up to two results [3]. Consequently, the operators in a matrix algorithm have the following characteristics:

1. Scalar, or primitive, operators (such as add, multiply, rotation, and sine) are basic unary, binary or ternary operations that can produce up to two outputs and whose computation time is independent of the data. In practice, scalar operators produce a single result, except in cases such as rotation of a pair of elements, which produce two outputs.
2. Vector operators have up to two vector operands and produce up to two vector results. The same primitive operator is applied to each element of the vector operands to produce the vector results. An additional scalar operand is common (broadcast) to all the instances of the primitive operator. Figure 8.8(a) shows the dependency graph of a vector operator.
3. Matrix operators have up to one matrix operand, one vector operand common to rows of the matrix operator, and a second vector operand common to columns of the matrix operator. A matrix operator produces a matrix result by applying the same primitive operator to each element of the matrix operand (and associated elements from the vector operands). Figure 8.8(b) illustrates the dependency graph of a matrix operator.

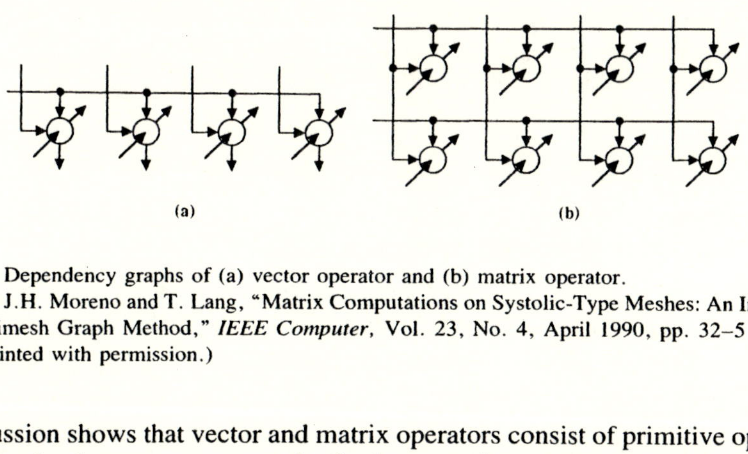

Figure 8.8 Dependency graphs of (a) vector operator and (b) matrix operator.
Source: [1] J.H. Moreno and T. Lang, "Matrix Computations on Systolic-Type Meshes: An Introduction to the Multimesh Graph Method," *IEEE Computer*, Vol. 23, No. 4, April 1990, pp. 32–51. (© 1990 IEEE. Reprinted with permission.)

This discussion shows that vector and matrix operators consist of primitive operations tied together by the common operands. Such operands correspond to the broadcasting of data throughout the elements of the vector or matrix. To implement these operators in mesh arrays we must eliminate the broadcast data, which we do by applying a transformation (replacing data broadcasting by transmitted data).

Because primitive operators can be unary, binary, or ternary, matrix and vector operators are not required to have all the operands indicated above. For example, the addition of a constant to each element of a vector does not require a second vector operand. However, the properties of matrix and vector operators rule out performing operations with two matrices or three vectors (that is, adding two matrices or rotating elements of two vectors by corresponding angles contained in a third vector). For our purposes, such cases correspond to sets of scalar operations.

The limitations on the input and output to and from the operators in a matrix algorithm, as described, arise from the objective of representing the regularized algorithms as MMGs. As a result, an operator cannot have more than three operands. Moreover, because cells are capable only of nearest-neighbor data transfers and the array has external I/O only at the boundaries, we must transfer broadcasted data through the cells (as transmitted data); such data appear as output from nodes in the MMGs. Consequently, the maximum number of input operands is three, and the combined number of computed results and transmitted data from a node also is limited to three.

The canonical form of matrix algorithms shown in Figure 8.7 and described in the previous paragraphs is quite general; for instance, it is directly applicable to algorithms such as matrix decompositions (LU, Cholesky, and QR), transitive closure, Faddeev algorithm, and $\mathbf{BA}^{-1}$. Other representations also are allowed with the MMG method, as long as they have the type of operators listed earlier and are amenable to the transformational process. This canonical form excludes branches or loops with a data-dependent range. Such properties are not present in matrix algorithms of interest, in particular those algorithms suitable for parallel implementation, so this limitation is minor. The only usual case of a data-dependent loop arises when testing for some termination (convergence) condition. However, that case can be handled by treating the algorithm as a loop with a sufficiently large range.

8.2.4 The Regularization Stage

As indicated earlier, the regularization stage produces the three-dimensional graph, called an MMG, with the following characteristics:

- unidirectional and nearest-neighbor dependency;
- edges parallel to axes of the three-dimensional space;
- nodes at integer values of the axes.

As will be shown later, this regularized form is suitable for mapping onto arrays because it preserves all the information present in the algorithm. In addition, it makes it easy to incorporate the characteristics of an array and the cells, to allocate data and operations, and to determine the schedule of operations, I/O, and the data transfer.

Some researchers have regarded the dependency graph of a matrix algorithm as multidimensional instead of three-dimensional. Such a conclusion was reached by representing in a graph the index dependencies of the algorithm. In other words, the dimensionality of the graph was defined by the number of indices appearing in an algorithm. In those approaches, every variable must have all indices, so that each instance of a variable is associated with a point in the multidimensional index space [13–15]. In contrast, the dimensionality of the data-dependency graph in the MMG method is determined by the maximum of three inputs and three outputs per primitive node. A comparison between data dependency and index dependency is available [3, Chapter 7].

The regularization stage in the MMG method is performed in two steps. First, obtain the *fully parallel* (data-dependency) *graph* (FPG) of the algorithm. The graph is obtained by tracing the execution of the algorithm (the outer loop and the loop body). This means that the algorithm is executed symbolically, tracking which variables are used and when; then, operations are allocated to nodes of a graph and data references to its edges. In other words, the FPG corresponds to an unfolded data-flow graph. Note that the FPG of a small problem (for example, $n = 4, 5$) is sufficient to capture the features of an algorithm, because the FPG consists of several subgraphs with the same dependency structure but perhaps different sizes.

As an example, Figure 8.9 depicts the LU decomposition algorithm. The outer loop body of this algorithm consists of two vector operators, with a single output each, followed by a matrix operator that uses the output from both vector operators. The symbolic execution of this algorithm also is depicted in Figure 8.9, which leads to the FPG shown in Figure 8.10. As indicated above, iterations of the outermost loop in the algorithm lead to subgraphs with the same structure, as depicted in the figure.

In the second step, transform the FPG into a three-dimensional MMG, like those in Figure 8.11. To do this, transform the FPG to remove properties not allowed in MMGs. The complete set of such properties consists of only

- data broadcasting, which is replaced by transmitted data;
- bidirectional data flow, which is eliminated by moving dependent operations to one side of the data source;
- nonregular dependencies, which are regularized by adding delay nodes in the nonregular part.

```
For i3 = 1 to n do
  u[i3, i3] = 1/a[i3, i3]
  For i2 = (i3 + 1) to n do
    u[i3, i2] = a[i3, i2]
  For i1 = (i3 + 1) to n to
    l[i1, i3] = a[i1, i3] * u[i3, i3]
  For i1 = (i3 + 1) to n to
    For i2 = (i3 + 1) to n to
      a[i1, i2] = a[i1, i2] - l[i1, i3] * u[i3, i2]
```

$$
\begin{aligned}
u[1,1] &\leftarrow 1/a[1,1] \\
u[1,2] &\leftarrow a[1,2] \\
u[1,3] &\leftarrow a[1,3] \\
&\vdots \\
l[2,1] &\leftarrow a[2,1] * u[1,1] \\
l[3,1] &\leftarrow a[3,1] * u[1,1] \\
&\vdots \\
a[2,2] &\leftarrow a[2,2] - l[2,1] * u[1,2] \\
a[2,3] &\leftarrow a[2,3] - l[2,1] * u[1,3] \\
&\vdots
\end{aligned}
$$

Figure 8.9 Symbolic execution of the LU decomposition algorithm.

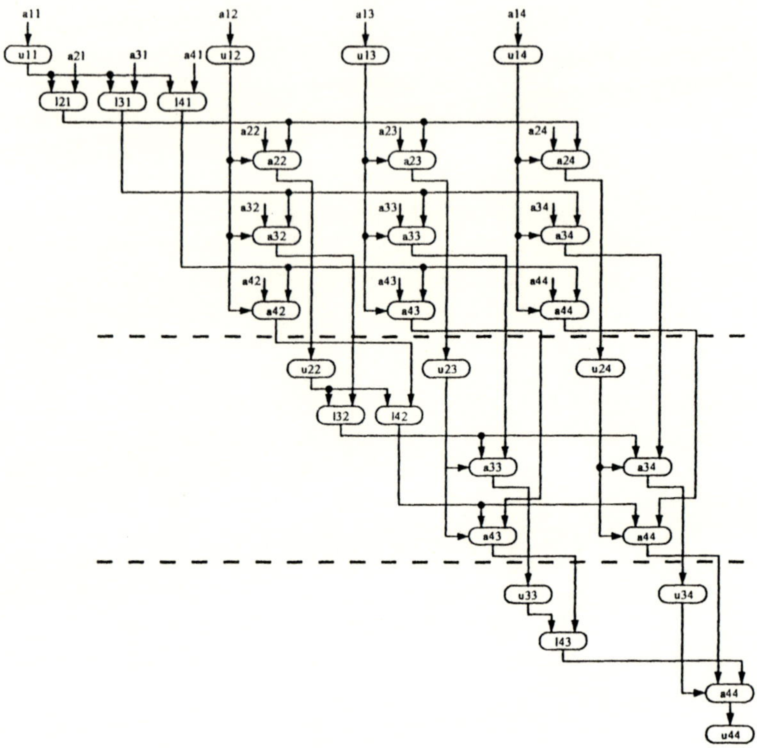

Figure 8.10 Fully parallel graph of the LU decomposition for $n = 4$.

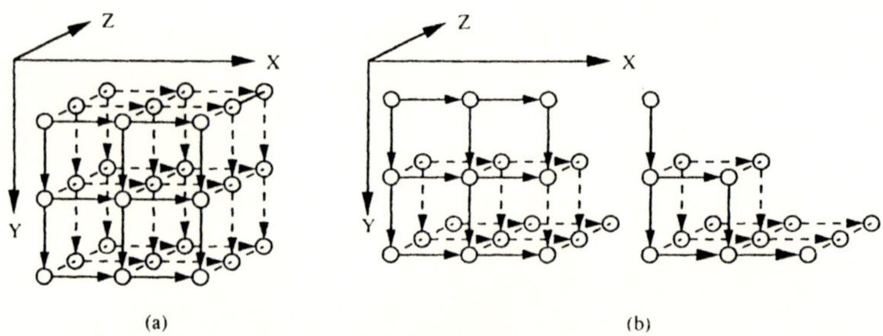

Figure 8.11 Examples of MMGs: (a) complete; (b) incomplete.
Source: [1] J.H. Moreno and T. Lang, "Matrix Computations on Systolic-Type Meshes: An Introduction to the Multimesh Graph Method," *IEEE Computer*, Vol. 23, No. 4, April 1990, pp. 32–51. (© 1990 IEEE. Reprinted with permission.)

A detailed discussion of these transformations is available [3]. In the LU decomposition algorithm, broadcasted data corresponds to value u_{ii}, which are used to compute the values of l_{ji}, and values u_{ij} and l_{ji}, which are used to compute the updated values a_{ij}. The FPG first is transformed by replacing data broadcasting with transmitted data, as shown in Figure 8.12. In this figure, thick lines represent transmitted data, and thin lines correspond to computed (dependent) data. The graph then is transformed into an MMG by allocating each subgraph (each iteration of the outermost loop) to a different plane in a three-dimensional space, as shown in Figure 8.13.

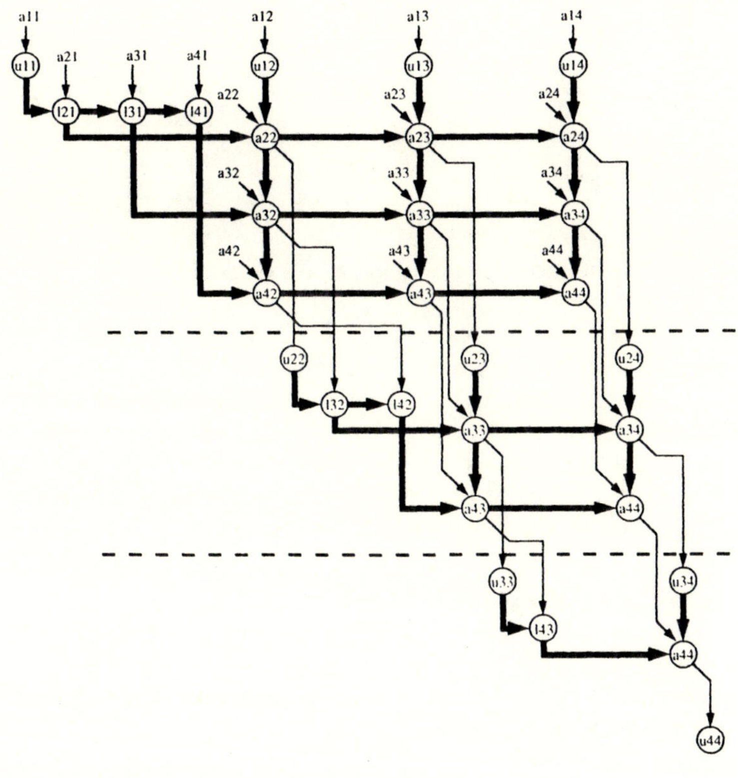

Figure 8.12 FPG without broadcasting for the LU decomposition algorithm (n = 4).

8.2.5 The Mapping Stage

As stated earlier, mapping an MMG onto an array consists of determining the assignment of operations to cells, the assignment of operands and results to storage

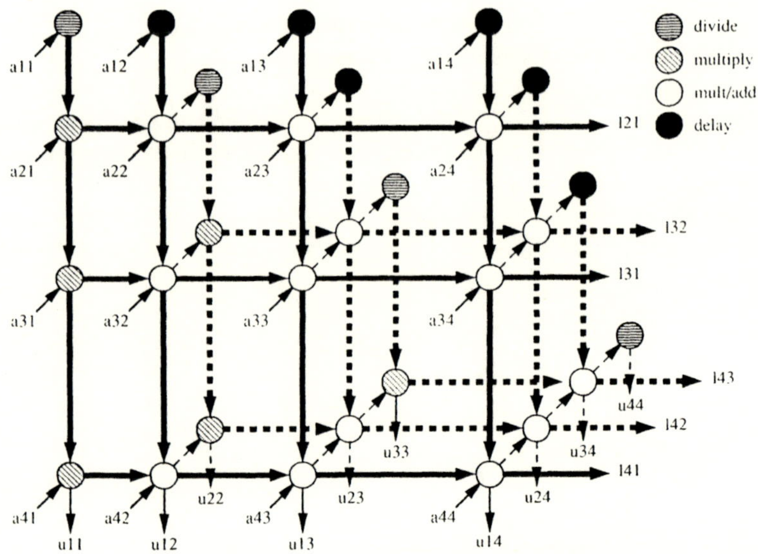

Figure 8.13 MMG for the LU decomposition algorithm ($n = 4$).

modules, the schedule of operations and data transfers, and the I/O schedule. Consequently, the mapping process is influenced by the characteristics of the target architecture. As these characteristics may have a large number of alternatives, the method has to be adapted to each particular case. Moreover, the characteristics may influence the process in conflicting ways, making it necessary to analyze different alternatives and select the most appropriate one. The issues to consider in the adaptation include

- the type, number, and organization of functional units per cell;
- the data paths to access memory, neighboring cells, and the cell's rf;
- the I/O paths to internal cells;
- the organization and size of the rf (used to store intermediate results or operands that are accessed several times).

Mapping the MMG onto a local-access array is performed as follows:

1. Determine the assignment of operations and data to cells by coalescing the MMG onto a *G-graph* that has the same topology and number of *G-nodes* as cells in the target array (linear or two dimensional).
2. Schedule the order of execution of the primitive operations that compose a *G*-node.

These two steps are described in detail in Section 8.2.7.

8.2.6 The Performance Evaluation Measures

The main performance measures used to evaluate different mappings of an algorithm onto a class-specific, local-access mesh array are computation time (t), utilization of PEs (U), and speed-up (S). These measures are computed as follows:

$$t_K(n) = t_b$$

$$U_K(n) = \frac{N_{\text{total}}(n)}{KFt_K(n)}$$

$$S_K(n) = \frac{t_1(n)}{t_K(n)} = K\frac{U_K(n)}{U_1(n)}$$

where

- n = the size of the matrices,
- K = the number of cells,
- t_b = the computation time of the largest partition,
- N_{total} = the total number of floating-point operations required by the algorithm,
- F = the cell computing capacity (number of operations per cell per time step),
- t_1 = the computation time using one cell.

Because we are interested in the performance achieved with the mapping method, speed-up is measured with respect to a single cell of the same type used in the array.

8.2.7 Mapping the Regularized Algorithm

Collapsing the MMG onto a G-graph is accomplished by dividing the MMG into partitions and coalescing each partition into a G-node (Figure 8.14). These partitions extend along one complete axis of the MMG. Partitions parallel to axes of the three-dimensional space lead to simpler and more efficient implementations, so that the direction of coalescing can be limited to three alternatives, one along each axis. Partitions for linear arrays consist of groups of meshes from the MMG, whereas for two-dimensional arrays the partitions are composed of sets of submeshes (portions of meshes), as depicted in Figure 8.14.

Approaches to Coalescing

As can be inferred from Figure 8.13, the MMG might have uneven distribution of primitive nodes throughout meshes. For example, as is the case for LU decompo-

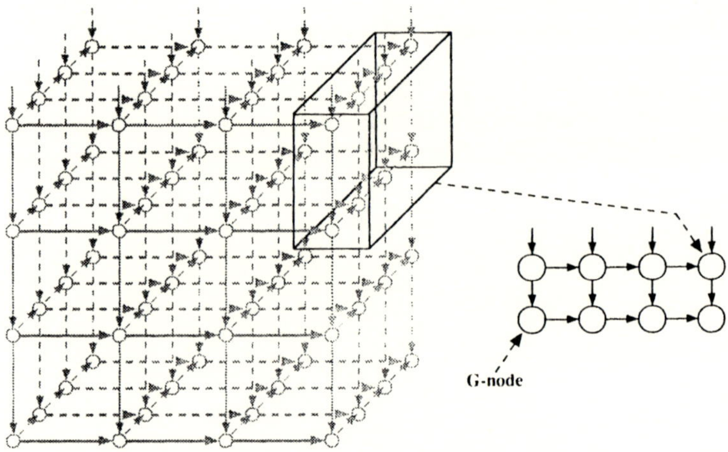

Figure 8.14 Collapsing an MMG onto a *G*-graph.
Source: [1] J.H. Moreno and T. Lang, "Matrix Computations on Systolic-Type Meshes: An Introduction to the Multimesh Graph Method," *IEEE Computer*, Vol. 23, No. 4, April 1990, pp. 32–51. (© 1990 IEEE. Reprinted with permission.)

sition, there might be fewer nodes at the left of the graph than at the right (or *vice versa*); the same might be true from top to bottom. Moreover, the MMG might have uneven distribution of computational requirements throughout nodes. For example, a matrix algorithm might have many "lightweight" nodes that compute multiplies-adds (a basic operation), and a few "heavyweight" nodes that require a complex operation (such as division, square root, or computation of a rotation angle). We refer to these characteristics as the uneven *density* of the MMG.

The MMG method permits the selection of uniform and nonuniform sized partitions when coalescing the MMG, as follows:

- *Uniform partitions* means that the MMG is divided into partitions with the same number of neighbor (sub)meshes. This approach, which is equivalent to "block" methods [16], has the advantage of simplicity, but may lead to an unbalanced distribution of load in the array and, consequently, inefficient utilization of cells due to the uneven density of the MMG.
- *Nonuniform partitions* mean that the MMG is divided into partitions with variable numbers of neighbor (sub)meshes. The size of the partitions is selected in such a way that the total computational load assigned to each partition is roughly the same.

The nonuniform partitions require additional work to determine an adequate size for each, but this approach leads to better utilization of cells than uniform partitions because it provides better load balancing. We suggest a heuristic partitioning technique, where meshes from the MMG are allocated to a partition until the load accumulated on the partition is close to an ideal average (the total load divided by the number of cells).

For a linear array, the heuristic strategy is illustrated in Figure 8.15 and is described by the following algorithm. Compute N_t = total load in execution of the algorithm. Then compute

$$N_{\text{avg}} = \frac{N_t}{\#\ \text{cells}} = \text{ideal average load per cell}$$

For $i = 1, \ldots, n$, compute n_i = load in mesh i of MMG. Set $i = 0$.

For $k = 1, \ldots, \#$ cells repeat $i = i + 1$, and allocate mesh i of the MMG to G-node k until the load in G-node k is as near as possible to N_{avg}.

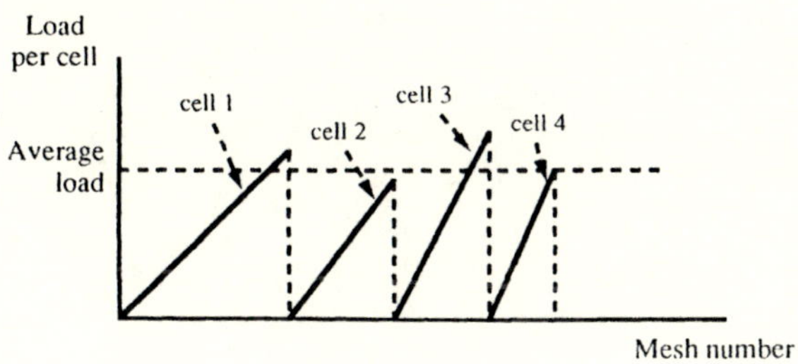

Figure 8.15 Selecting partition size in nonuniform partitioning.

In this approach, the load placed on a G-node is computed assuming that a mesh both is and is not allocated to the G-node, and the resulting loads are compared to the ideal average. The final selection is determined by the smallest difference, in a greedy strategy (select the smallest difference with respect to the ideal average for each G-node, regardless of what is best overall).

This scheme might lead to having too large (or too small) a load left for the last G-node (that is, allocating mostly below—or above—average loads to the different G-nodes). This effect can be eliminated by carrying an accumulated error that is used in making each decision. Thus, the successive selection of below average loads will add more weight to the selection of an above average load for the next G-node.

Uniform and nonuniform partitioning will be illustrated later by using LU decomposition and a linear array. Other alternative ways to allocate partitions from the MMG to cells ("scattered" or interleaved, for example) are discussed in [3]. However, those alternatives produce more complex schedules, bidirectional communications, and less efficient performance than what can be achieved with these schemes.

The Schedule of Operations

We discuss now how operations are scheduled for execution. For this purpose, let us consider a *complete* MMG (see Figure 8.16); that is, an MMG with nodes in every intersection of integer values in the three-dimensional space, with three input and three output edges in every node, and where all nodes have unit computation time. Moreover, this MMG has two flows of broadcasted (transmitted) data, parallel to the x and y axes, and one flow of dependent (computed) data. Input and output to and from the algorithm appear at the boundaries of the MMG, parallel to each axis.

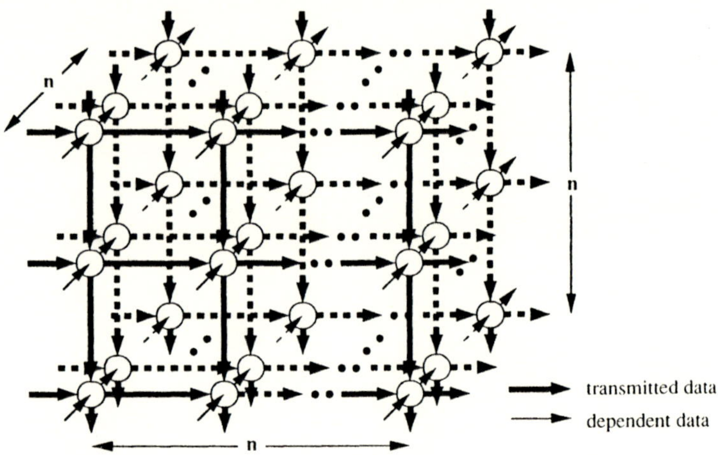

Figure 8.16 A complete MMG.
Source: [1] J.H. Moreno and T. Lang, "Matrix Computations on Systolic-Type Meshes: An Introduction to the Multimesh Graph Method," *IEEE Computer,* Vol. 23, No. 4, April 1990, pp. 32–51. (© 1990 IEEE. Reprinted with permission.)

We consider only the simple cases where nodes are scheduled by traversing them parallel to one axis in the three-dimensional space, because these schedules lead to more efficient mapping [3]. Consequently, the schedule can take any of three possible directions. The selection of a particular direction depends on whether cells are in a pipeline and on the memory bandwidth. If cells are not in a pipeline, a schedule may be obtained by traversing nodes parallel to the z-axis (that is, all nodes in one path parallel to the z-axis are scheduled before moving to another path); in this case, the value computed in a node is used immediately for another operation.

By contrast, a schedule parallel to the z-axis is not suitable if cells are in a pipeline. In such a case, the next node in a path along the z-axis is dependent on the one being executed, and the required datum is not yet available. To solve this problem, the schedule must follow the flow of transmitted data [3]. Due to the bandwidth and storage considerations discussed next, we suggest a schedule of operations by *prisms* of the MMG, as depicted in Figure 8.17. The proposed schedule traverses the q nodes within the prism in the topmost path parallel to the x-axis, then q nodes in the path immediately below it, and so on until traversing the pq nodes in the outermost mesh of the prism. Nodes in inner meshes of the prism are executed in the same order, once nodes in the outermost mesh have been executed.

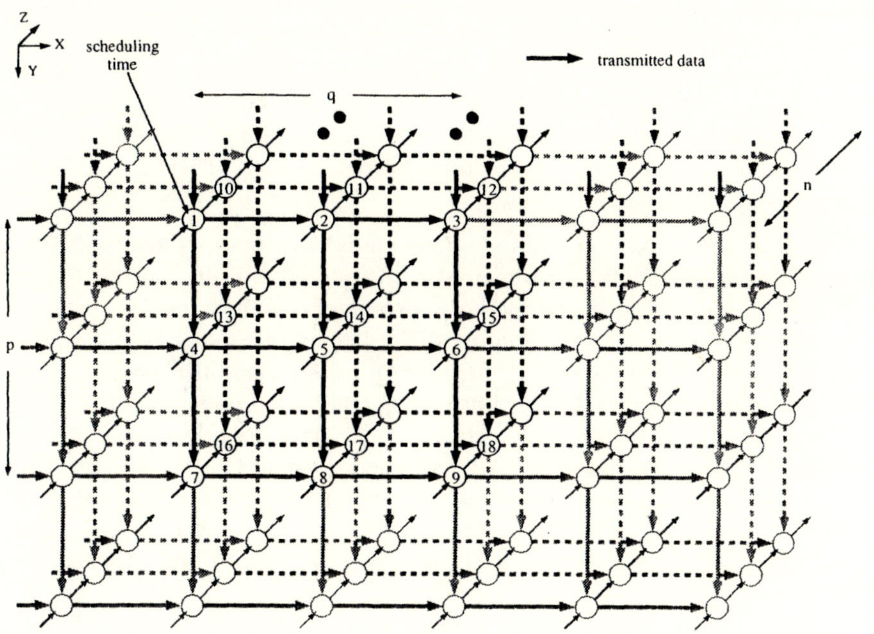

Figure 8.17 Prism in the complete MMG.

As a result of this schedule, the length of the pipeline (S) is related to the size of the base of the prisms because there must be at least as many successive independent operations as length of the pipeline; that is, $S \leq q$.

The size of the prisms also is related to the cell bandwidth for reading the transmitted data. In particular, the schedule shown in Figure 8.17 indicates that a value flowing parallel to the x-axis is read and used during q time steps, so that no other value from that direction is needed until q time steps later. The situation is similar for data flowing parallel to the y-axis, although the traffic there is in bursts of q values every pq time steps.

The Allocation of Data

Data must be stored and flow through the array. We will assume that the communication between cells consists of messages, and that the data being transferred into a cell are stored in memory before being used, independent of the execution of operations (direct memory access). Consequently, input data to the portion of the MMG executed in a cell always is stored in memory.

Consider an inner prism in the portion of a complete MMG allocated to a cell. The execution of primitive nodes requires cells capable of transferring data to or from three places in parallel (for the data flowing in each direction). The z-flow requires one read and one write access per time step, so that a suitable storage location is a register file. Because one result is produced in every cycle along the z-axis, the single write capability of the register file is used up completely.

Flows x and y hold the transmitted data and are only read. We suggest the following alternatives for these two sources of data, which differ in their bandwidth and storage requirements:

1. Use a dual-port memory with single-cycle access time, so that the values of x and y are read directly from memory. In such a case, the register file (of size pq) is used only for z. The memory also can be implemented as two independent parallel modules, one holding the x data and the other holding the y data. In addition, one module (say, the one with the x data) also holds the z inputs.
2. Use a standard memory (one port), a register file composed of two separate banks (each capable of one read and one write access per cycle), and one additional dedicated register. Allocate the z flow to one register bank (of size pq), the y flow to the other bank (of size q), and the x flow to the single dedicated register. In this case, values of x are read from memory, stored in the dedicated register, and reused from there q times; values of y also are read from memory, stored in their own register bank, and used from there p times during pq cycles before being replaced. Input values of z are read from memory, whereas computed z values are stored in and read from the corresponding register bank.

In the second case, the size of the base of the prisms ($p \times q$) is related to the total size of the register banks within a cell (R) as follows:

$$R = q + pq = q(p + 1)$$

The allocation and use of data in each cell are depicted in Figure 8.18.

Because all input data are stored in memory, the schedule and allocation of data as described lead to access conflicts when executing nodes in the outermost mesh of the prism (two or more data need to be read from the same memory module). Consequently, executing those nodes implies accessing memory more than once,

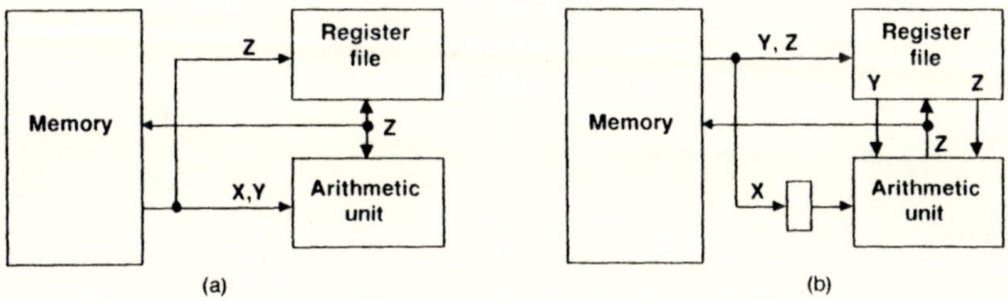

Figure 8.18 Data allocation in (a) dual-port memory and (b) dual register bank systems.

which requires more than one cycle. Because some data are later allocated to registers, a suitable approach consists of loading those registers in advance; this process can be regarded as adding delay nodes to the graph representing the data transfers.

Operation and Synchronization of Cells

The use of coalescing as a partitioning strategy allows cells to operate in a loosely coupled manner: input data are preloaded into the local memories, and during computation, each cell reads operands from its local storage and saves the results locally. Communication between cells is low, because it takes place only when a cell executes a node at the boundary of a partition of the MMG.

Synchronization between cells is necessary when an operation depends on data transferred from the neighbor cell upstream, which means data flowing parallel to the x-axis. Because the data correspond to a transmitted value, they can be transferred as soon as produced or used for the first time in the cell upstream. As a result, the transmitted data moves through cells at the rate of one cell per time step, so that the latency for data to propagate through the array is proportional to the length of the array (instead of the size of the matrices).

So far we have assumed that all nodes in the MMG have unit computation time, but operations in fact may have variable computation time. Because cells operate in a loosely coupled mode, each cell advances on its computation independent from other cells; if a cell requires data from a neighbor cell that is not yet available, then the cell must wait. However, because the dependency between cells arises from transmitted data and such data are transferred as soon as available, they usually arrive before they are needed and the problem of idle time waiting for data can be neglected.

8.3 EXAMPLE: MAPPING THE LU DECOMPOSITION ALGORITHM

We now illustrate the mapping process for a linear and a two-dimensional, class-specific, local-access array, using LU decomposition as an example. Note that the method and criteria described here are equally applicable to other structures (other linear or two-dimensional arrays) and to other algorithms.

8.3.1 The Target Architecture

The architectures used in this example consist of a linear and a two-dimensional array of processing cells. We assume the following characteristics of the cells, whose architecture is as depicted in Figure 8.18(b):

- a large local memory, single-cycle access time;
- a two-stage pipeline arithmetic unit consisting of a multiplier and an ALU connected in cascade;
- two register banks of size 4 and 16 words, respectively;
- one additional dedicated register;
- internal program memory.

The arithmetic pipeline allows the initiation of one multiply and one ALU operation per time step. Complex operations (such as divide, square root, rotation) are implemented as sequences of basic operations [17]. For example, we assume that a divide is implemented in the equivalent of eight multiply operations. Consequently, primitive operations in an algorithm can have different computation time in the target arrays. All operations and data transfers are performed on single-precision floating-point data.

8.3.2 The LU Decomposition Algorithm

The LU decomposition algorithm was given in Figure 8.9, and its MMG was shown in Figure 8.13. For an $n \times n$ matrix, the computational load of this algorithm (the number of operations excluding data transfers such as $u[k, j] = a[k, j]$) is as follows:

$$N_{\text{div}} = n$$

$$N_{\text{mult}} = \sum_{l=1}^{n} (l - 1) = n(n - 1)/2$$

$$N_{\text{mult/add}} = \sum_{l=1}^{n} (l - 1)^2 = n(n - 1)(2n - 1)/6$$

$$N_{\text{total}}(n) = 8n + \frac{n(n - 1)}{2} + \frac{2n(n - 1)(2n - 1)}{6} = n(4n^2 - 3n + 47)/6 \text{ ops}$$

Numbering meshes from right to left along the x-axis in Figure 8.13, the number of operations in mesh i is

$$\text{Divides } D_x^i = 1$$

$$\text{Multiplies } M_x^i = (i - 1)$$

$$\text{Multiply/adds } MA_x^i = \sum_{j=i}^{n-1} j = [n(n-1) - i(i-1)]/2$$

$$\text{Delays } \delta_x^i = 2n - i$$

$$\text{Total} = N_x^i = 8 + (i-1) + [n(n-1) - i(i-1)]/2 + (2n - i)$$

$$= 7 + n(n+3)/2 - i(i-1)/2$$

The expression for δ_x^i includes delay nodes in an outermost mesh (not shown in Figure 8.13) corresponding to loading the input data into the registers. These expressions allow computing the load allocated to cells by summing over the range of (sub)meshes assigned to each cell.

Computing the LU Decomposition in a Single Cell

Before mapping the LU decomposition onto target arrays, we look into the performance that can be achieved when computing the whole algorithm in a single cell. This result gives a measure of the cell's suitability for the specific algorithm. The schedule of primitive operations, described in Section 8.2.7, introduces no degradation in a single cell due to dependencies in the pipeline; as a result, a new operation is initiated every time step. However, some time steps are lost while executing the delay nodes.

From the expressions given earlier for number of operations in the algorithm and because a multiply/add is initiated in a single cycle, the total number of time steps required to execute LU decomposition in a single cell is

$$t_1(n) = 8n + \frac{n(n-1)}{2} + \frac{n(n-1)(2n-1)}{6} + \left[n^2 + \frac{n(n+1)}{2}\right]$$

$$= n(2n^2 + 9n + 49)/6$$

Consequently, mapping a problem of size $n = 200$ onto a single cell has*

$$t_1(200) = 2{,}728{,}300 \text{ time steps}$$

$$U_1(200) = \frac{N_{\text{total}}(200)}{K \cdot F \cdot t_1(200)} = \frac{5{,}314{,}900}{1 \cdot 2 \cdot 2{,}728{,}300} = 0.97$$

That is, executing the LU decomposition algorithm in one cell of the target arrays with the schedule and distribution of data proposed here produces almost unitary utilization. Loss of utilization arises only when computing simple multiply operations and divides, because no other operation is executed in the ALU at the same time, and when loading data into the registers. However, as determined from the MMG, these operations are few compared with the number of multiplications and additions that can be pipelined in the single cell.

8.3.3 Mapping onto a Linear Array

Uniform Coalescing

Consider first performing uniform coalescing onto a linear array. As indicated before, the graph can be collapsed in three possible ways; namely, by coalescing partitions from the MMG parallel to each axis in the three-dimensional space. A detailed discussion of the three alternatives is given in [3]; here, we show only the partitioning along the x-axis, as depicted in Figure 8.19. In this case, uniform coalescing consists of allocating n/K columns of the input matrix to each cell of the linear array. Moreover, each cell produces part of the resulting matrix, in a columnwise manner, so that cells to the right generate more results than cells to the left. (In case n/K is not an integer value, we allocate $\lfloor n/K \rfloor$ meshes of the graph to part of the PEs [the leftmost ones], and $\lceil n/K \rceil$ meshes to the remaining ones.)

Scheduling operations by prisms, as described in Figure 8.17, produces little degradation due to pipeline latency, so that we can neglect it. Consequently, the computation time of each cell is determined by the computing load allocated to it. Figure 8.20 gives the number of cycles required by each cell for a matrix of size 200×200 and sixteen cells, which also is shown graphically in Figure 8.21. The total time required to compute the algorithm is given by the busiest cell; namely, PE_{16}. Note that load balancing is not good, due to the uneven density of the MMG.

*This analysis does not consider time steps lost due to control of program flow (loops and branches). Those losses depend on the instruction set of the cells and not on the data paths.

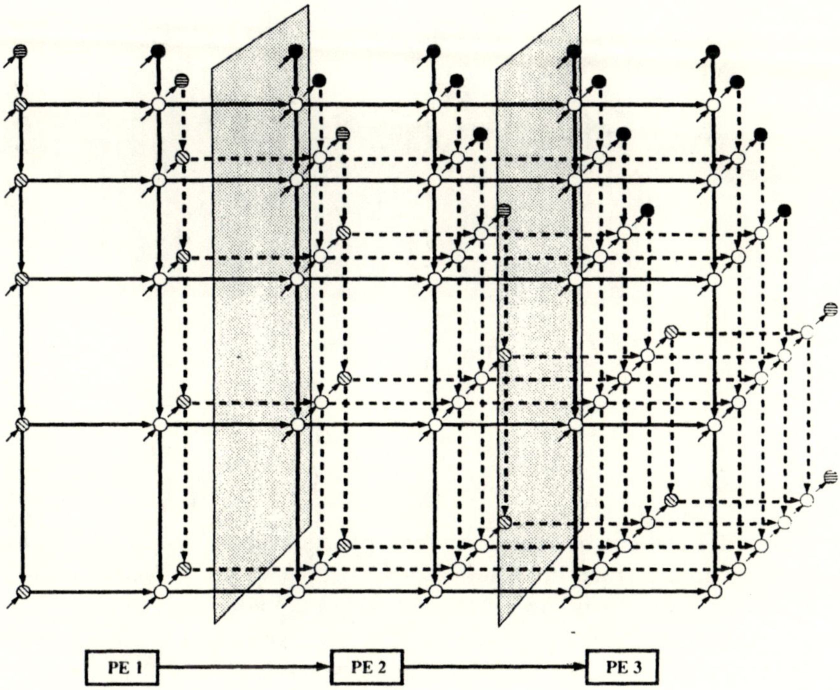

Figure 8.19 Uniform coalescing in a linear array.

PE	Cycles	PE	Cycles
1	20,527	9	195,014
2	52,130	10	207,110
3	81,536	11	217,478
4	108,745	12	226,118
5	133,757	13	233,030
6	156,572	14	238,214
7	177,190	15	241,670
8	195,611	16	243,398

Figure 8.20 Load per cell in a linear array, uniform coalescing ($n = 200$, $K = 16$).

The resulting performance estimates are as follows:

$$t_{16}(200) = 243{,}398 \text{ time steps}$$

$$U_{16}(200) = 0.68$$

$$S_{16}(200) = 11.2$$

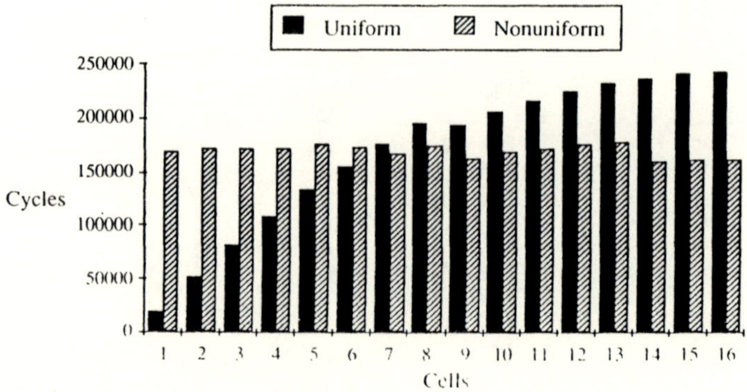

Figure 8.21 Load per cell in a linear array.

Nonuniform Coalescing

Consider now the case of assigning to cells partitions of variable size. Using the heuristic procedure described in Section 8.2.7 together with the expressions for number of operations per mesh in the MMG, we obtain the allocation of meshes to cells shown in Figure 8.22 and the corresponding cycles required. The load per cell also is depicted graphically in Figure 8.21. The computation time of the algorithm is given by PE_{13}, with 179,079 time steps. The performance estimates obtained are

$$t_{16}(200) = 179{,}079 \text{ times steps}$$

$$U_{16}(200) = 0.93$$

$$S_{16}(200) = 15.2$$

These estimates show that nonuniform coalescing mapping produces utilization U =

PE	Cycles	Meshes	PE	Cycles	Meshes
1	169,207	200 - 160	9	164,013	69 - 61
2	173,223	159 - 141	10	168,873	60 - 52
3	172,795	140 - 126	11	173,004	51 - 43
4	172,627	125 - 113	12	176,406	42 - 34
5	176,198	112 - 101	13	179,079	33 - 25
6	174,207	100 - 90	14	160,836	24 - 17
7	167,750	89 - 80	15	161,860	16 - 9
8	175,650	79 - 70	16	162,372	8 - 1

Figure 8.22 Load per cell in a linear array, nonuniform coalescing (n = 200, K = 16).

0.93, almost the maximum possible for the LU decomposition algorithm in the target array (determined by the mapping onto a single cell).

Note that, as with any heuristic strategy, these results may not necessarily represent optimal partitioning in terms of load balancing. However, the results are obtained with a single execution of a simple greedy algorithm, without the need for enumeration and comparison of alternative partition sizes. Moreover, the heuristic strategy is good because it results in utilization very close to the optimal value.

8.3.4 Mapping onto a Two-Dimensional Array

Uniform Coalescing

Consider now mapping the LU decomposition onto a two-dimensional array. Uniform partitioning consists of coalescing the MMG onto a two-dimensional G-graph where all G-nodes correspond to portions from the MMG of the same size. The results obtained with this approach for a problem of size 200×200 in an array of 16 cells (4×4) are given in Figure 8.23 and depicted graphically in Figure 8.24. Note that due to the uneven density of the MMG, load balancing is not good, and

PE	Cycles	PE	Cycles	PE	Cycles	PE	Cycles
(1,1)	45,775	(1,2)	66,250	(1,3)	66,250	(1,4)	66,250
(2,1)	66,250	(2,2)	170,775	(2,3)	191,250	(2,4)	191,250
(3,1)	66,250	(3,2)	191,250	(3,3)	295,775	(3,4)	316,250
(4,1)	66,250	(4,2)	191,250	(3,4)	316,250	(4,4)	420,775

Figure 8.23 Load per cell in a two-dimensional array, uniform coalescing (n = 200, K = 16).

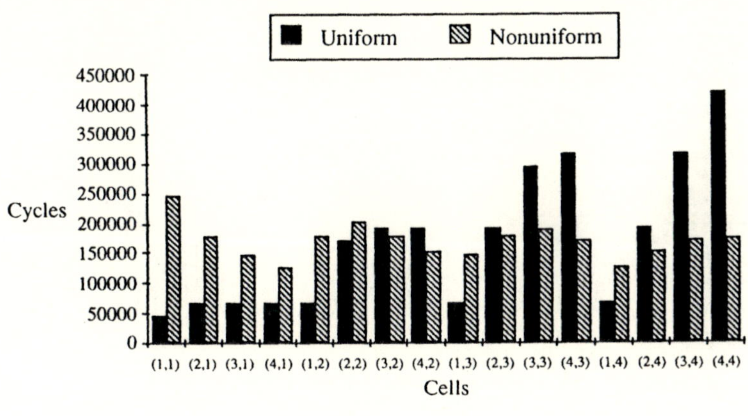

Figure 8.24 Load per cell in a two-dimensional array.

even worse than in uniform coalescing for the linear array. The corresponding performance estimates are

$$t_{16}(200) = 420{,}775 \text{ time steps}$$

$$U_{16}(200) = 0.39$$

$$S_{16}(200) = 6.5$$

Nonuniform Coalescing

Two-dimensional arrays require a more complex strategy to determine the size of nonuniform partitions than the one used for linear arrays. This complexity is a result of the uneven density of the MMG as well as of the need to obtain a simple communication pattern between cells. This last requirement leads to the allocation of neighbor portions of the same width to all G-nodes in the same vertical position in the G-graph, as well as portions of the same height to all G-nodes in the same horizontal position, as depicted in Figure 8.25. For example, if the portion mapped to a G-node $(\sqrt{K}, \sqrt{K})$ has the size (s_1, s_2, n), then the width of the portions for G-nodes $(i, \sqrt{K})$ is s_1, and the height of the portions for G-nodes $(\sqrt{K}, j)$ is s_2. However, the density of the MMG is not constant, so that, if the load allocated to G-node $(\sqrt{K}, \sqrt{K})$ is close to the ideal average, then the load allocated to at least one G-node $(i, \sqrt{K})$ or $(\sqrt{K}, j)$ is not. Similar situations occur for any other G-node.

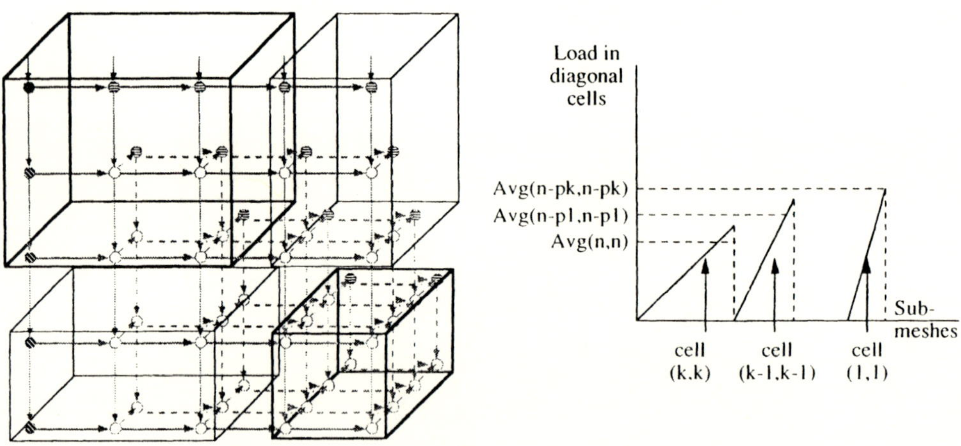

Figure 8.25 Selecting a partition size in nonuniform partitioning for a two-dimensional array.

We propose a heuristic partitioning strategy for two-dimensional arrays that first determines the size of diagonal *G*-nodes. Once these diagonal *G*-nodes have been determined, the size of off-diagonal *G*-nodes is determined from the intersection of the width and height of the portions collapsed onto the corresponding diagonal nodes, as shown in Figure 8.25. For the diagonal *G*-nodes, we propose the following approach, which also is illustrated in Figure 8.25:

- Compute the ideal average load per *G*-node for the entire problem. Allocate submeshes from the MMG to *G*-node $(\sqrt{K}, \sqrt{K})$ until its load is higher than the ideal average.
- Exclude from the MMG all nodes in vertical or horizontal paths that intersect with nodes allocated to *G*-node $(\sqrt{K}, \sqrt{K})$.
- Compute a new average load per *G*-node for the remaining part of the MMG assuming an array of $(\sqrt{K} - 1, \sqrt{K} - 1)$ cells. Allocate submeshes from the remaining MMG to *G*-node $(\sqrt{K} - 1, \sqrt{K} - 1)$ until its load is higher than the new average.
- Repeat the last two steps until reaching *G*-node (1, 1), reducing each time the effective size of the MMG.

The results obtained from applying this technique to LU decomposition are depicted in Figures 8.26 and 8.24; the corresponding estimates of performance are

$$t_{16}(200) = 247{,}509 \text{ time steps}$$

$$U_{16}(200) = 0.67$$

$$S_{16}(200) = 11.0$$

Note that, in contrast to the linear case, diagonal cells are always allocated a load higher than the ideal average. Moreover, this average increases from PE$(\sqrt{K}, \sqrt{K})$ to PE(1, 1). These are consequences of the uneven density of the MMG. The

PE	Cycles	Submeshes	PE	Cycles	Submeshes
(1,1)	247,509	(1,1) - (89,89)	(1,3)	147,384	(1,134) - (89,169)
(2,1)	180,136	(90,1) - (133,89)	(2,3)	178,200	(90,134) - (133,169)
(3,1)	147,384	(134,1) - (169,89)	(3,3)	190,122	(134,134) - (169,169)
(4,1)	126,914	(170,1) - (200,89)	(3,4)	170,190	(170,134) - (200,169)
(1,2)	180,136	(1,90) - (89,133)	(1,4)	126,914	(1,170) - (89,200)
(2,2)	203,918	(90,90) - (133,133)	(2,4)	153,450	(90,170) - (133,200)
(3,2)	178,200	(134,90) - (169,133)	(3,4)	170,190	(134,170) - (169,200)
(4,2)	153,450	(170,90) - (200,133)	(4,4)	174,003	(170,170) - (200,200)

Figure 8.26 Load per cell in two-dimensional array, nonuniform coalescing (n = 200, K = 16).

heuristics could be improved by computing the ideal average load as the average load in the diagonal G-nodes obtained above, and using this new average in a new allocation of submeshes to G-nodes.

From the results presented in Figures 8.20–8.26, we can conclude that a linear array with the same number of cells exhibits better performance than a two-dimensional array. Better load balancing is achieved in the linear case than in the two-dimensional array, as a consequence of the uneven density of the MMG. Note that with the simple heuristic technique to determine the size of nonuniform partitions, the linear array reaches 93 percent utilization; in contrast, the more complex strategy for the two-dimensional array leads to 67 percent utilization. Although we could improve the heuristics for two-dimensional arrays, it will never lead to better load balancing than in the linear case, due to the characteristics of the MMG. In addition, nonuniform partitioning leads to better utilization than uniform (block) methods.

REFERENCES

1. Moreno, J.H., and T. Lang, "Matrix Computations on Systolic-Type Meshes: An Introduction to the Multimesh Graph Method," *IEEE Computer*, Vol. 23, No. 4, April 1990, pp. 32–51.
2. Kung, H.T., "Why Systolic Arrays?" *IEEE Computer*, Vol. 15, No. 1, Jan. 1982, pp. 37–46.
3. Moreno, J.H., *Matrix Computations on Mesh Arrays*, Ph.D. thesis, Computer Science Department, University of California Los Angeles, June 1989.
4. Fortes, J.A.B., and B.W. Wah, eds., *Special Issue on Systolic Arrays*, IEEE Computer Society Press, Los Alamitos, CA, July 1987.
5. Kung, S.Y., *VLSI Array Processors*, Prentice-Hall, Englewood Cliffs, NJ, 1988.
6. Speiser, J.M., and H. Whitehouse, "A Review of Signal Processing with Systolic Arrays," in *SPIE Real-Time Signal Processing VI*, Bellingham, WA, August 1983, pp. 2–6.
7. McCanny, J., J. McWhirter, and E. Swartzlander, eds., *Systolic Array Processors*, Prentice-Hall, Englewood Cliffs, NJ, 1989.
8. Bromley, K., S.Y. Kung, and E. Schwatzlander, eds., *Proc. Int. Conf. Systolic Arrays*, IEEE Computer Society Press, Los Alamitos, CA, May 1988.
9. Fortes, J.A.B., K. Fu, and B.W. Wah, "Systematic Approaches to the Design of Algorithmically Specified Systolic Arrays," in V. Milutinović, ed., *Computer Architecture*, Elsevier North Holland, New York, 1988, pp. 454–494.
10. Kung, H.T., "Let's Design Algorithms for VLSI Systems," Caltech Conference on VLSI, January 1979, pp. 65–90.
11. Moreno, J.H., and T. Lang, "On Partitioning the Faddeev Algorithm," *Proc. Int. Conf. Systolic Arrays*, IEEE Computer Society Press, Los Alamitos, CA, May 1988, pp. 125–134.
12. Navarro, J.J., J.M. Llaberia, and M. Valero, "Partitioning: An Essential Step in Mapping Algorithms into Systolic Array Processors," *IEEE Computer*, Vol. 20, No. 7, July 1987, pp. 77–89.
13. Rao, S.K., and T. Kailath, "Regular Iterative Algorithms and Their Implementation on Processor Arrays," *Proc. IEEE*, Vol. 76, No. 3, March 1988, pp. 259–269.
14. Moldovan, D.I., "On the Design of Algorithms for VLSI Systolic Arrays," *Proc. IEEE*, Vol. 71, No. 1, January 1983, pp. 113–120.
15. Quinton, P., "Automatic Synthesis of Systolic Arrays from Uniform Recurrent Equations," *Proc.*

11th Annual Symp. Computer Architecture, IEEE Computer Society Press, Los Alamitos, CA, June 1984, pp. 208–214.

16. Fox, G., *et al., Solving Problems in Concurrent Processors, Vol. 1,* Prentice-Hall, Englewood Cliffs, NJ, 1987, pp. 167–186, 363–424.
17. Hwang, K., *Computer Arithmetic: Principles, Architecture and Design,* John Wiley & Sons, New York, 1979.

Chapter 9
Recent Developments

The subjects treated briefly in this last chapter can be related to the previous material in the following manner. By removing the requirement of stationarity in narrowband processing in Chapters 3 and 4, we need a signal processing approach that estimates instantaneous frequency; for example, the Wigner-Ville distribution. By no longer assuming plane-wave propagation, we must correct for wavefront curvature; the result is a focused beamformer capable of estimating range. By relaxing the requirement that the noise be Gaussian distributed, processing tools such as polyspectra are available to discriminate between Gaussian and non-Gaussian noise. Finally, by including normal modes in the propagation model, we can passively estimate range and depth with a vertical array of sensors.

9.1 WIGNER-VILLE DISTRIBUTION

In those sections of Chapters 3 and 4 dealing with narrowband processing we assumed that the parameters of the signal are constant (at least over the observation period). This assumption may be invalid if the mechanism generating the signal introduces small short-term fluctuations (as would result from unbalanced rotational machinery) or if one or more of the signal parameters is intentionally varied with time (the linear FM pulse in Chapter 5). Processing such a signal with a length-N DFT might result in output that is "smeared" over several output bins, if the signal frequency varies appreciably during N samples. Various distributions have been devised that permit representation of a signal simultaneously in time and frequency. A comprehensive treatment of such *time-frequency distributions* (TFDs) is found in a tutorial paper by Cohen [1].

A particular TFD that has been applied to the narrowband sonar problem is the *Wigner-Ville distribution* (WVD). (The interested reader can find references to the WVD in [1], and especially papers by Nuttall [2–4].) A continuous-variable form of the WVD is [5, eq. 9]

$$W(t, f) = \int_{-\infty}^{\infty} s(t + \tau/2)s^*(t - \tau/2) \exp(-j2\pi f\tau) \, d\tau \tag{9.1}$$

where $s(t)$ is a complex-valued time function. The cited references show that the first-order moments of the WVD provide estimates of the instantaneous frequency and time delay. Integrating the WVD with respect to time yields the power spectral density at frequency f, and the instantaneous power is obtained by integrating with respect to frequency. Therefore, the WVD is useful in analyzing nonstationary phase-modulated signals.

Example 9.1

We consider the continuous-variable linear FM pulse and illustrate qualitatively the results of calculating the WVD. The signal input is

$$s(t) = \exp[j2\pi(f_0t + mt^2/2)] \tag{9.2}$$

Substituting (9.2) into (9.1) yields

$$\begin{aligned} W(t, f) = \int_{-\infty}^{\infty} & \exp\{j2\pi[f_0(t + \tau/2) + m(t + \tau/2)^2/2]\} \\ & \cdot \exp\{-j2\pi[f_0(t - \tau/2) + m(t - \tau/2)^2/2]\} \\ & \cdot \exp(-j2\pi f\tau) \, d\tau \end{aligned} \tag{9.3}$$

After expanding each of the exponential factors in (9.3) and simplifying, we obtain

$$W(t, f) = \int_{-\infty}^{\infty} \exp[j2\pi(f_0 + mt - f)\tau] \, d\tau \tag{9.4}$$

Expression (9.4) is the continuous-variable delta function:

$$W(t, f) = \delta(f_0 + mt - f) \tag{9.5}$$

which has a peak at $f = f_0 + mt$. Therefore, for a given value of time t, the location of the peak of the WVD provides an estimate of the instantaneous frequency of the LFM pulse. This instantaneous frequency is shown in Figure 5.7. (End of example.)

We discussed the uncertainty function of a discrete-time signal in Section 5.2.1. A slightly different formulation is given by Yen [5, eq. 23] and Van Trees [6, eq. 28], where

$$A(\zeta, \tau) = \int_{-\infty}^{\infty} s(t + \tau/2)s^*(t - \tau/2) \exp(-j2\pi\zeta\tau)\, dt \tag{9.6}$$

Expression (9.6) is called the *ambiguity function* by Yen and the *two-dimensional correlation function* by Van Trees, where τ is replaced by $-\tau$. Yen relates the uncertainty function of $s(t)$ in (9.6) to the WVD (9.1) by the expression

$$A(\zeta, \tau) = \int_{-\infty}^{\infty}\int_{-\infty}^{\infty} W(t, f) \exp[-j(2\pi\zeta t - 2\pi f\tau)]\, dt\, df$$

Thus, the WVD is the two-dimensional Fourier transform of the uncertainty function (9.6) of $s(t)$.

Discrete-variable approximations to (9.1) have been developed by various researchers. For example, the following *smoothed pseudo-Wigner distribution* (SPWD) is used by Yu and Cheng [7, eq. 60]:

$$W_s(n, f) = \sum_{k=-L}^{L} w(k)R(k, n) \exp(-j2\pi kf) \tag{9.7}$$

where $W_s(n, f)$ denotes the SPWD. The sequence $w(k)$ is a symmetric real window function, and $R(k, n)$ is an estimate of the autocorrelation function defined by

$$R(k, n) = \sum_{m=-M+1}^{M-1} g(m)z(n + m + k)z^*(n + m - k)$$

where $g(m)$ is a symmetric real window function, and $z(n) = s(n) + j\hat{s}(n)$ is the analytic signal representation of the input sequence, $s(n)$. The implementation of (9.7) is shown in Figure 9.1.

Consider the discrete-variable LFM pulse given in (5.20) and repeated here:

$$s(n) = \begin{cases} \exp\{j2\pi[f_0 n/f_s + mn^2/(2f_s^2)]\}, & -N/2 \leq n \leq N/2 - 1 \\ 0, & \text{otherwise} \end{cases} \tag{9.8}$$

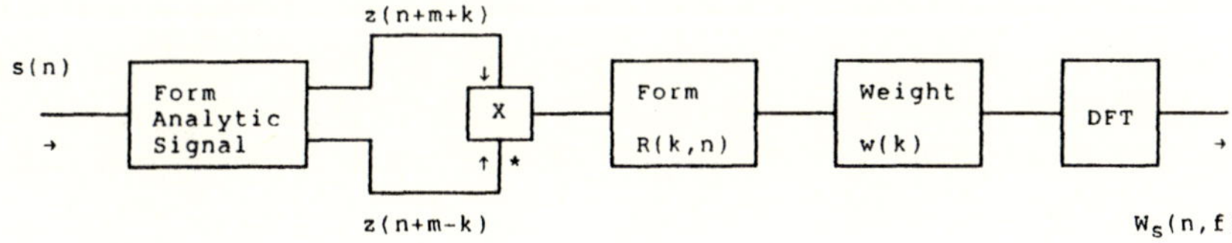

Figure 9.1 Implementation of the smoothed pseudo-Wigner distribution.

A sketch of the SPWD for the LFM pulse (9.8) is provided in Figure 9.2. The information given in this figure is a discrete-variable version of the idealized response of the WVD to an LFM pulse (9.5). Other approaches to obtaining smoothed distributions are discussed in [1, pp. 961–962].

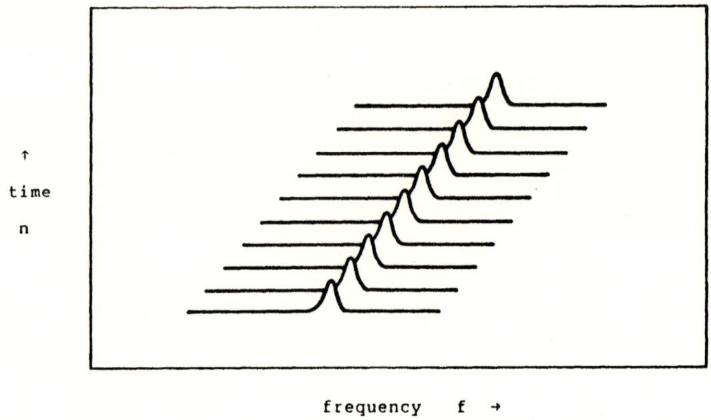

Figure 9.2 Sketch of the output of the smoothed pseudo-Wigner distribution for the LFM pulse.

Kay and Boudreaux-Bartels [8] show that the optimal detector for the LFM pulse in additive complex white Gaussian noise can be implemented by integrating the WVD along the straight line in the time-frequency domain that corresponds to the instantaneous frequency. (Recall that, in a detection problem, the values of signal parameters, such as the sweep rate of the LFM pulse, are known.)

If the arrival time and sweep rate of the LFM pulse are unknown, estimates of these quantities can be obtained by integrating the WVD along all possible straight lines in the time-frequency domain [8]. The maximum output of this integration process then is compared to a threshold. If the threshold is exceeded, a detection is declared. The LFM pulse parameters are estimated from the position and slope in the time-frequency plane of the line yielding the maximum output.

Three recent developments concerning the WVD are presented next. First, the WVD often generates "cross terms" when the input consists of two or more signal components. Choi and Williams [9 and 1, pp. 950–952] describe a technique by which these spurious terms can be minimized. Second, it has been generally believed that, to form the WVD from discrete samples, it is necessary to sample the band-limited signal at $f_s \geq 4f_{\max}$ to avoid aliasing [1, p. 960]. Here $f_{\max}$ is the highest frequency in the signal. Nuttall [4] provides an alias-free method for calculating the WVD from a signal sampled at the Nyquist frequency of $2f_{\max}$. Third, the implementation of the SPWD given in Figure 9.1 provides a processor for nonstationary signals that has the resolution of the DFT that appears in this figure. A high-resolution processor for nonstationary signals can be obtained by replacing the DFT in Figure 9.1 with one of the high-resolution spectral estimation techniques discussed by Marple [10].

9.2 FOCUSED BEAMFORMER

The results obtained for an array of processors in Chapter 2 for the conventional beamformer, as well as in Chapter 3 for the detection and estimation of broadband signals, are based on the assumption of plane-wave propagation. That is, we assumed that the array is in the far field relative to the source. When this assumption is violated, the curvature of the wavefront presents the opportunity for passive estimation of both the bearing and the range of a source.

To illustrate the effect of wavefront curvature, we consider a three-sensor array, with a sensor separation of d. The geometry is shown in Figure 9.3, where the distance from the source to the ith sensor is denoted by p_i, $i = 1, 2, 3$. We find from Figure 9.3 that the differences in path lengths, relative to sensor 1, are given by

$$p_1 - p_2 = p_1 - (p_1^2 + 2dp_1 \cos\theta - d^2)^{1/2} \tag{9.9}$$
$$p_1 - p_3 = p_1 - (p_1^2 + 4dp_1 \cos\theta - 4d^2)^{1/2}$$

We rewrite (9.9) in the form

$$p_1 - p_2 = p_1 - p_1[1 + 2d(\cos\theta)/p_1 - d^2/(p_1^2)]^{1/2} \tag{9.10}$$
$$p_1 - p_3 = p_1 - p_1[1 + 4d(\cos\theta)/p_1 - 4d^2/(p_1^2)]^{1/2}$$

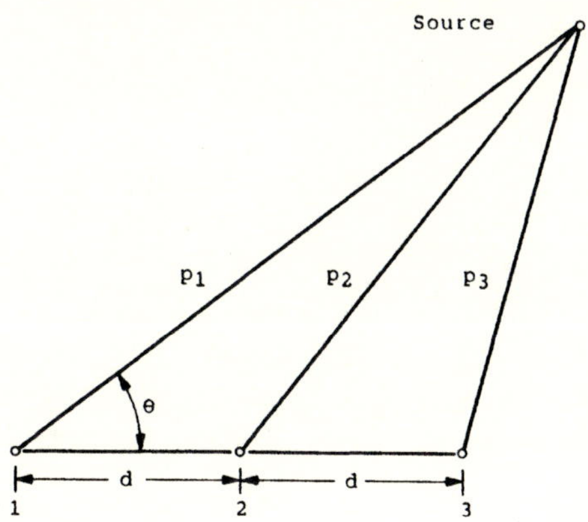

Figure 9.3 Wavefront curvature geometry.

Assuming that $p_1 \gg d$, we can apply the approximation $(1 + x)^{1/2} \approx 1 + x/2$ for $x \ll 1$, to (9.10). This yields

$$p_1 - p_2 \approx p_1 - p_1[1 + d(\cos\theta)/p_1 - d^2/(2p_1^2)]$$
$$p_1 - p_3 \approx p_1 - p_1[1 + 2d(\cos\theta)/p_1 - 2d^2/(p_1^2)]$$

or, equivalently,

$$p_1 - p_2 \approx -d\cos\theta + d^2/(2p_1) \tag{9.11}$$
$$p_1 - p_3 \approx -2d\cos\theta + 2d^2/p_1$$

To coherently combine the sensor outputs, we would delay the outputs of sensors 2 and 3 by $(p_1 - p_2)/c$ and $(p_1 - p_3)/c$, respectively, where c is the speed of sound. The terms $d(\cos\theta)/c$ and $2d(\cos\theta)/c$ from (9.11) correspond to the plane-wave case (Section 2.2.2) whereas the terms $d^2/(2cp_1)$ and $2d^2/(cp_1)$, correct for wavefront curvature. Note that (9.11) is approximately correct for a "semidistant" source. That is, $p_1 \gg d$ but d^2 and p_1 are comparable so that wavefront curvature cannot be ignored.

We next extend (9.11) to the case of an arbitrary three-dimensional array. We use the coordinate system in Figure 2.3 where again $\mathbf{r}_i$ is the position vector of the ith sensor, and $\mathbf{u}$ is the unit direction vector of the source. In addition, the location of the source is denoted by the vector $\mathbf{r}$. The direct-path propagation distance from the source to the ith sensor is

$$p_i = \|\mathbf{r} - \mathbf{r}_i\| \tag{9.12}$$

where $\|\cdot\|$ is the Euclidean norm defined by (7.26). Expanding (9.12) yields

$$p_i = \|\mathbf{r}\|\left(1 - \frac{2\mathbf{r}_i^T\mathbf{r}}{\|\mathbf{r}\|^2} + \frac{\|\mathbf{r}_i\|^2}{\|\mathbf{r}\|^2}\right)^{1/2} \tag{9.13}$$

Assuming that $\|\mathbf{r}\| \gg \|\mathbf{r}_i\|$, (9.13) is approximately equal to

$$p_i \approx \|\mathbf{r}\| - \frac{\mathbf{r}_i^T\mathbf{r}}{\|\mathbf{r}\|} + \frac{\|\mathbf{r}_i\|^2}{2\|\mathbf{r}\|} \tag{9.14}$$

We recognize $\mathbf{r}/\|\mathbf{r}\|$ as the unit direction vector, $\mathbf{u}$. Thus the path length in (9.14) from the source to the ith sensor is

$$p_i \approx \|\mathbf{r}\| - \mathbf{r}_i \cdot \mathbf{u} + \frac{\|\mathbf{r}_i\|^2}{2\|\mathbf{r}\|} \tag{9.15}$$

By dividing (9.15) by c, we find that $\|\mathbf{r}\|/c$ is a bulk delay, common to all sensors, that disappears when a set of relative delays is calculated for the array. The term $\mathbf{r}_i \cdot \mathbf{u}/c$ is the delay associated with the plane-wave assumption (Section 2.2.1) and the term $\|\mathbf{r}_i\|^2/(2c\|\mathbf{r}\|)$ is the correction for wavefront curvature.

Maximum Likelihood Estimation of Position

We presented the Neyman-Pearson broadband detector for an array of sensors in Figure 3.8. The sensor filters for the kth frequency

$$\mathbf{h}^+(k) = \mathbf{p}^+(k)\mathbf{Q}^{-1}(k)$$

were derived in Section 3.2.2 by using a knowledge of the signal direction, to calculate $\mathbf{p}(k)$, and of the normalized noise spatial correlation matrix, $\mathbf{Q}(k)$. Instead of assuming *a priori* knowledge of signal direction, we can calculate a $\mathbf{p}(k)$ vector for each of a set of hypothesized range-bearing pairs in the vicinity of the array. Carter [11, 12] has shown that calculating the output of the processor in Figure 9.4 for this set of range-bearing pairs and searching for the largest output provides the ML estimate of source position.

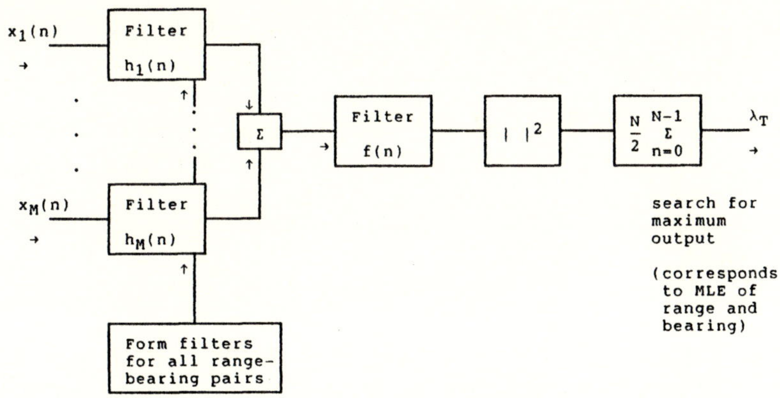

Figure 9.4 Maximum-likelihood estimator for range and bearing.
Source: After [12] G.C. Carter, "Coherence and Time Delay Estimation," *Proc. IEEE,* Vol. 75, No. 2, February 1987, pp. 236–255.

If we have control over the locations of the sensors of a linear array, we can improve estimation performance by appropriately specifying the geometry of the array. Carter [11, 12] shows that, for a linear array of M sensors and a noise field uncorrelated from sensor to sensor, the best array geometry for estimating bearing consists of grouping $M/2$ sensors at each end of the array. (This is the result obtained by MacDonald and Schultheiss in [13].) Placing $M/2$ sensors at the center of the array and $M/4$ sensors at each end minimizes the variance of the range estimate. Finally, if we are interested in minimizing the error in position (that is, both bearing and range), the appropriate array geometry is one in which $M/3$ sensors are located at the center of the array and $M/3$ sensors are placed at each end.

Example 9.2

To assess the computational impact of implementing a wavefront-curvature, or focused, beamformer, we consider a hypothetical array consisting of $M = 100$ sensors. It is desired that $N_B = 50$ values of bearing be investigated, each with $N_R = 15$ values of range. That is, we wish to form $N_B \times N_R = 750$ range-bearing pairs in Figure 9.4. Assuming a sampling frequency of $f_s = 10{,}000$ Hz, we find that the implementation of a simple delay-and-sum beamformer requires

$$(N_R \times N_B) \text{ range-bearing pairs} \times M \text{ sensors} \times f_s \text{ samples/sensor/s} \times 1 \text{ flop/sample} = 750 \text{ Mflops}$$

This result would increase if beamformer weighting were used. The computational requirement of 750 Mflops is within the capabilities of modern sonar systems.

9.3 BISPECTRUM

We let $x(n)$ be a sample sequence of a real stationary zero-mean random process. The power spectral density, or spectrum, of $x(n)$ has the form (Section 1.1.2)

$$S(f) = \sum_m R_{xx}(m) \exp(-j2\pi f m/f_s)$$

where $R_{xx}(m) = E\{x(n)x(n + m)\}$ is the autocorrelation sequence of $x(n)$. It is also of interest to consider higher order spectra. In particular, the bispectrum is defined as [14, eq. 5]

$$B(f_1, f_2) = \sum_m \sum_n C_{xxx}(m, n) \exp[-j2\pi(f_1 m + f_2 n)]$$

where the third-order cumulant, $C_{xxx}(m, n)$, is given by

$$C_{xxx}(m, n) = E\{x(p)x(p + m)x(p + n)\} \tag{9.16}$$

As discussed in the paper by Nikias and Raghuveer [15], the bispectrum finds applications in sonar signal processing in the following areas. First, if $x(n)$ is a sample sequence of a stationary zero-mean Gaussian random process, the third-order cumulant (9.16) is zero for all (m, n); and therefore the bispectrum is zero. It thus is possible to detect departures from a Gaussian distribution by using the bispectrum.

Hinich, Marandino, and Sullivan [14] describe work in which the bispectra of ship-radiated noise and background ocean noise are compared. Background ocean noise measurements were obtained by forming a broadside beam with a towed array whereas the noise of the towing platform was observed with a forward-looking endfire beam. Data from the broadside beam, which consists primarily of Gaussian noise, did not contain significant bispectral components. The bispectrum of data from the endfire beam contained a number of components, however, that the authors attribute to the non-Gaussian characteristics of noise sources on the towing platform. Thus, the bispectrum offers the possibility of differentiating between the noise radiated by a ship and background ocean noise. The conventional spectra of these two broadband noise sources often have different shapes, but reliable differentiation between background noise and background noise added to ship-radiated noise in general is difficult for low values of SNR.

A second application of the bispectrum involves what Nikias and Raghuveer [15] refer to as *quadratic phase coupling*. Such coupling arises if interaction between a pair of narrowband components produces another component at the sum or difference frequency. We consider the two sequences [15, eq. 3.11]:

$$\begin{aligned} y_1(n) &= \cos(2\pi f_1 n/f_s + \phi_1) + \cos(2\pi f_2 n/f_s + \phi_2) \\ &\quad + \cos[2\pi(f_1 + f_2)n/f_s + \phi_3] \end{aligned} \tag{9.17}$$

$$\begin{aligned} y_2(n) &= \cos(2\pi f_1 n/f_s + \phi_1) + \cos(2\pi f_2 n/f_s + \phi_2) \\ &\quad + \cos[2\pi(f_1 + f_2)n/f_s + \phi_1 + \phi_2] \end{aligned} \tag{9.18}$$

where ϕ_1, ϕ_2, and ϕ_3 are samples of a uniformly distributed independent random process. These samples satisfy $0 \leq \phi_i \leq 2\pi$, $i = 1, 2, 3$, and $\phi_3 \neq \phi_1 + \phi_2$. Equation (9.17) contains a narrowband component with a frequency equal to the sum of the frequencies of the first two components. Because $\phi_3 \neq \phi_1 + \phi_2$, the components in (9.17) are not phase coupled. The sequence in (9.18) is an example of phase coupling because both the frequency and the phase of the third term in (9.18) are equal to the sums of the frequencies and the phases, respectively, of the first two terms. Note that the third term in (9.18) appears if nonlinear effects cause formation of the product of $\cos(2\pi f_1 n/f_s + \phi_1)$ and $\cos(2\pi f_2 n/f_s + \phi_2)$.

Both $y_1(n)$ and $y_2(n)$ have identical spectra (because phase information is not retained), so the conventional spectrum cannot be used to differentiate between the three unrelated narrowband components in (9.17) and the harmonic set in (9.18).

The bispectrum, however, provides such a differentiation because the bispectrum of (9.17) is zero whereas that of (9.18) is not.

9.4 CONCLUSION

Numerous other areas of study in the field of sonar signal processing currently are being pursued. The three topics presented in this chapter are intended to indicate the variety of research being conducted in universities, government laboratories, and private industry. Another such topic is the use of normal modes to passively obtain estimates of the range and depth of a source. Normal mode theory is discussed in Burdic [16, Ch. 4], Clay and Medwin [17, Ch. 9], and Urick [18, Chs. 5 and 6]; the first four modes are depicted in Figure 9.5. The shallow-water sound pressure field can be described at each frequency as the sum of the contributions of a set of normal modes. Wilson, Koch, and Vidmar [19] and Yang [20] discuss the signal processing required to obtain an estimate of the range and depth of the source by measuring the pressure field with a vertical array of sensors. In particular, an FFT-based matched-mode processing scheme is presented in [19]. It is interesting to note that QR decomposition (Section 7.2.2) is used to solve a least squares problem that arises in the process of implementing this algorithm.

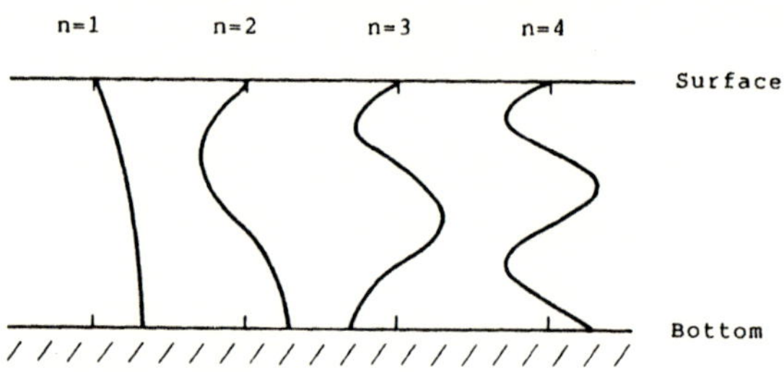

Figure 9.5 Sound pressure *versus* depth for the first four normal modes.
Source: After [17] C.S. Clay and H. Medwin, *Acoustical Oceanography: Principles and Applications*, John Wiley & Sons, New York, 1977.

It is our hope that the treatment of sonar signal processing provided in the first eight chapters of this book will serve as a basis for the exploration of new advances in this exciting field.

REFERENCES

1. Cohen, L., "Time-Frequency Distributions—A Review," *Proc. IEEE*, Vol. 77, No. 7, July 1989, pp. 941–981.
2. Nuttall, A.H., "Wigner Distribution Function: Relation to Short-Term Spectral Estimation,

Smoothing, and Performance in Noise," *NUSC Technical Report* No. 8225, Naval Underwater Systems Center, New London, CT, February 16, 1988.

3. Nuttall, A.H., "The Wigner Distribution Function with Minimum Spread," *NUSC Technical Report* No. 8317, Naval Underwater Systems Center, New London, CT, June 1, 1988.
4. Nuttall, A.H., "Alias-Free Wigner Distribution Function and Complex Ambiguity Function for Discrete-Time Samples," *NUSC Technical Report* No. 8533, Naval Underwater Systems Center, New London, CT, April 14, 1989.
5. Yen, N., "Time and Frequency Representation of Acoustic Signals by Means of the Wigner Distribution Function: Implementation and Interpretation," *J. Acoustical Society of America*, Vol. 81, No. 6, June 1987, pp. 1841–1850.
6. Van Trees, H.L., "Optimum Signal Design and Processing for Reverberation-Limited Environments," *IEEE Trans. on Military Electronics*, Vol. MIL-19, No. 3-4, July–October 1965, pp. 212–229.
7. Yu, K.-B., and S. Cheng, "Signal Synthesis from Pseudo-Wigner Distribution and Applications," *IEEE Trans. on Acoustics, Speech, and Signal Processing*, Vol. ASSP-35, No. 9, September 1987, pp. 1289–1302.
8. Kay, S., and G. F. Boudreaux-Bartels, "On the Optimality of the Wigner Distribution for Detection," *Proc. Int. Conf. Acoustics, Speech, and Signal Processing*, Tampa, FL, March 26–29, 1985, pp. 1017–1020.
9. Choi, H.-I., and W.J. Williams, "Improved Time-Frequency Representation of Multicomponent Signals Using Exponential Kernels," *IEEE Trans. on Acoustics, Speech, and Signal Processing*, Vol. ASSP-37, No. 6, June 1989, pp. 862–871.
10. Marple, S.L., Jr., *Digital Spectral Analysis with Applications*, Prentice-Hall, Englewood Cliffs, NJ, 1987.
11. Carter, G.C., "Variance Bounds for Passively Locating an Acoustic Source with a Symmetric Line Array," *J. Acoustical Society of America*, Vol. 62, No. 4, October 1977, pp. 922–926.
12. Carter, G.C., "Coherence and Time Delay Estimation," *Proc. IEEE*, Vol. 75, No. 2, February 1987, pp. 236–255.
13. MacDonald, V.H., and P.M. Schultheiss, "Optimum Passive Bearing Estimation in a Spatially Incoherent Noise Environment," *J. Acoustical Society of America*, Vol. 46, No. 1 (Part 1), 1969, pp. 37–43.
14. Hinich, M.J., D. Marandino, and E.J. Sullivan, "Bispectrum of Ship-Radiated Noise," *J. Acoustical Society of America*, Vol. 85, No. 4, April 1989, pp. 1512–1517.
15. Nikias, C.L., and M.R. Raghuveer, "Bispectrum Estimation: A Digital Signal Processing Framework," *Proc. IEEE*, Vol. 75, No. 7, July 1987, pp. 869–891.
16. Burdic, W.S., *Underwater Acoustic System Analysis*, Prentice-Hall, Englewood Cliffs, NJ, 1984.
17. Clay, C.S., and H. Medwin, *Acoustical Oceanography: Principles and Applications*, John Wiley & Sons, New York, 1977.
18. Urick, R.J., *Principles of Underwater Sound*, McGraw-Hill, New York, 1975.
19. Wilson, G.R., R.A. Koch, and P.J. Vidmar, "Matched Mode Localization," *J. Acoustical Society of America*, Vol. 84, No. 1, July 1988, pp. 310–320.
20. Yang, T.C., "A Method of Range and Depth Estimation by Modal Decomposition," *J. Acoustical Society of America*, Vol. 82, No. 5, November 1987, pp. 1736–1745.

Glossary

Acronym, Symbol, or Abbreviation	*Definition*
A/D	analog-to-digital
AGC	automatic gain control
ALI	automatic line integration
ALU	arithmetic logic unit
AR	autoregressive
ARMA	autoregressive-moving average
BTR	bearing-time record
β_{eff}	effective noise bandwidth
CPA	closest point of approach
CRLB	Cramer-Rao lower bound
CRT	cathode ray tube
CW	continuous wave
cpsd	cross power spectral density
DFT	discrete Fourier transform
DHT	discrete Hartley transform
DI	directivity index
DIF	decimation-in-frequency
DIT	decimation-in-time
DI_t	directivity index (transmitter array)
DSP	digital signal processing
DT	detection threshold
d	S/N at detector output (10 $\log_{10} d$ is the detection index)
dB	decibel
EE	echo excess
ENBW	equivalent noise bandwidth
ESD	energy spectral density
EV	eigenvector
erfc	complementary error function
FFT	fast Fourier transform

Acronym, Symbol, or Abbreviation	*Definition*
FHT	fast Hartley transform
FIM	Fisher information matrix
FIR	finite impulse response
FM	frequency modulation
FPG	fully parallel data-dependency graph
FRAZ	frequency-azimuth (display)
HF	high frequency
IDFT	inverse discrete Fourier transform
IFFT	inverse fast Fourier transform
IIR	infinite impulse response
I/O	input-output
ITA	intermediate-time average
K_I	number of independent samples (for integration)
LF	low frequency
LFA	low-frequency array
LFM	linear FM
LMS	least mean square
LTA	long-time average
LTI	long-time integration
Mbyte	megabyte
MDS	minimum detectable signal
MFA	medium-frequency array
MFLOPS	1 million floating point operations per second
ML	maximum likelihood
MLE	maximum-likelihood estimator
MMG	multimesh dependency graph
MRA	maximum response axis
MUSIC	multiple signal classification
NL	noise level
NP	Neyman-Pearson
ODN	ownship doppler nullification
OTA	order-truncate-average
Pa	Pascal
P_D	probability of detection
PE	processing element
P_{fa}	probability of false alarm
PL	processing, or implementation, loss
PRN	pseudorandom noise
pdf	probability density function
psd	power spectral density
rf	register file
RL	reverberation level
ROC	receiver operating characteristic
SE	signal excess

Acronym, Symbol, or Abbreviation	*Definition*
S/H	sample-and-hold
SL	scalloping loss
SL	source level
SL_{av}	average scalloping loss
SLD	square-law detector
S/N	signal-to-noise ratio
SPWD	smoothed pseudo-Wigner distribution
S/QN	signal-to-quantization noise ratio
SSA	sector-space average
STA	short-time average
STI	short-time integration
SVD	singular value decomposition
sup	supremum
TFD	time-frequency distribution
TL	transmission loss
TS	target strength
TVG	time-varying gain
VLSI	very large scale integration
WVD	Wigner-Ville distribution

Index

Printed in the United States
21123LVS00003BE/2

9 780890 064535